INTUITIVE
BIOSTATISTICS

INTUITIVE
BIOSTATISTICS

Harvey Motulsky, M.D.

Department of Pharmacology
University of California, San Diego
  and
President, GraphPad Software, Inc.

New York   Oxford
OXFORD UNIVERSITY PRESS
1995

Oxford University Press

Oxford   New York
Athens   Auckland   Bangkok   Bombay
Calcutta   Cape Town   Dar es Salaam   Delhi
Florence   Hong Kong   Istanbul   Karachi
Kuala Lumpur   Madras   Madrid   Melbourne
Mexico City   Nairobi   Paris   Singapore
Taipei   Tokyo   Toronto

and associated companies in
Berlin   Ibadan

Library of Congress Cataloging-in-Publication Data
Motulsky, Harvey.
Intuitive biostatistics / Harvey Motulsky.
p.   cm. Includes index.
ISBN 0-19-508606-6.—ISBN 0-19-508607-4 (pbk.)
1. Medicine—Research—Statistical methods.   I. Title.
R853.S7M68   1995
610′.01′5195—dc20   95-8166

9 8 7 6 5 4 3 2 1

Printed in the United States of America
on acid-free paper

PREFACE

APPROACH

This book provides a nonmathematical introduction to statistics for medical students, physicians, graduate students, and researchers in the health sciences. So do plenty of other books, but this one has a unique approach.

- Explanations rather than mathematical proofs. To those with appropriate training and inclination, mathematical notation is a wonderful way to say things clearly and concisely. Why read a page of explanation when you can read two equations? But for many (perhaps most) students and scientists, mathematical notation is confusing and threatening. This book explains concepts in words, with few mathematical proofs or equations (except for those provided in reference sections marked with a # symbol).
- Emphasis on interpreting results rather than analyzing data. Statistical methods will only be *used* by those who collect data. Statistical results will be *interpreted* by everyone who reads published papers. This book emphasizes the interpretation of published results, although it also explains how to analyze data. For example, I explain how to interpret P values long before I present any tests that compute P values. In some cases, I discuss statistical methods without showing the equations needed to analyze data with those methods.
- Emphasis on confidence intervals. Statistical calculations generate both P values and confidence intervals. While most books emphasize the calculation of P values, I emphasize confidence intervals. Although P values and confidence intervals are related and are calculated together, I present confidence intervals first, and delay presenting P values until Chapter 10.
- Examples from the clinical literature. In a few places, I've included simple examples with fake data. But most of the examples are from recent medical literature.* To focus the discussion on basic statistical understanding, I have sometimes simplified the findings a bit, without (I hope) obscuring the essence of the results.

---

*These papers were not selected because they are particularly good or particularly bad. They are just a sampling of papers from good journals that I happened to stumble across when searching for examples (so the authors don't need to be particularly proud or embarrassed to see their work included).

• Explanation of Bayesian thinking. Bayesian thinking helps one interpret P values, lab results, genetic counseling, and linkage analysis. Whereas most introductory books ignore Bayesian analysis, I discuss it in reasonable detail in Part IV.

TOPICS COVERED

In choosing topics to include in this book I've chosen breadth over depth. This is because so many statistical methods are commonly used in the biomedical literature. Flip through any medical or scientific journal and you'll soon find use of a statistical technique not mentioned in most introductory books. To guide those who read those papers, I included many topics omitted from other books: relative risk and odds ratios, prediction intervals, nonparametric tests, survival curves, multiple comparisons, the design of clinical trials, computing the power of a test, nonlinear regression, interpretation of lab tests (sensitivity, specificity, etc.). I also briefly introduce multiple regression, logistic regression, proportional hazards regression, randomization tests, and lod scores. Analysis of variance is given less emphasis than usual.

CHAPTERS TO SKIP

As statistics books go, this one is pretty short. But I realize that it still is more than most people want to read about statistics. If you just want to learn the main ideas of statistics, with no detail, read Chapters 1 through 5, 10 through 13, and 19.

This book is for anyone who reads papers in the biomedical literature, not just for people who read clinical studies. Basic scientists may want to skip Chapters 6, 9, 20, 21, 32, and 33 which deal with topics uncommonly encountered in basic research. The other chapters are applicable to both clinicians and basic scientists.

ANALYZING DATA WITH COMPUTER PROGRAMS

We are lucky to live in an era where personal computers are readily available. Although this book gives the equations for many statistical tests, most people will rely on a computer program instead. Unfortunately, most statistics programs are designed for statisticians and are too complicated and too expensive for the average student or scientist. That's why my company, GraphPad Software, created GraphPad InStat, an inexpensive and extremely easy statistical program available for DOS and Macintosh computers.* (See Appendix 2 for details about this program.) Although this book shows sample output from InStat, you do not need InStat to follow the examples in this book or to work the problems.

Although spreadsheet programs were originally developed to perform financial calculations, current versions are very versatile and adept at statistical computation. See Appendix 3 to learn how to use Microsoft Excel to perform statistical calculations.

---

*By the time you read this, a Windows version may be available.

REFERENCES AND ACKNOWLEDGMENTS

I have organized this book in a unique way, but none of the ideas are particularly original. All of the statistical methods are standard, and have been discussed in many text books. Rather than give the original reference for each method, I have listed text book references in Appendix 1.

I would like to thank everyone who reviewed various sections of the book in draft form and gave valuable comments, including Jan Agosti, Cedric Garland, Ed Jackson, Arno Motulsky, Paige Searle, and Christopher Sempos. I especially want to thank Harry Frank, whose lengthy comments improved this book considerably. This book would be very different if it weren't for his repeated lengthy reviews. I also want to thank all the students who helped me shape this book over the last five years. Of course, any errors are my own responsibility. Please email comments and suggestions to HMotulsky@graphpad.com.

New York   H.M.
January 1995

CONTENTS IN BRIEF

Contents, xi

1. Introduction to Statistics, 3

PART I.   CONFIDENCE INTERVALS, 9

2. Confidence Interval of a Proportion, 11

3. The Standard Deviation, 22

4. The Gaussian Distribution, 31

5. The Confidence Interval of a Mean, 39

6. Survival Curves, 53

PART II.   COMPARING GROUPS WITH CONFIDENCE INTERVALS, 61

7. Confidence Interval of a Difference Between Means, 63

8. Confidence Interval of the Difference or Ratio of Two Proportions:
   Prospective Studies, 70

9. Confidence Interval of the Ratio of Two Proportions:
   Case-Control Studies, 81

PART III.   INTRODUCTION TO P VALUES, 91

10. What Is a P Value?, 93

11. Statistical Significance and Hypothesis Testing, 106

12. Interpreting Significant and Not Significant P Values, 113

13. Multiple Comparisons, 118

PART IV.   BAYESIAN LOGIC, 127

14. Interpreting Lab Tests: Introduction to Bayesian Thinking, 129

15. Bayes and Statistical Significance, 140

16. Bayes' Theorem in Genetics, 149

PART V.   CORRELATION AND REGRESSION, 153

17. Correlation, 155

18. An Introduction to Regression, 165

19. Simple Linear Regression, 167

PART VI.   DESIGNING CLINICAL STUDIES, 181

20. The Design of Clinical Trials, 183

21. Clinical Trials where N = 1, 192

22. Choosing an Appropriate Sample Size, 195

PART VII.   COMMON STATISTICAL TESTS, 205

23. Comparing Two Groups: Unpaired t Test, 207

24. Comparing Two Means: The Randomization and Mann-Whitney Tests, 217

25. Comparing Two Paired Groups: Paired t and Wilcoxon Tests, 225

26. Comparing Observed and Expected Counts, 230

27. Comparing Two Proportions, 233

PART VIII.   INTRODUCTION TO ADVANCED STATISTICAL TESTS, 243

28. The Confidence Interval of Counted Variables, 245

29. Further Analyses of Contingency Tables, 250

30. Comparing Three or More Means: Analysis of Variance, 255

31. Multiple Regression, 263

32. Logistic Regression, 268

33. Comparing Survival Curves, 272

34. Using Nonlinear Regression to Fit Curves, 277

35. Combining Probabilities, 284

PART IX.   OVERVIEWS, 291

36. Adjusting for Confounding Variables, 293

37. Choosing a Test, 297

38. The Big Picture, 303

PART X.   APPENDICES, 307

Index, 383

CONTENTS

1. **Introduction to Statistics, 3**

   Why do we need statistical calculations?, 3
   Many kinds of data can be analyzed without statistical analysis, 4
   Statistical calculations extrapolate from sample to population, 4
   What statistical calculations can do, 5
   What statistical calculations cannot do, 6
   Why is it hard to learn statistics?, 7
   Arrangement of this book, 8

PART I. CONFIDENCE INTERVALS, 9

2. **Confidence Interval of a Proportion, 11**

   Proportions versus measurements, 11
   The binomial distribution: From population to sample, 11
   The confidence interval of a proportion: From sample to population, 12
   What exactly does it mean to say that you are ''95% sure''?, 13
   Assumptions, 14
   Obtaining the confidence interval of a proportion from a table, 16
   The special cases of 0 and 100 percent, 17
   Example, 17
   Calculating the confidence interval of a proportion, 18
   The binomial equation, 19
   How the confidence intervals are derived, 19
   Summary, 20
   Objectives, 20
   Problems, 21

3. **The Standard Deviation, 22**

   Source of variability, 22
   Displaying variability with histograms, 23
   The mean and median, 24
   Quantifying variability with percentiles, 25
   Quantifying variability with the variance and standard deviation, 26
   N or N − 1? The sample SD versus the population SD, 27

Calculating the sample standard deviation with a calculator, 28
Interpreting the standard deviation, 28
Coefficient of variation (CV), 29
Summary, 29
Objectives, 29
Problems, 30

4. The Gaussian Distribution, 31

Probability distributions, 31
The Gaussian distribution, 31
Using the Gaussian distribution to make inferences about the population, 33
The prediction interval, 35
Normal limits and the "normal" distribution, 36
Summary, 37
Objectives, 38
Problems, 38

5. The Confidence Interval of a Mean, 39

Interpreting a confidence interval of a mean, 39
Assumptions that must be true to interpret the 95% CI of a mean, 40
Calculating the confidence interval of a mean, 41
The central limit theorem, 42
The standard error of the mean, 44
The t distribution, 44
The Gaussian assumption and statistical inference, 46
Confidence interval of a proportion revisited, 47
Error bars, 48
Summary, 49
Objectives, 50
Problems, 50

6. Survival Curves, 53

A simple survival curve, 53
Censored survival data, 54
Creating a survival curve, 55
Confidence interval of a survival curve, 55
Median survival, 57
Assumptions, 58
Problems with survival studies, 58
Summary, 59
Objectives, 59
Problems, 59

PART II.   COMPARING GROUPS WITH CONFIDENCE INTERVALS, 61

7. Confidence Interval of a Difference Between Means, 63

Interpreting the 95% CI for the difference between two means, 63
Calculating the 95% CI for the difference between means of unpaired groups, 65

Why are paired subjects analyzed differently?, 66
How to calculate the 95% CI of the mean difference of paired subjects, 67
Summary, 68
Objectives, 68
Problems, 68

8. Confidence Interval of the Difference or Ratio of Two Proportions:
Prospective Studies, 70

Cross-sectional, prospective, and retrospective studies, 70
An example of a clinical trial, 71
Difference between two proportions, 72
Relative risk, 73
Assumptions, 73
How the relative risk can be misleading, 74
Probabilities versus odds, 75
The odds ratio, 76
Relative risks from survival studies, 76
What is a contingency table?, 77
Calculating confidence intervals, 77
Summary, 78
Objectives, 79
Problems, 79

9. Confidence Interval of the Ratio of Two Proportions:
Case-Control Studies, 81

What is a case-control study?, 81
Why can't you calculate the relative risk from case-control data?, 82
How to interpret the odds ratio, 82
Why the odds ratio from a case-control study approximates the real relative
risk, 83
Advantages and disadvantages of case-control studies, 85
Assumptions in case-control studies, 86
Matched pairs, 87
Calculating the 95% CI of an odds ratio, 88
Summary, 88
Objectives, 89
Problems, 89

PART III.   INTRODUCTION TO P VALUES, 91

10. What Is a P Value?, 93

Introduction to P values, 93
A simple example: Blood pressure in medical students, 93
Other null hypotheses, 97
Common misinterpretations of P values, 97
One-tailed versus two-tailed P values, 97
Example 10.1 Comparing two proportions from an experimental study, 99
Example 10.2 Comparing two proportions from a case-control study, 100

Example 10.3 Comparing two means with the t test, 101
Example 10.4 Comparing means of paired samples, 102
Example 10.5 Comparing two survival curves with the log-rank test, 103
Summary, 104
Objectives, 104
Problems, 105

11. Statistical Significance and Hypothesis Testing, 106

Statistical hypothesis testing, 106
The advantages and disadvantages of using the phrase *statistically significant*, 107
An analogy: Innocent until proven guilty, 108
Type I and Type II errors, 109
Choosing an appropriate value for α, 109
The relationship between α and P values, 110
The relationship between α and confidence intervals, 111
Statistical significance versus scientific importance, 111
Summary, 111
Objectives, 112
Problems, 112

12. Interpreting Significant and Not Significant P Values, 113

The term *significant*, 113
Extremely significant results, 113
Borderline P values, 114
The term *not significant*, 114
Interpreting *not significant* results with confidence intervals, 114
Interpreting *not significant* P values using power analyses, 116
Summary, 117

13. Multiple Comparisons, 118

Coincidences, 118
Multiple independent questions, 119
Multiple groups, 120
Multiple measurements to answer one question, 123
Multiple subgroups, 123
Multiple comparisons and data dredging, 124
Summary, 125
Problem, 126

PART IV. BAYESIAN LOGIC, 127

14. Interpreting Lab Tests: Introduction to Bayesian Thinking, 129

The accuracy of a qualitative lab test, 129
The accuracy of a quantitative lab test, 130
The predictive value of a test result, 132
Calculating the predictive value of a positive or negative test, 133
Bayes' Theorem, 136

A review of probability and odds, 136
Bayes' equation, 137
Some additional complexities, 138
Summary, 138
Objectives, 139
Problems, 139

15. Bayes and Statistical Significance, 140

Type I errors and false positives, 140
Type II errors and false negatives, 140
Probability of obtaining a false-positive lab result: Probability that a significant
    result will occur by chance, 142
The predictive value of significant results: Bayes and P values, 142
The controversy regarding Bayesian statistics, 145
Applying Bayesian thinking informally, 145
Multiple comparisons, 146
Summary, 146
Objectives, 147
Problems, 148

16. Bayes' Theorem in Genetics, 149

Bayes' theorem in genetic counseling, 149
Bayes and genetic linkage, 150
Problems, 152

PART V.   CORRELATION AND REGRESSION, 153

17. Correlation, 155

Introducing the correlation coefficient, 155
Interpreting r, 157
Interpreting r^2, 158
Assumptions, 158
Outliers, 159
Spearman rank correlation, 160
Don't combine two populations in correlation, 161
Calculating the correlation coefficient, 161
The 95% CI of a correlation coefficient, 162
Calculating the Spearman correlation coefficient, 162
Calculating a P value from correlation coefficients, 162
Summary, 163
Objectives, 163
Problems, 164

18. An Introduction to Regression, 165

What is a model?, 165
Why bother with models?, 165
Different kinds of regression, 166

19. Simple Linear Regression, 167

An example of linear regression, 167
Comparison of linear regression and correlation, 169
The linear regression model, 170
The assumptions of linear regression, 171
Linear regression as a least squares method, 172
The meaning of r^2, 173
Maximum likelihood, 173
Graphing residuals, 174
Using the regression line as a standard curve to find new values of Y from X, 175
The regression fallacy, 175
Calculating linear regression, 176
Summary, 179
Objectives, 179
Problems, 179

PART VI.   DESIGNING CLINICAL STUDIES, 181

20. The Design of Clinical Trials, 183

Designing the study, 184
The need for a study protocol, 187
What is in the study protocol, 188
Human subject committees and informed consent, 189
Ethics of placebos, 189
How is the population defined?, 190
Reviewing data from clinical trials, 190

21. Clinical Trials where N = 1, 192

Example, 193

22. Choosing an Appropriate Sample Size, 195

Confidence intervals, 195
Hypothesis testing, 197
Interpreting a statement regarding sample size and power, 201
Sequential studies, 202
Summary, 203
Objectives, 203
Problems, 203

PART VII.   COMMON STATISTICAL TESTS, 205

23. Comparing Two Groups: Unpaired t Test, 207

Some notes on t, 208
Obtaining the P value from t, using a table, 208
The t distribution and the meaning of the P value, 209
Assumptions of a t test, 210
The relationship between confidence intervals and hypothesis testing, 211

Calculating the power of a t test, 213
Example 23.2, 215
Problems, 216

24. Comparing Two Means: The Randomization and Mann-Whitney Tests, 217

Calculating the exact randomization test, 218
Large samples: The approximate randomization test, 219
The relationship between the randomization test and the t test, 220
Mann-Whitney test, 221
Performing the Mann-Whitney test, 222
Assumptions of the Mann-Whitney test, 223
When to use nonparametric tests, 224
Problems, 224

25. Comparing Two Paired Groups: Paired t and Wilcoxon Tests, 225

When to use special tests for paired data, 225
Calculating the paired t test, 225
Assumptions of paired t test, 227
"Ratio" t tests, 227
The Wilcoxon Signed Rank Sum test, 228
Problems, 229

26. Comparing Observed and Expected Counts, 230

Analyzing counted data, 230
The Yates' continuity correction, 231
Where does the equation come from?, 232
Problems, 232

27. Comparing Two Proportions, 233

Fisher's exact test, 233
Chi-square test for 2×2 contingency tables, 233
How to calculate the chi-square test for a 2×2 contingency table, 234
Assumptions, 235
Choosing between chi-square and Fisher's test, 236
Calculating power, 236
Problems, 241

PART VIII.   INTRODUCTION TO ADVANCED STATISTICAL TESTS, 243

28. The Confidence Interval of Counted Variables, 245

The Poisson distribution, 245
Understanding the confidence interval of a count, 246
Calculating the confidence interval of a count, 247
Changing scales, 247
Objectives, 248
Problems, 248

29. Further Analyses of Contingency Tables, 250

McNemar's chi-square test for paired observations, 250
Chi-square test with large tables (more than two rows or columns), 251
Chi-square test for trend, 253
Mantel-Haenszel chi-square test, 253

30. Comparing Three or More Means: Analysis of Variance, 255

What's wrong with multiple t tests?, 255
One-way ANOVA, 255
An example, 256
Assumptions of ANOVA, 258
Mathematical model of one-way ANOVA, 258
Multiple comparison post tests, 258
Repeated-measures ANOVA, 260
Nonparametric ANOVA, 260
Two-way ANOVA, 261
Perspective on ANOVA, 261

31. Multiple Regression, 263

The uses of multiple regression, 263
The multiple regression model, 264
Assumptions of multiple regression, 264
Interpreting the results of multiple regression, 265
Choosing which X variables to include in a model, 266
The term *multivariate statistics*, 267

32. Logistic Regression, 268

Introduction to logistic regression, 268
How logistic regression works, 269
Assumptions of logistic regression, 270
Interpreting results from logistic regression, 270

33. Comparing Survival Curves, 272

Comparing two survival curves, 272
Assumptions of the log-rank test, 273
A potential trap: Comparing survival of responders versus nonresponders, 273
Will Rogers' phenomenon, 274
Multiple regression with survival data: Proportional hazards regression, 274
How proportional hazards regression works, 275
Interpreting the results of proportional hazards regression, 276

34. Using Nonlinear Regression to Fit Curves, 277

The goals of curve fitting, 277
An example, 277
What's wrong with transforming curved data into straight lines?, 279
Using a nonlinear regression program, 279

The results of nonlinear regression, 281
Polynomial regression, 283

35. Combining Probabilities, 284

Propagation of errors, 284
Example of error propagation, 286
Decision Analysis, 286
Meta-analysis, 288

PART IX.   OVERVIEWS, 291

36. Adjusting for Confounding Variables, 293

What is a confounding variable?, 293
Designing studies to avoid confounding variables, 293
The need to correct for confounding variables, 294
Statistical tests that correct for confounding variables, 295
Interpreting results, 296

37. Choosing a Test, 297

Review of available statistical tests, 297
Review of nonparametric tests, 297
Choosing between parametric and nonparametric tests: The easy cases, 297
Choosing between parametric and nonparametric tests: The hard cases, 299
Choosing between parametric and nonparametric tests: Does it matter?, 300
One- or two-sided P value?, 300
Paired or unpaired test?, 301
Fisher's test or the chi-square test?, 301
Regression or correlation?, 302

38. The Big Picture, 303

Look at the data!, 303
Beware of very large and very small samples, 303
Beware of multiple comparisons, 303
Don't focus on averages: Outliers may be important, 304
Non-Gaussian distributions are normal, 304
Garbage in, garbage out, 304
Confidence limits are as informative as P values (maybe more so), 304
Statistically significant does not mean scientifically important, 304
P < 0.05 is not sacred, 305
Don't overinterpret not significant results, 305
Don't ignore pairing, 305
Correlation or association does not imply causation, 305
Distinguish between studies designed to generate a hypothesis and studies
    designed to test one, 305
Distinguish between studies that measure an important outcome and studies that
    measure a proxy or surrogate outcome, 306
Published P values are optimistic (too low), 306
Confidence intervals are optimistic (too narrow), 306

PART X.   APPENDICES, 307

Appendix 1. References, 309

Appendix 2. GraphPad InStat and GraphPad Prism, 311

Appendix 3. Analyzing Data With a Spreadsheet Program or Statistics
Program, 313

Appendix 4. Answers to Problems, 316

Appendix 5. Statistical Tables, 360

Index, 383

INTUITIVE
BIOSTATISTICS

Introduction to Statistics

There is something fascinating about science. One gets such a
wholesale return of conjecture out of a trifling investment of fact.
Mark Twain (Life on the Mississippi, 1850)

This is a book for "consumers" of statistics. The goals are to teach you enough
statistics to

1. Understand the statistical portions of most articles in medical journals.
2. Avoid being bamboozled by statistical nonsense.
3. Do simple statistical calculations yourself, especially those that help you interpret
   published literature.
4. Use a simple statistics computer program to analyze data.
5. Be able to refer to a more advanced statistics text or communicate with a statistical
   consultant (without an interpreter).

Many statistical books read like cookbooks; they contain the recipes for many
statistical tests, and their goal (often unstated) is to train "statistical chefs" able to
whip up a P value on moment's notice. This book is based on the assumption that
statistical tests are best calculated by computer programs or by experts. This book,
therefore, will not teach you to be a chef, but rather to become an educated connoisseur
or critic who can appreciate and criticize what the chef has created. But just as you
must learn a bit about the differences between broiling, boiling, baking, and basting
to become a connoisseur of fine food, you must learn a bit about probability distributions
and null hypotheses to become an educated consumer of the biomedical literature.
Hopefully this book will make it relatively painless.

WHY DO WE NEED STATISTICAL CALCULATIONS?

When analyzing data, your goal is simple: You wish to make the strongest possible
conclusions from limited amounts of data. To do this, you need to overcome two
problems:

- Important differences are often obscured by biological variability and/or experimental
  imprecision, making it difficult to distinguish real differences from random variation.
- The human brain excels at finding patterns and relationships, but tends to overgeneralize. For example, a 3-year-old girl recently told her buddy, "You can't become a

doctor; only girls can become doctors.'' To her this made sense, as the only three doctors she knew were women. This inclination to overgeneralize does not seem to go away as you get older, and scientists have the same urge. Statistical rigor prevents you from making this kind of error.

MANY KINDS OF DATA CAN BE ANALYZED WITHOUT STATISTICAL ANALYSIS

Statistical calculations are most helpful when you are looking for fairly small differences in the face of considerable biological variability and imprecise measurements. Basic scientists asking fundamental questions can often reduce biological variability by using inbred animals or cloned cells in controlled environments. Even so, there will still be scatter among replicate data points. If you only care about differences that are large compared with the scatter, the conclusions from such studies can be obvious without statistical analysis. In such experimental systems, effects small enough to require statistical analysis are often not interesting enough to pursue.

   If you are lucky enough to be studying such a system, you may heed the following aphorisms:

> If you need statistics to analyze your experiment, then you've done the wrong experiment.

> If your data speak for themselves, don't interrupt!

   Most scientists are not so lucky. In many areas of biology, and especially in clinical research, the investigator is faced with enormous biological variability, is not able to control all relevant variables, and is interested in small effects (say 20% change). With such data, it is difficult to distinguish the signal you are looking for from the noise created by biological variability and imprecise measurements. Statistical calculations are necessary to make sense out of such data.

STATISTICAL CALCULATIONS EXTRAPOLATE FROM SAMPLE TO POPULATION

Statistical calculations allow you to make general conclusions from limited amounts of data. You can extrapolate from your data to a more general case. Statisticians say that you extrapolate from a *sample* to a *population*. The distinction between sample and population is key to understanding much of statistics. Here are four different contexts where the terms are used.

* *Quality control.* The terms *sample* and *population* make the most sense in the context of quality control where the sample is randomly selected from the overall population. For example, a factory makes lots of items (the population), but randomly selects a few items to test (the sample). These results obtained from the sample are used to make inferences about the entire population.
* *Political polls.* A random sample of voters (the sample) is polled, and the results are used to make conclusions about the entire population of voters.

- *Clinical studies.* The sample of patients studied is rarely a random sample of the larger population. However, the patients included in the study are representative of other similar patients, and the extrapolation from sample to population is still useful. There is often room for disagreement about the precise definition of the population. Is the *population* all such patients that come to that particular medical center, or all that come to a big city teaching hospital, or all such patients in the country, or all such patients in the world? While the population may be defined rather vaguely, it still is clear we wish to use the sample data to make conclusions about a larger group.
- *Laboratory experiments.* Extending the terms *sample* and *population* to laboratory experiments is a bit awkward. The data from the experiment(s) you actually performed is the sample. If you were to repeat the experiment, you'd have a different sample. The data from all the experiments you could have performed is the population. From the sample data you want to make inferences about the ideal situation.

In biomedical research, we usually assume that the population is infinite, or at least very large compared with our sample. All the methods in this book are based on that assumption. If the population has a defined size, and you have sampled a substantial fraction of the population (>10% or so), then you need to use special methods that are not presented in this book.

WHAT STATISTICAL CALCULATIONS CAN DO

Statistical reasoning uses three general approaches:

Statistical Estimation

The simplest example is calculating the mean of a sample. Although the calculation is exact, the mean you calculate from a sample is only an estimate of the population mean. This is called a *point estimate.* How good is the estimate? As we will see in Chapter 5, it depends on the sample size and scatter. Statistical calculations combine these to generate an interval estimate (a range of values), known as a *confidence interval* for the population mean. If you assume that your sample is randomly selected from (or at least representative of) the entire population, then you can be 95% sure that the mean of the population lies somewhere within the 95% confidence interval, and you can be 99% sure that the mean lies within the 99% confidence interval. Similarly, it is possible to calculate confidence intervals for proportions, for the difference or ratio of two proportions or two means, and for many other values.

Statistical Hypothesis Testing

Statistical hypothesis testing helps you decide whether an observed difference is likely to be caused by chance. Various techniques can be used to answer this question: If there is no difference between two (or more) populations, what is the probability of randomly selecting samples with a difference as large or larger than actually observed? The answer is a probability termed the *P value.* If the P value is small, you conclude that the difference is statistically *significant* and unlikely to be due to chance.

Statistical Modeling

Statistical modeling tests how well experimental data fit a mathematical model constructed from physical, chemical, genetic, or physiological principles. The most common form of statistical modeling is linear regression. These calculations determine "the best" straight line through a particular set of data points. More sophisticated modeling methods can fit curves through data points.

WHAT STATISTICAL CALCULATIONS CANNOT DO

In theory, here is how you should apply statistical analysis to a simple experiment:

1. Define a population you are interested in.
2. Randomly select a sample of subjects to study.
3. Randomly select half the subjects to receive one treatment, and give the other half another treatment.
4. Measure a single variable in each subject.
5. From the data you have measured in the samples, use statistical techniques to make inferences about the distribution of the variable in the population and about the effect of the treatment.

When applying statistical analysis to real data, scientists confront several problems that limit the validity of statistical reasoning. For example, consider how you would design a study to test whether a new drug is effective in treating patients infected with the human immunodeficiency virus (HIV).

- The population you really care about is all patients in the world, now and in the future, who are infected with HIV. Because you can't access that population, you choose to study a more limited population: HIV patients aged 20 to 40 living in San Francisco who come to a university clinic. You may also exclude from the population patients who are too sick, who are taking other experimental drugs, who have taken experimental vaccines, or who are unable to cooperate with the experimental protocol. Even though the population you are working with is defined narrowly, you hope to extrapolate your findings to the wider population of HIV-infected patients.
- Randomly sampling patients from the defined population is not practical, so instead you simply attempt to enroll all patients who come to morning clinic during two particular months. This is termed a *convenience sample*. The validity of statistical calculations depends on the assumption that the results obtained from this convenience sample are similar to those you would have obtained had you randomly sampled subjects from the population.
- The variable you really want to measure is survival time, so you can ask whether the drug increases life span. But HIV kills slowly, so it will take a long time to accumulate enough data. As an alternative (or first step), you choose to measure the number of helper (CD4) lymphocytes. Patients infected with the HIV have low numbers of CD4 lymphocytes, so you can ask whether the drug increases CD4 cell number (or delays the reduction in CD4 cell count). To save time and expense, you have switched from an important variable (survival) to a proxy variable (CD4 cell count).

- Statistical calculations are based on the assumption that the measurements are made correctly. In our HIV example, statistical calculations would not be helpful if the antibody used to identify CD4 cells was not really selective for those cells.
- Statistical calculations are most often used to analyze one variable measured in a single experiment, or a series of similar experiments. But scientists usually draw general conclusions by combining evidence generated by different kinds of experiments. To assess the effectiveness of a drug to combat HIV, you might want to look at several measures of effectiveness: reduction in CD4 cell count, prolongation of life, increased quality of life, and reduction in medical costs. In addition to measuring how well the drug works, you also want to quantify the number and severity of side effects. Although your conclusion must be based on all these data, statistical methods are not very helpful in blending different kinds of data. You must use clinical or scientific judgment, as well as common sense.

In summary, statistical reasoning can not help you overcome these common problems:

- The population you really care about is more diverse than the population from which your data were sampled.
- You collect data from a "convenience sample" rather than a random sample.
- The measured variable is a proxy for another variable you really care about.
- Your measurements may be made or recorded incorrectly, and assays may not always measure exactly the right thing.
- You need to combine different kinds of measurements to reach an overall conclusion.

You must use scientific and clinical judgment, common sense, and sometimes a leap of faith to overcome these problems. Statistical calculations are an important part of data analysis, but interpreting data also requires a great deal of judgment. That's what makes research challenging. This is a book about statistics, so we will focus on the statistical analysis of data. Understanding the statistical calculations is only a small part of evaluating clinical and biological research.

WHY IS IT HARD TO LEARN STATISTICS?

Five factors make it difficult for many students to learn statistics:

- The terminology is deceptive. Statistics gives special meaning to many ordinary words. To understand statistics, you have to understand that the statistical meaning of terms such as *significant, error,* and *hypothesis* are distinct from the ordinary uses of these words. As you read this book, pay special attention to the statistical terms that sound like words you already know.
- Many people seem to believe that statistical calculations are magical and can reach conclusions that are much stronger than is actually possible. The phrase *statistically significant* is seductive and is often misinterpreted.
- Statistics requires mastering abstract concepts. It is not easy to think about theoretical concepts such as populations, probability distributions, and null hypotheses.
- Statistics is at the interface of mathematics and science. To really grasp the concepts of statistics, you need to be able to think about it from both angles. This book

emphasizes the scientific angle and avoids math. If you think like a mathematician, you may prefer a text that uses a mathematical approach.

• The derivation of many statistical tests involves difficult math. Unless you study more advanced books, you must take much of statistics on faith. However, you can learn to *use* statistical tests and interpret the results even if you don't fully understand how they work. This situation is common in science, as few scientists really understand all the tools they use. You can interpret results from a pH meter (measures acidity) or a scintillation counter (measures radioactivity), even if you don't understand *exactly* how they work. You only need to know enough about how the instruments work so that you can avoid using them in inappropriate situations. Similarly, you can calculate statistical tests and interpret the results even if you don't understand how the equations were derived, as long as you know enough to use the statistical tests appropriately.

ARRANGEMENT OF THIS BOOK

Parts I through V present the basic principles of statistics. To make it easier to learn, I have separated the chapters that explain confidence intervals from those that explain P values. In practice, the two approaches are used in parallel. Basic scientists who don't care to learn about clinical studies may skip Chapters 6 (survival curves) and 9 (case-control studies) without loss of continuity.

Part VI describes the design of clinical studies and discusses how to determine sample size. Basic scientists who don't care to learn about clinical studies can skip this entire part. However, Chapter 22 (sample size) is of interest to all. Part VII explains the most common statistical tests. Even if you use a computer program to calculate the tests, reading these chapters will help you understand how the tests work. The tests mentioned in this section are described in detail.

Part VIII gives an overview of more advanced statistical tests. These tests are not described in detail, but the chapters provide enough information so that you can be an intelligent consumer of papers that use these tests. The chapters in this section do not follow a logical sequence, so you can pick and choose the topics that interest you. The only exception is that you should read Chapter 31 (multiple regression) before Chapters 32 (logistic regression) or the parts of Chapter 33 (comparing survival curves) dealing with proportional hazards regression.

The statistical principles and tests discussed in this book are widely used, and I do not give detailed references. For more information, refer to the general textbook references listed in Appendix 1.

I

CONFIDENCE INTERVALS

Statistical analysis of data leads to two kinds of results: confidence intervals and P values. The two give complementary information and are often calculated in tandem. For the purposes of clarity and simplicity, this book presents confidence intervals first and then presents P values. Confidence intervals let you state a result with *margin of error*. This section explains what this means and how to calculate confidence intervals.

2

Confidence Interval of a Proportion

PROPORTIONS VERSUS MEASUREMENTS

The results of experiments can be expressed in different ways. In this chapter we will consider only results expressed as a proportion or fraction. Here are some examples: the proportion of patients who become infected after a procedure, the proportion of patients with myocardial infarction who develop heart failure, the proportion of students who pass a course, the proportion of voters who vote for a particular candidate. Later we will discuss other kinds of variables, including measurements and survival times.

THE BINOMIAL DISTRIBUTION: FROM POPULATION TO SAMPLE

If you flip a coin fairly, there is a 50% probability (or chance) that it will land on heads and a 50% probability that it will land on tails. This means that, in the long run, a coin will land on heads about as often as it lands on tails. But in any particular series of tosses, you may not see the coin land on heads exactly half the time. You may even see all heads or all tails.

Mathematicians have developed equations, known as the *binomial distribution,* to calculate the likelihood of observing any particular outcome when you know the proportion in the overall population. Using the binomial distribution, you can answer questions such as these:

- If you flip a coin 10 times, what is the probability of getting exactly 7 heads?
- If you flip a coin 10 times, what is the probability of getting 7 or more heads?
- If 5% of patients undergoing an operation get infected, what is the chance that 10 or more of the next 30 patients will be infected?
- If a couple's chance of passing a genetic disease to each child is 25%, what is the chance that their first three children will all be unaffected?
- If 40% of voters are Democrats, what is the chance that a random sample of 500 voters will include more than 45% Democrats?

Perhaps you've seen the equations that help you answer these kinds of questions, and recall that there are lots of factorials. If you're interested, the equation is presented at the end of this chapter.

The binomial distribution is not immediately useful when analyzing data because it works in the wrong direction. The theory starts with a known probability (i.e., 50% of coin flips are heads) and calculates the likelihood of any particular result in a sample. When analyzing data, we need to work in the opposite direction. We don't know the overall probability. That's what we are trying to find out. We do know the proportion observed in a single sample and wish to make inferences about the overall probability.

The binomial distribution can still be useful, but it must be turned backwards to generate confidence intervals. I show you how to do this at the end of the chapter. For now, accept the fact that it can be done and concentrate on interpreting the results.

THE CONFIDENCE INTERVAL OF A PROPORTION: FROM SAMPLE TO POPULATION

Let's start with an example. Out of 14 patients you have treated with a particular drug, three suffered from a particular side effect. The proportion is 3/14, which equals 0.2143. What can you say about the probability of complications in the entire population of patients who will be treated with this drug?

There are two issues to think about. First, you must think about whether the 14 patients are representative of the entire population of patients who will receive the drug. Perhaps these patients were selected in such a way as to make them more (or less) likely than other patients to develop the side effect. Statistical calculations can't help you answer that question, and we'll assume that the sample adequately represents the population. The second issue is random sampling, sometimes referred to as *margin of error.* Just by chance, your sample of 14 patients may have had an especially high or an especially low rate of side effects. The overall proportion of side effects in the population is unlikely to equal exactly 0.2143.

Here is a second example. You polled 100 randomly selected voters just before an election, and only 33 said they would vote for your candidate. What can you say about the proportion of *all* voters who will vote for your candidate? Again, there are two issues to deal with. First, you need to think about whether your sample is really representative of the population of voters, and whether people tell the pollsters the truth about how they will vote. Statistical calculations cannot help you grapple with those issues. We'll assume that the sample is perfectly representative of the population of voters and that every person will vote as they said they would on the poll. Second, you need to think about sampling error. Just by chance, your sample may contain a smaller or larger fraction of people voting for your candidate than does the overall population.

Since we only know the proportion in one sample, there is no way to be sure about the proportion in the population. The best we can do is calculate a range of values that bracket the true population proportion. How wide does this range of values have to be? In the overall population, the fraction of patients with side effects could be as low as 0.000001% (or lower) or as high as 99.99999% (or higher). Those values are exceedingly unlikely but not absolutely impossible. If you want to be 100% sure that your range includes the true population value, the range has to include these possibilities. Such a wide range is not helpful. To create a narrower and more useful range, you must accept the possibility that the interval will not include the true population value.

Scientists usually accept a 5% chance that the range will not include the true population value. The range or interval is called the *95% confidence interval,* abbreviated *95% CI.* You can be 95% sure that the 95% CI includes the true population value. It makes sense that the margin of error depends on the sample size, so that the confidence interval is wider in the first example (14 subjects) than in the second (100 subjects). Before continuing, you should think about these two examples and write down your intuitive estimate of the 95% CIs. Do it now, before reading the answer in the next paragraph.

Later in this chapter you'll learn how to calculate the confidence interval. But it is easier to use an appropriate computer program to calculate the 95% CIs instantly. All examples in this book were calculated with the simple program GraphPad InStat (see Appendix 2), but many other programs can perform these calculations. Here are the results. For the first example, the 95% CI extends from 0.05 to 0.51. For the second example, the 95% CI extends from 0.24 to 0.42. How good were your guesses? Many people tend to imagine that the interval is narrower than it actually is.

What does this mean? Assuming that our samples were randomly chosen from the entire populations, we can be 95% sure that the range of values includes the true population proportion. Note that there is no uncertainty about what we observed in the sample. We are absolutely sure that 21.4% of our subjects suffered from the side effect and that 33.0% of the people polled said they would vote for our candidate. Calculation of a confidence interval cannot overcome any mistakes that were made in tabulating those numbers. What we don't know is the proportion in the entire population. However, we can be 95% sure that it lies within the calculated interval.

The term *confidence interval,* abbreviated CI, refers to the range of values. The correct syntax is to express the CI as 5% to 51%, as 0.05 to 0.51, or as [0.05,0.51]. It is considered to be bad form to express the CI as 5%–51% or as 28% ± 23%. The two ends of the CI are called the *confidence limits.*

WHAT EXACTLY DOES IT MEAN TO SAY THAT YOU ARE "95% SURE"?

When you only have measured one sample, you don't know the value of the population proportion. It either lies within the 95% CI you calculated or it doesn't. There is no way for you to know. If you were to calculate a 95% CI from many samples, the population proportion will be included in the CI in 95% of the samples, but will be outside of the CI the other 5% of the time. More precisely, you can be 95% certain that the 95% CI calculated from your sample includes the population proportion.

Figure 2.1 illustrates the meaning of CIs. Here we assume that the proportion of voters in the overall population who will vote for your candidate equals 0.28 (shown as the horizontal dotted line). We created 50 samples, each with 100 subjects and calculated the 95% CI for each. Each 95% CI is shown as a vertical line extending from the lower confidence limit to the upper confidence limit. The value of the observed proportion in each sample is shown as a small hatch mark in the middle of each CI. The first line (on the left) corresponds to our example. The other 49 lines represent results that could have been obtained by random sampling from the same population.

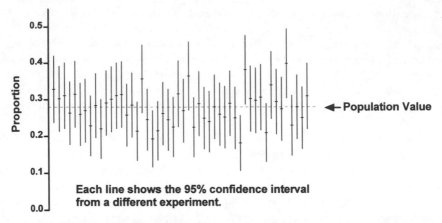

Figure 2.1. The meaning of a confidence interval. In this example, we know that the true proportion of "success" in the population equals 0.28. This is shown as a horizontal dotted line. Using a computer program that can generate random numbers, we randomly selected 50 samples from this population. Each vertical line shows the 95% CI for the proportion "success" calculated from that one sample. With most of the samples, the observed proportion is close to the true proportion, but in some samples there is a big discrepancy. In four of the samples (for example, the fifth from the right), the 95% CI does not include the true value. If you were to collect data from many samples, you'd expect to see this 5% of the time.

In most of the samples, the population value is near the middle of the 95% CI. In some of the samples, the population value is near one of the ends of the 95% CI. And in four of the samples, the population value lies outside the 95% CI. In the long run, 1 out of 20 95% CIs will not include the population value.

Figure 2.1 is useful for understanding CIs, but you cannot create such a figure when you analyze data. When you analyze data, you don't know the actual population value, as you only have results from a single sample. There is no way you can know whether the 95% CI you calculate includes the population value. All you know is that, in the long run, 95% of such intervals will contain the population value and 5% will not. Of course, every CI you calculate will contain the sample proportion you obtained. What you can't know for sure is whether the interval also contains the population proportion.

Note that the CI is not always symmetrical around the sample proportion. In the first example, it extends further to the right than the left. With larger sample sizes, the 95% CI becomes narrower and more symmetrical.

ASSUMPTIONS

The interpretation of the CI depends on the following assumptions:

• Random (or representative) sample
• Independent observations
• Accurate assessment
• Assessing an event you really care about

Random (or Representative) Sample

The 95% CI is based on the assumption that your sample was randomly selected from the population. In many cases, this assumption is not true. You can still interpret the CI as long as you assume that your sample is representative of the population.

This assumption would be violated if the 14 patients included in the sample were the first 14 patients to be given the drug. If so, they are likely to be sicker than future patients and perhaps more likely to get a side effect. The assumption would be violated in the election example if the sample was not randomly selected from the population of voters. In fact, this mistake was made in the Roosevelt-Landon U.S. Presidential Election. To find subjects, the pollsters relied heavily on phone books and automobile registration lists. In 1936, Republicans were far more likely than Democrats to own a phone or a car, and therefore the poll selected too many Republicans. The poll predicted that Landon would win by a large margin, but that didn't happen.

The reputable polling organizations no longer make this type of mistake and go to a lot of trouble to ensure that their samples are representative of the entire population. However, many so-called polls are performed in which television viewers are invited to call in their opinion. In the United States, this usually is a 900 phone number for which the caller pays for the privilege of being polled! Clearly, the self-selected "sample" tabulated by such "polls" is not representative of any population, so the data mean nothing. For example, in June 1994 the football star O.J. Simpson was arrested for allegedly murdering his ex-wife, and the events surrounding the arrest were given a tremendous amount of television coverage. One news show performed a telephone "poll" of its viewers asking whether the press was giving too much coverage. Clearly the results are meaningless, as people who really thought that there was too much coverage probably weren't watching the show.

Independent Observations

The 95% CI is only valid when all subjects are sampled from the same population and each has been selected independently of the others. Selecting one member of the population should not change the chance of selecting anyone else. This assumption would be violated if some patients were given the drug twice and included in the sample more than once (maybe we only had 12 patients but two were double counted). The assumption would also be violated if several of the patients came from one family. The propensity for some drug reactions is familial, so observations from two patients from one family are not independent observations. In the election sample, the assumption would be violated if the pollsters polled both husband and wife in each family, or if some voters were polled more than once or if half the subjects were sampled from one city and half from another.

Accurate Assessment

The 95% CI is only valid when the number of subjects in each category is tabulated correctly. This assumption would be violated in our first example if one of the patients actually had taken a different drug, or if one of the "drug side effects" was actually caused by something else. The assumption would be violated in the election example if the pollster recorded some of the opinions incorrectly.

Assessing an Event You Really Care About

The 95% CI allows you to extrapolate from the sample to the population for the event that you tabulated. But sometimes you really care about a different event. In our drug reaction example, our interpretation of the results must depend on the severity of the drug reactions. If we included all possible reactions, no matter how mild, the results won't tell us what we really care about: the proportion of patients who develop severe (life-threatening) drug reactions.

In the voting example, we assessed our sample's response on a poll on a particular date, so the 95% CI gives us the margin of error for how the population would respond on that poll on that date. We wish to extrapolate to election results in the future but can do so only by making an additional assumption—that people will vote as they said they would. This assumption was violated in a classic mistake in the polling prior to the Dewey versus Truman Presidential Election in the United States in 1948. Polls of many thousand voters showed that Dewey would win by a large margin. Because the CI was so narrow, the pollsters were very confident. Newspapers were so sure of the results that they prematurely printed the erroneous headline ''Dewey Beats Truman.'' In fact, Truman won. Why was the poll wrong? The polls were performed in September and early October, and the election was held in November. Many voters changed their mind in the interim period. The 95% CI correctly presented the margin of error in September but was inappropriately used to predict voting results 2 months later.

OBTAINING THE CONFIDENCE INTERVAL OF A PROPORTION FROM A TABLE

The width of the CI depends on both the value of the proportion and the size of the sample. If you make the sample larger, the CI becomes narrower. If the sample proportion is close to 50%, the CI is wider than if the proportion is far from 50%.

The CI is determined from the binomial distribution using calculations described later in the chapter. The answers have been tabulated in Table 2.1 shown here, which is abridged from Table A5.1 in the Appendix. Find the column corresponding to the numerator and the row corresponding to the denominator, and read the 95% CI.

Consider another example. You have performed a procedure 15 times with a single adverse incident. Without any other knowledge of the procedure, what is the

Table 2.1. Confidence Interval of a Proportion

Denominator	Numerator				
	0	1	2	3	4
10	0.00 to 0.31	<0.01 to 0.45	0.03 to 0.56	0.07 to 0.65	0.12 to 0.74
11	0.00 to 0.28	<0.01 to 0.41	0.02 to 0.52	0.06 to 0.61	0.11 to 0.69
12	0.00 to 0.26	<0.01 to 0.38	0.02 to 0.48	0.05 to 0.57	0.10 to 0.65
13	0.00 to 0.25	<0.01 to 0.36	0.02 to 0.45	0.05 to 0.54	0.09 to 0.61
14	0.00 to 0.23	<0.01 to 0.34	0.02 to 0.43	0.05 to 0.51	0.08 to 0.58
15	0.00 to 0.22	<0.01 to 0.32	0.02 to 0.40	0.04 to 0.48	0.08 to 0.55

95% CI for the average rate of adverse incidents? The answer is the 95% CI of the observed proportion 1/15, which extends from <0.01 to 0.32. With 95% confidence, the true proportion of complications may be less than 1% or as high as 32%. Most people are surprised by how wide the CIs are.

THE SPECIAL CASES OF 0 AND 100 PERCENT*

The 95% CI allows for a 2.5% chance that the population proportion is higher than the upper confidence limit and a 2.5% chance that the population proportion is lower than the lower confidence limit. If you observed 0 successes out of N trials, then you know that there is no possibility that the population proportion is less than the observed proportion. You only need an upper confidence limit, as the lower limit must be exactly 0. A similar problem occurs when you observed N successes out of N trials. You only need a lower confidence limit, as the upper limit must be 100%. Because the uncertainty only goes in one direction, the ''95%'' confidence interval really gives you 97.5% confidence.

EXAMPLE

In order to better counsel the parents of premature babies, M.C. Allen et al. investigated the survival of premature infants.† They retrospectively studied all premature babies born at 22 to 25 weeks gestation at the Johns Hopkins Hospital during a 3-year period. The investigators separately tabulated deaths for infants by their gestational age. Of 29 infants born at 22 weeks gestation, none survived 6 months. Of 39 infants born at 25 weeks gestation, 31 survived for at least 6 months.

    The investigators presented these data without CI, but you can calculate them. It only makes sense to calculate a CI when the sample is representative of a larger population about which you wish to make inferences. It is reasonable to think that these data from several years at one hospital are representative of data from other years at other hospitals, at least at big-city university hospitals in the United States. If you aren't willing to make that assumption, you shouldn't calculate a CI. But the data wouldn't be worth collecting if the investigators didn't think that the results would be similar in other hospitals in later years.

    For the infants born at 25 weeks gestation, we want to determine the 95% CI of 31/39. These values are not on Table A5.1, so you'll need to calculate the CI by computer or by hand. If you use the InStat program, you'll find that the 95% CI ranges from 63% to 91%. (If you use Equation 2.1, which follows, you'll calculate an approximate interval: 67% to 92%). This means that if the true proportion of surviving infants was any less than 63%, there is less than a 2.5% chance of observing such a large proportion just by chance. It also means that if the true proportion were any greater than 91%, the chance observing such a small proportion just by chance is less

---

*This section is more advanced than the rest. You may skip it without loss of continuity.

†MC Allen, PK Donohue, AE Dusman. The limit of viability—Neonatal outcome of infants born at 22 to 25 weeks gestation. *N Engl J Med* 329:1597–1601, 1993.

than 2.5%. That leaves us with a 95% chance (100% − 2.5% − 2.5%) that the true proportion is between 63% and 91%.

For the infants born at 22 weeks gestation, we want to determine the CI of 0/29. You shouldn't use Equation 2.2, because the numerator is too small. Instead use a computer program, or Table A5.1 in the Appendix. The 95% CI extends from 0% to 11.9%. We can be 95% sure that the overall proportion of surviving infants is somewhere in this range. (Since we observed 0%, the CI really only goes in one direction, so we really can be 97.5% sure rather than 95% sure.) Even though no babies born at 22 weeks gestational age survived in our sample, our CI includes the possibility that the overall survival rate is as high as 11.9%. This means that if the overall survival rate in the population was any value greater than 11.9%, there would be less than a 2.5% chance of observing 0 survivors in a sample of 29.

These CIs only account for sampling variability. When you try to extrapolate these results to results you expect to see in your hospital, you also need to account for the different populations served by different hospitals and the different methods used to care for premature infants. The true CIs, which are impossible to calculate, are almost certainly wider than the ones you calculate.

CALCULATING THE CONFIDENCE INTERVAL OF A PROPORTION*

The calculations used to determine the exact CIs shown in the table are quite complex and should be left to computer programs. But it is easy to calculate an approximate CI as long as the numbers are not too small.

Equation 2.1 is a reasonable approximation for calculating the 95% CI of a proportion p assessed in a sample with N subjects. The confidence limits this equation calculates are not completely accurate, but it is a reasonable approximation if at least five subjects had each outcome (in other words, the numerator of the proportion is 5 or greater and the denominator is at least 5 greater than the numerator).

Approximate 95% CI of proportion:

$$\left(p - 1.96 \sqrt{\frac{p(1 - p)}{N}}\right) \text{ to } \left(p + 1.96 \sqrt{\frac{p(1 - p)}{N}}\right) \tag{2.1}$$

Beware of the variable p. The p used here is not the same as a P value (which we will discuss extensively in later chapters). The use of the letter p for both purposes is potentially confusing. This book uses an upper case P for P values and a lower case p for proportions, but not all books follow this convention.

The approximation can be used with the election example given at the beginning of the chapter. The sample proportion is 33/100 or 33.0%. The number of subjects

*This section contains the equations you need to calculate statistics yourself. You may skip it without loss of continuity.

with each outcome is greater than 5 (33 said they would vote for one candidate, and 67 said they would vote for the other), so the approximation can be used. The 95% CI ranges from 0.24 to 0.42. Because the sample is large, these values match the values determined by the exact methods to two decimal places. With smaller samples, this approximate method is less accurate.

You may calculate CIs for any degree of confidence. By convention, 95% CIs are presented most commonly. If you want to be more confident that your interval contains the population value, you must make the interval wider. If you are willing to be less confident, then the interval can be narrower. To generate a 90% CI, substitute the number 1.65 for 1.96 in Equation 2.1. To generate a 99% CI, substitute 2.58. You'll learn where these numbers come from in Chapter 4.

THE BINOMIAL EQUATION*

Assume that in the overall population, the proportion of "successes" is p. In a sample of N subjects, what is the chance that observing exactly R successes? The answer is calculated by Equation 2.2:

$$\text{Probability of R successes in N trials} = \left(\frac{N!}{R!(N-R)!}\right) p^R (1-p)^{N-R}. \quad (2.2)$$

The exclamation point denotes factorial. For example, $3! = 3 \times 2 \times 1 = 6$. The term on the right $[p^R(1-p)^{N-R}]$ is the probability of obtaining a particular sequence of "successes" and "failures." That term is very small. The term on the left takes into account that there are many different sequences of successes and failures that lead to the same proportion success.

Equation 2.2 calculates the probability of observing exactly R successes in N trials. Most likely, you really want to know the chance of observing *at least* S successes in N trials. To do this, reevaluate Equation 2.2 for $R = S, R = S + 1, R = S + 2...$ up to $R = N$ and sum all the resulting probabilities.

The word *success* is used quite generally to denote one of the possible outcomes. You could just as well use the word *failure* or *outcome A*. Equation 2.2 uses p and (1 − p) symmetrically, so it doesn't matter which outcome you label *success* and which outcome you label *failure*.

HOW THE CONFIDENCE INTERVALS ARE DERIVED

You can calculate and interpret CIs without knowing how they are derived. But it is nice to have a feel for what's really going on. I'll try to give the flavor of the logic using our second example. Recall that we observed that 33 out of 100 subjects polled said they would vote a certain way. The 95% CI is 0.24 to 0.42.

---

*This section contains the equations you need to calculate statistics yourself. You may skip it without loss of continuity.

The binomial equation works from population to sample. It lets us answer this kind of question. Given a hypothetical proportion in the population, what is the chance of observing some particular proportion (or higher) in a sample of defined size? The equations are a bit messy, but it is not surprising that probability theory can answer those kinds of questions.

To find the upper 95% confidence limit (U), we need to pose this question: If the population proportion equals U, what is the probability that a sample proportion (N = 100) will equal 0.33 or less? We set the answer to 2.5% and then solve for U. To find the lower 95% confidence limit (L), we pose the opposite question. If the population proportion equals L, what is the probability that a sample proportion (N = 100) will equal 0.33 or more? We set the answer to 2.5% and then solve for L. Solving those equations for U and L is not easy. Fortunately, the answers have been tabulated: L = 0.24 and U = 0.42. We've set up the problem to allow for a 2.5% chance of being wrong in each direction. Subtracting these from 100% leaves you with 95% confidence that the population proportion is between L and U, between 0.24 and 0.42.

So the trick is to use mathematical theory that makes predictions about the samples from hypothetical populations, and twist it around so that it can make inferences about the population from the sample. It's tricky to understand and it's amazing that it works. Fortunately you can calculate statistical analyses and interpret statistical results without having a solid understanding of the mathematical theory.

SUMMARY

Many types of data can be reduced to either/or categories and can be expressed as a proportion of events or subjects that fall into each category. The distribution of such events is described by the binomial distribution.

The challenge of statistics is to start with an observation in one sample and make generalizations about the overall population. One way to do this is to express the results as a confidence interval. Having observed a proportion in your sample, you can calculate a range of values that you can be 95% sure contains the true population proportion. This is the 95% confidence interval. You can calculate the interval for any degree of confidence you want, but 95% is standard. If you want to be more confident, you must make the interval wider.

OBJECTIVES

1. You must be familiar with the following terms:
   - Binomial distribution
   - Cumulative binomal distribution
   - Confidence interval
   - Confidence limit
   - Sample
   - Population
   - Random sample
   - Independence

2. When you see a proportion reported in a scientific paper (or elsewhere), you should be able to

- Define the population.
- Calculate (or determine from a table) the 95% CI of the proportion.
- Interpret the 95% CI in the context of the study.
- State the assumptions that must be true for the interpretation to be valid.

PROBLEMS

1. Of the first 100 patients to undergo a new operation, 6 die. Can you calculate the 95% CI for the probability of dying with this procedure? If so, calculate the interval. If not, what information do you need? What assumptions must you make?
2. A new drug is tested in 100 patients and lowers blood pressure by an average of 6%. Can you calculate the 95% CI for the fractional lowering of blood pressure by this drug? If so, calculate the interval. If not, what information do you need? What is the CI for the fractional lowering of blood pressure? What assumptions must you make?
3. You use a hemocytometer to determine the viability of cells stained with Trypan Blue. You count 94 unstained cells (viable) and 6 stained cells (indicating that they are not viable). Can you calculate the 95% CI for the fraction of stained (dead) cells? If so, calculate the interval. If not, what information do you need? What assumptions must you make?
4. In 1989, 20 out of 125 second-year medical students in San Diego failed to pass the biostatistics course (until they came back for an oral exam). Can you calculate the 95% CI for the probability of passing the course? If so, calculate the interval. If not, what information do you need? What assumptions must you make?
5. Ross Perot won 19% of the vote in the 1992 Presidential Election in the United States. Can you calculate the 95% CI for the fraction of voters who voted for him? If so, calculate the interval. If not, what information do you need? What assumptions must you make?
6. In your city (population = 1 million) last year a rare disease had an incidence of 25/10,000 population. Can you calculate the 95% CI for the incidence rate? If so, calculate the interval. If not, what information do you need? What assumptions must you make?
7. Is it possible for the lower end of a CI for a proportion to be negative? To be zero?
8. Is it possible to calculate a 100% CI?

3

The Standard Deviation

The previous chapter dealt with data that can be expressed as a fraction or proportion. However, the results of many experiments are expressed as measurements, for example, blood pressure, enzyme activity, IQ score, blood hemoglobin, or oxygen saturation. Working with these kinds of variables is more difficult than proportions, because you must deal with variability within samples as well as differences between groups.

SOURCE OF VARIABILITY

When you measure a variable in a number of subjects, the results will rarely be the same in all subjects. Scatter comes from three sources:

- *Imprecision or experimental error.* Many statistics books (especially those designed for engineers) implicitly assume that most variability is due to imprecision. In medical studies, imprecision is often a small source of variability.
- *Biological variability.* People (and animals, even cells) are different from one another, and these differences are important! Moreover, people (and animals) vary over time due to circadian variations, aging, and alterations in activity, mood, and diet. In biological and clinical studies, much or most of the scatter is often due to biological variation.
- *Blunders.* Mistakes and glitches (lousy pipetting, mislabeled tubes, transposed digits, voltage spikes, etc.) also contribute to variability.

Many statisticians use the word *error* when referring to all of these sources of variability. Note that this use of the word *error* is quite different from the everyday use of the word to mean *mistake*. I prefer the term *scatter* to *error*.

Another term you should know is *bias*. Biased measurements are systematically wrong. Bias can be caused by any factor that consistently alters the results: the proverbial thumb on the scale, miscalibrated pipettes, bugs in computer programs, the placebo effect, and so on. As used in statistics, the word *bias* refers to anything that leads to systematic errors, not only the preconceived notions of the experimenter. Bias will not usually contribute to scatter, as it will increase (or decrease) all values. Statistical calculations cannot detect or correct bias (unless you do additional experiments), and statistical analysis of biased data will rarely be helpful. As computer programmers say: Garbage in, garbage out.

DISPLAYING VARIABILITY WITH HISTOGRAMS

To discuss the distribution of data, we will work with an artificial example. One hundred medical students worked as pairs, and each student measured the systolic blood pressure of his or her partner. It is difficult to make much sense from a list of 100 numbers, so instead we display the results on a histogram as in Figure 3.1. A histogram is a bar graph. Systolic blood pressure is shown on the horizontal axis, measured in mmHg (same as torr). The base of each bar spans a range of blood pressure values, and the height of each bar is the number of individuals who have blood pressures that fall within that range.

To create a histogram you must decide how wide to make each bar. To make Figure 3.1, we set the width of each bar to 5 mmHg. The tallest bar in the graph shows that 15 subjects had blood pressure values between 117.5 and 122.5 mmHg. If the bars are too wide, the histogram will show too little detail, as shown in the left panel of Figure 3.2. If the bars are too narrow, the histogram will show too much detail and it is hard to interpret (right half of Figure 3.2). Some histograms plot relative frequencies (proportions or percentages) on the Y axis rather than the actual number of observations.

Before continuing with our analysis of the numbers, we should think a bit about what the numbers really tell us (or what they would tell us if they weren't made up). Blood pressure is affected by many variables. In serious studies of blood pressure, investigators go to a great deal of trouble to make sure that the measurements are as consistent and unbiased as possible.

This study of students has a number of flaws. Most important,

- The recorded values may differ substantially from arterial pressure. Every measurement was made by a different inexperienced person. Measuring blood pressure with

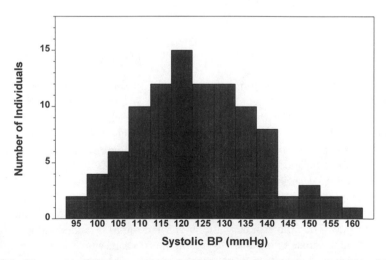

Figure 3.1. Histogram of blood pressures of 100 students. Each bar spans a width of 5 mmHg on the horizontal axis. The height of each bar represents the number of students whose blood pressure lies within that range.

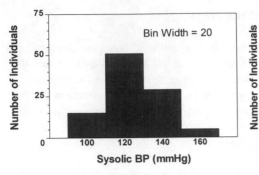
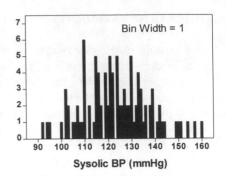

Figure 3.2. More histograms of blood pressure of 100 students. In the left panel, each bar spans a width of 20 mmHg. Because there are so few bars, the graph is not very informative. In the right panel, each bar has a width of 1 mmHg. Because there are so many bars, the graph contains too much detail, and it is hard to see the distribution of values. Histograms are generally most useful when they contain approximately 10 to 20 bars.

a cuff is somewhat subjective, as it requires noting the position of a bouncing column of mercury at the instant that a faint sound appears (systolic) or disappears (diastolic). It is also important to place the cuff correctly (at the level of the heart). Experience is required.

- The values cannot be compared to values for resting supine blood pressure measured in other studies. Having a pressure measured by a friend for the first time is stressful and may elevate the pressure.

I won't belabor the flaws of this example, except to repeat: Garbage in, garbage out! Statistical analysis of flawed data is unlikely to be useful. Designing good studies is difficult—much harder than statistical analysis.

THE MEAN AND MEDIAN

One way to describe the distribution of the data is to calculate the mean or average. You probably already know how to do that: Add up all the numbers and divide by the number of observations. For these 100 blood pressures, the mean is 123.4 mmHg. If you think visually, you should think of the mean as the "center of gravity" of the histogram. If you printed the bars of the histogram of Figure 3.1 on a sheet of wood or plastic, it would balance on a fulcrum placed at X = 123.4, the mean.

Another way to describe the central tendency of the data is to calculate the median. To do this you must rank the values in order and find the middle one. Since there are an even number of data points in this sample, the median lies between the 50th and 51st ranked value. The 50th ranked value is 122 and the 51st ranked value is 124. We take the average of those two values to obtain the median, which is 123 mmHg. You are probably already familiar with the term *percentile*. The median is the value at the 50th percentile because 50% of the values fall below the median.

In this example, the mean and median are almost the same. This is not always true. What would happen if the largest value (160 mmHg) were mistakenly entered into the computer as 16,000 mmHg? The median would be unchanged at 123 mmHG. Changing that one value does not change the value of the 50th and 51st ranked values, so does not change the median. The mean, however, increases to 281 mmHg. In this case, there would be one value greater than the mean and 99 values below the mean. There are always just as many values below the median as above it. The number of data points above and below the mean will not be equal if the histogram is not symmetrical.

QUANTIFYING VARIABILITY WITH PERCENTILES

In addition to quantifying the middle of the distribution with the mean or median, you also want to quantify the scatter of the distribution. One approach is to report the lowest and highest values. For this example the data range from 92 to 160 mmHg. Another approach is to report the 25th and 75th percentiles. For this example, the 25th percentile is 115 mmHg. This means that 25% of the data points fall below 115. The 75th percentile is 133 mmHg. This means that 75% of the values lie below 133 mmHg. The difference between the 75th and 25th percentiles is called the *interquartile range,* which equals 18 mmHg.

The interquartile range can be depicted graphically as a box and whisker plot, as shown in Figure 3.3. The box extends from the 25th to 75th percentiles with a horizontal line at the median (50th percentile). The whiskers indicate the smallest and largest values. (Whiskers can be defined in other ways as well.)

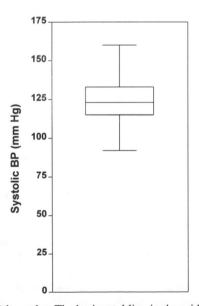

Figure 3.3. A box-and-whiskers plot. The horizontal line in the middle shows the median (50th percentile) of the sample. The top and bottom of the box show the 75th and 25th percentiles, respectively. In this graph, the top and bottom of the whiskers show the maximum and minimum values (other box-and-whiskers graphs define the whiskers in different ways). From this graph you can instantly see that the median is about 125 mmHg, and half the sample have systolic blood pressures between about 115 and 135 mmHg.

QUANTIFYING VARIABILITY WITH THE VARIANCE AND STANDARD DEVIATION

One problem with the interquartile range is that it tells you nothing about the values that lie below the 25th percentile or above the 75th percentile. We'd like a measure of scatter that somehow includes all the values. One way to quantify the scatter of all the values is to calculate the average of the deviations of the values from the mean. But it turns out that this is not helpful—the positive deviations balance out the negative deviations so the average deviation is always 0. Another thought would be to take the average of the absolute values of the deviations. This seems reasonable, but it turns out to be pretty much a statistical dead end that does not facilitate other inferences. Instead statisticians quantify scatter as the average of the square of the deviations of each value from the mean. This value is termed the *variance*. The variance of a population (abbreviated σ^2) is defined by Equation 3.1, where μ is the population mean, N is the number of data points, and Y_i is an individual value.

$$\text{Population variance} = \sigma^2 = \frac{\sum_{i=1}^{i=N}(Y_i - \mu)^2}{N}. \tag{3.1}$$

You may wonder why the deviations are squared. Why not just sum the absolute values of the deviations? This is difficult to explain intuitively. It just falls out of the math and makes statistical equations work. Basically, the idea is that one large deviation contributes more scatter to the data than do two deviations half that size. More practically, defining the variance in this way facilitates calculations of confidence intervals (CIs) and P values, as you will see in later chapters.

The variance is expressed in the units of the data squared. Thinking about the square of units is difficult, and so the variance is not an intuitive way to quantify scatter. The square root of the variance is called the *standard deviation*, abbreviated SD or σ (Equation 3.2):

$$\text{Population SD} = \sigma = \sqrt{\frac{\sum_{i=1}^{i=N}(Y_i - \mu)^2}{N}} \tag{3.2}$$

Caution! Don't use this equation until you read the next section.

The SD is expressed in the same units as the data. This makes it much easier to think about than the variance. Even so, you may not find it easy to think about the square root of the average of the squares of the deviations! You'll find it easier to think about SDs after you read the next chapter. We will discuss the interpretation of the SD later in this chapter and in the next chapter.

Returning to our blood pressure example, we encounter a problem when we try to use Equations 3.1 and 3.2 to figure out the variance and SD. The problem is that those equations apply only to the entire population, but we know only about a sample of 100 subjects. To use Equations 3.1 or 3.2, you need to know the value of μ, the population mean, but you don't know it. Read on to see how the equations need to be adjusted to deal with samples.

N OR N − 1? THE SAMPLE SD VERSUS THE POPULATION SD

Equations 3.1 and 3.2 assume that you have made measurements on the entire population. As we have already discussed, this is rarely the case. The whole point of statistical
calculations is to make inferences about the entire population from measurements of
a sample. This introduces an additional complexity to the calculation of the variance
and SD. To calculate the SD using Equation 3.2, you need to calculate the deviation
of each value from the population mean. But you don't know the population mean.
All you know is the sample mean. Except for the rare cases where the sample mean
happens to equal the population mean, values are always closer (on average) to their
sample mean than to the overall population mean. The sum of the squares of the
deviations from the sample mean is therefore smaller than the sum of squares of the
deviations from the population mean, and Equation 3.2 gives too small a value for the
SD. This problem is eliminated by reducing the denominator to N−1, rather than N.
Calculate the variance and SD from a sample of data using Equation 3.3:

$$\text{Sample variance} = s^2 = \frac{\sum_{i=1}^{N}(Y_i - m)^2}{N-1}. \tag{3.3}$$

$$\text{Sample SD} = s = \sqrt{\frac{\sum_{i=1}^{N}(Y_i - m)^2}{N-1}}.$$

Note that we switched from Greek (μ, σ) to Roman (m, s) letters when switching
from discussions of population mean and SD to discussions of sample mean and SD.
This book hides a lot of the mathematical detail and sometimes glosses over the
distinction. If you refer to more mathematical books, pay attention to the difference
between Greek and Roman letters.

> If the difference between N and N−1 ever matters to you, then you are probably up to
> no good anyway—e.g. trying to substantiate a questionable hypothesis with marginal data.
>
> W.H. Press et al., *Numerical Recipes*

Here is another way to understand why the denominator is N−1 rather than N.
When we calculate the sample mean m, we take the sum of all Y values and divide
by the number of values N. Why divide by N? You learned to calculate a mean so
long ago that you probably never thought about it. The mean is technically defined as
the sum divided by degrees of freedom. The sample mean has N degrees of freedom
because each of the N observations is free to assume any value. Knowing some of the
values does not tell you anything about the remaining values. The sample variance is
the mean of the square of the deviations of the values from the sample mean. This
mean has only N−1 degrees of freedom. Why? It is because you must calculate the
sample mean m before you can calculate the sample variance and SD. Once you know
the sample mean and N−1 values, you can calculate the value of the remaining (Nth)
value with certainty. The Nth value is absolutely determined from the sample mean
and the other N−1 values. Only N−1 of the values are free to assume any value.

Therefore, we calculate the average of the squared deviations by dividing by $N-1$, and say that the sample variance has $N-1$ *degrees of freedom,* abbreviated df.

We'll discuss degrees of freedom again in later chapters. Many people find the concept of degrees of freedom to be quite confusing. Fortunately, being confused about degrees of freedom is not a big handicap! You can calculate statistical tests and interpret the results with only a vague understanding of degrees of freedom.

Returning to the blood pressure example, the sample SD is 14.0 mmHg. We'll discuss how to interpret this value soon. The sample variance is 196.8 mmHg^2. It is rare to see variances reported in the biomedical literature* but common to see SDs.

The term *sample SD* is a bit misleading. The sample SD is the best possible estimate of the SD of the entire population, as determined from one particular sample. In clinical and experimental science, one should routinely calculate the sample SD.

CALCULATING THE SAMPLE STANDARD DEVIATION WITH A CALCULATOR

Many calculators can calculate the SD. You have to first press a button to place the calculator into statistics or SD mode. Then enter each value and press the button labeled "enter" or "M+." After you have entered all values, press the appropriate button to calculate the SD. As we have seen, there are two ways to calculate the SD. Some calculators can compute either the sample or the population SD, depending on which button you press. The button for a sample SD is often labeled σ_{N-1}. Other calculators (especially those designed for use in business rather than science), compute only the population SD (denominator = N) rather than the sample SD (denominator = $N-1$).

To test your calculator, here is a simple example. Calculate the sample SD of these values: 120, 80, 90, 110, and 95. The mean is 99.0, and the sample SD is 16.0. If your calculator reports that the SD equals 14.3, then it is attempting to calculate the population SD. If you haven't collected data for the entire population, this is an invalid calculation that underestimates the SD. If your calculator computes the SD incorrectly, you may correct it with Equation 3.4:

$$\text{Sample SD} = \text{Invalid "population" SD} \cdot \sqrt{\frac{N}{N-1}} \qquad (3.4)$$

INTERPRETING THE STANDARD DEVIATION

We will discuss the interpretation of the SD in the next chapter. For now, you can use the following rule of thumb: Somewhat more than half of the observations in a population usually lie within 1 SD on either side of the mean. With our blood pressure example, we can say that most likely somewhat more than half of the observations will be within 14 mmHg of the mean of 123 mmHg. In other words, most of the

---

*However, it is common to see results analyzed with a method known as analysis of variance (ANOVA), which is explained in Chapter 30.

observations probably lie between 109 and 137 mmHg. This definition is unsatisfying because the words *somewhat* and *usually* are so vague. We'll interpret the SD more rigorously in the next chapter.

COEFFICIENT OF VARIATION (CV)*

The SD can be normalized to the coefficient of variation (CV) using Equation 3.5:

$$CV\% = 100 \cdot \frac{SD}{mean}. \tag{3.5}$$

Because the SD and mean are expressed in the same units, the CV is a unitless fraction. Often the CV is expressed as a percentage. If you change the units in which you express the data, you do not change the CV. In our example of 100 blood pressure measurements, the mean blood pressure is 123.4 mmHg and the SD is 14.0 mmHg. The CV is 11.4%. The CV is useful for comparing scatter of variables measured in different units.

SUMMARY

Many kinds of data are expressed as measurements. You can display the scatter of measurements in a sample on a histogram. The center of the distribution can be described by the mean or median. The spread or scatter of the data can be described by the range, the interquartile range, the variance, the SD, or the coefficient of variation.

OBJECTIVES

1. You must be familiar with the following terms:
   - Error
   - Bias
   - Histogram
   - Mean
   - Median
   - Standard deviation
   - Variance
   - Coefficient of variation
2 Given a list of measurements, you should be able to
   - Define the population the values came from.
   - Determine whether it is reasonable to think that the data are representative of that population.
   - Draw a histogram of the distribution of values, and a box-and-whiskers plot.
   - Calculate the mean, median, standard deviation, and coefficient of variation.

*This section is more advanced than the rest. You may skip it without loss of continuity.

PROBLEMS

1. Estimate the SD of the age of the students in a typical medical school class (just a rough estimate will suffice).
2. Estimate the SD of the number of hours slept each night by adults.
3. The following cholesterol levels were measured in 10 people (mg/dl):
   260, 150, 160, 200, 210, 240, 220, 225, 210, 240
   Calculate the mean, median, variance, standard deviation, and coefficient of variation. Make sure you include units.
4. Add an 11th value (931) to the numbers in Question 3 and recalculate all values.
5. Can a SD be 0? Can it be negative?

4

The Gaussian Distribution

PROBABILITY DISTRIBUTIONS

As you saw in the last chapter, a histogram plots the distribution of values in a sample. In many situations, the histogram is bell shaped. The first two panels in Figure 4.1 show bell-shaped distributions of two samples. The second sample is larger, and the histogram was created with narrower bins. This makes it appear more smooth.

The third panel in the figure plots the distribution of values in the entire population. This is not a histogram because the population is infinitely large. What you see instead is a probability distribution. Although probability distributions are similar to histograms, the Y axes are different. In a histogram, the Y axis shows the number of observations in each bin. In a probability distribution, the Y axis can't show the number of observations in each bin because the population is infinitely large and there aren't any bins. Instead we represent probability as the *area under the curve*. The area under the entire curve represents the entire population, and the proportion of area that falls between two X values is the probability of observing a value in that interval. The Y axis is termed the *probability density,* a term that is difficult term to define precisely. Fortunately, you can understand probability distributions without rigorously defining the scale of the Y axis.

THE GAUSSIAN DISTRIBUTION

The symmetrical bell-shaped distribution is called a *Gaussian distribution.* You expect variables to approximate a Gaussian distribution when the variation is caused by many independent factors. For example, in a laboratory experiment with membrane fragments, variation between experiments might be caused by imprecise weighing of reagents, imprecise pipetting, the variable presence of chunks in the membrane suspension, and other factors. When you combine all these sources of variation, you are likely to see measurements that distribute with an approximately Gaussian distribution. In a clinical study, you expect to see an approximately Gaussian distribution when the variation is due to many independent genetic and environmental factors. When the variation is largely due to one factor (such as variation in a single gene), you expect to see bimodal or skewed distributions rather than Gaussian distributions.

You learned how to calculate the mean and standard deviation (SD) of a sample in the last chapter. Figure 4.2 shows the mean and SD of an ideal Gaussian distribution.

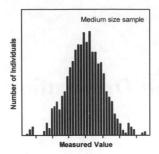 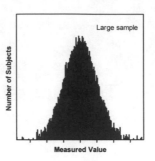 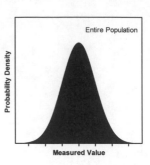

Figure 4.1. Histograms and probability distributions. The left panel shows a histogram for a medium-sized sample. The height of each bar denotes the number of subjects whose value lies within the span of its base. The middle panel shows a histogram for a larger sample. Each bar was made narrower. The right panel shows the distribution for the entire population. Since the population is infinite and there are no bars, the Y axis cannot denote numbers of subjects. Instead, it is called the *probability density*.

The mean, of course, is the center of the Gaussian distribution. The SD is a measure of the spread of the distribution. The area under the curve within 1 SD of the mean is shaded in Figure 4.3. The area under the entire curve represents the entire population. The shaded portion is about two thirds of the entire area. This means that about two thirds of a Gaussian population are within 1 SD of the mean.*

You can also see that the vast majority of the values lie within 2 SDs of the mean. More precisely, 95% of the values in a Gaussian distribution lie within 1.96 SDs of the mean. Areas under a portion of the Gaussian distribution are tabulated as the "z" distribution, where z is the number of standard deviations away from the mean.

For each value of z, the table shows four columns of probabilities. The first column shows the fraction of a Gaussian population that lies within z SDs of the mean. The last column shows the fraction that lies more than z SDs of the mean above the mean or more than z SDs below the mean. Thus the sum of the first column and the last column equals 100% for all values of z. Because the value in the last column includes components on both sides of the distribution, it is termed a *two-tailed probability*. This two-tailed probability (column four) is the sum of each tail, the probability that a value will be more than z SDs above the mean (the second column) and the probability that a value will be more than z SDs below the mean (the third column). The Gaussian distribution is symmetrical so the values in the second and third columns are identical and sum to the values in the fourth column. A longer version of this table is included in the Appendix as Table A5.2.

From Figure 4.3 we estimated that about two thirds of data in a Gaussian distribution lie within 1 SD of the mean. We can get a more exact value from this table. Look in the first column of the row for z = 1.0. The value is 68.27%. Look in the row for z = 2.0. Just over 95% of the population lies within 2 SDs of the mean. More precisely,

---

*I have deliberately avoided defining the scale of the Y axis or the units used to measure the area under the curve. We never care about area per se, but only about the fraction of the total area that lies under a defined portion of the curve. When you calculate this fraction the units cancel, so you can get by without understanding them precisely.

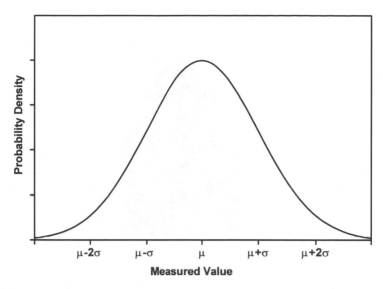

Figure 4.2. An ideal Gaussian distribution. The distribution is centered on the mean μ. The width of the distribution is determined by the standard deviation σ.

95% of the population lies within 1.96 SDs of the mean. You'll see that number 1.96 again, so you should remember what it means.

USING THE GAUSSIAN DISTRIBUTION TO MAKE INFERENCES ABOUT THE POPULATION

In the example of the previous chapter, we analyzed the blood pressure of 100 subjects. The sample mean was 123.4 mmHg and the sample SD was 14.0 mmHg. These values are also our best guesses for the mean and SD of the population.

Table 4.1. The Gaussian Distribution

	Fraction of the Population that is			
z	Within z SDs of the mean	More than z SDs above the mean	More than z SDs below the mean	More than z SDs above or below the mean
0.5	38.29%	30.85%	30.85%	61.71%
1.0	68.27%	15.87%	15.87%	31.73%
1.5	86.64%	6.68%	6.68%	13.36%
2.0	95.45%	2.28%	2.28%	4.55%
2.5	98.76%	0.62%	0.62%	1.24%
3.0	99.73%	0.13%	0.13%	0.27%
3.5	99.95%	0.02%	0.02%	0.05%

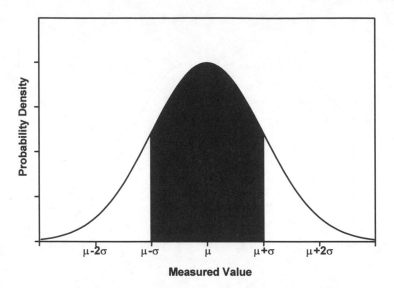

Figure 4.3. Interpreting probability distributions. The area under the entire curve represents the entire population. All values within 1 SD from the mean are shaded. The ratio of the shaded area to the total area is the fraction of the population whose value lies within 1 SD of the mean. You can see that the shaded area is about two thirds of the total area. If you measured it more exactly, you'd find that the shaded area equals 68.27% of the total area.

If you are willing to assume that the distribution of values in the population follows a Gaussian distribution, then you can plot your best guess for that distribution (Figure 4.4) and can make quantitative predictions about the population.

When reading research articles, you'll often see means and SDs without seeing the distribution of the values. When you read "mean = 123, SD = 14," you should have an intuitive sense of what that means if the population is Gaussian. If you have a talented right brain, you might be able to visualize Figure 4.4 in your head. If your left brain dominates, you should perform some very rough calculations in your head (two thirds of the values lie between 109 and 137). Either way, you should be able to interpret SDs in terms of the probable distribution of the data.

You can use the z table to answer questions such as this one: What fraction of the population has a systolic blood pressure greater than 140? Since the mean is 123.4 and the SD = 14.0, we are asking about deviations more than $(140.0 - 123.4)/14 = 1.2$ SDs from the mean. Setting $z = 1.2$, you can consult Table A5.2 in the Appendix and predict that 11.51% of the population has a blood pressure greater than 140.

More generally, you can calculate z for any value Y using Equation 4.1:

$$z = \frac{|Y - \text{mean}|}{\text{SD}}. \tag{4.1}$$

There are two problems with these calculations. One problem is that you don't know the mean and SD of the entire population. If your sample is small, this may introduce substantial error. The sample in this example had 100 subjects, so this is not a big problem. The second problem is that the population may not really be Gaussian.

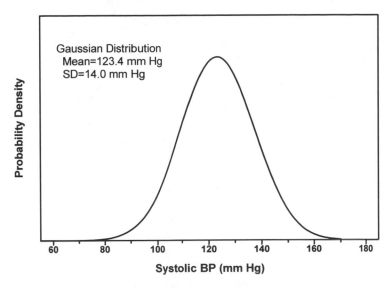

Figure 4.4. Estimating the distribution of values. If you knew only that the mean = 123.4 and the SD = 14 mmHg, what can you say about the distribution of values in the overall population? Before you can make any inferences, you need to make additional assumptions. If you assume that the population follows a Gaussian distribution (at least approximately), then this figure is your best guess.

This is a big problem. Even if the distribution only deviates a bit from a Gaussian distribution, the deviation will be most apparent in the tails of the distribution. But this example asks about the fraction of values in the tail. If the population is not really Gaussian, the calculation will give the wrong result.

THE PREDICTION INTERVAL*

You know that 95% of the values in a Gaussian population lie within 1.96 SDs of the population mean. What can you say about the distribution of values when you only know the mean and SD of one sample?

If your sample is large, then the mean and SD calculated from a sample are likely to be very close to the true population values and you expect 95% of the population to lie within 1.96 SDs of the sample mean. If your sample is small, you'll need to go more than 1.96 SDs in each direction to account for the likely discrepancies between the mean and SD you calculated from the sample and the true mean and SD in the overall population. To include 95% of the population, we need to create a range of values that goes more than 1.96 SDs in each direction of the sample mean. This range is called the *prediction interval*.

To calculate a prediction interval, you need to go K SDs from the mean, where K is obtained from Table 4.2 (the derivation of this table is beyond the scope of this book). For example, if your sample has five subjects (N = 5), then the prediction

*This section is more advanced than the rest. You may skip it without loss of continuity.

Table 4.2. 95% Prediction Interval

N	K	N	K
2	15.56	18	2.17
3	4.97	19	2.16
4	3.56	20	2.14
5	3.04	25	2.10
6	2.78	30	2.08
7	2.62	35	2.06
8	2.51	40	2.05
9	2.43	50	2.03
10	2.37	60	2.02
11	2.33	70	2.01
12	2.29	80	2.00
13	2.26	90	2.00
14	2.24	100	1.99
15	2.22	200	1.98
16	2.20	∞	1.96
17	2.18		

These values were calculated according to page 62 of
G. J. Hahn and W. Q. Meeker, *Statistical Intervals,* John
Wiley & Sons, 1991.

interval extends 3.04 SDs in each direction. With N = 10, the prediction interval extends 2.37 SDs in each direction.

Here is the precise definition of the 95% prediction interval: Each new observation has a 95% chance of being within the interval. On average, you expect that 95% of the entire population lies within the prediction interval, but there is a 50% chance that fewer than 95% of the values lie within the interval and a 50% chance that more than 95% of the values lie within the interval. The prediction interval is only valid when the population is distributed according to a Gaussian distribution.

NORMAL LIMITS AND THE "NORMAL" DISTRIBUTION

From our blood pressure data, what can we say about the "normal range" of blood pressures? Here is a simple, but flawed, approach that is often used:

> I am willing to assume that blood pressure in the population follows a Gaussian distribution. We know that 95% of the population lies within 1.96 SDs of the mean. We'll define the other 5% as abnormal. Using the sample mean and SD from our study, we can define the "normal range" as 123.4 ± (1.96 × 14.0) or 96 to 151 mm Hg.

There are a number of problems with this simple approach:

• It really doesn't make sense to define normal and abnormal just in terms of the distribution of values in the general population. What you really care about is consequences. We know that high blood pressure can cause atherosclerosis, heart disease, and strokes. So the questions we really want to answer are these: In which subjects is the blood pressure high enough to be dangerous? More precisely, we want to know which subjects have blood pressure high enough that the increased risk of heart disease and stroke is high enough to justify the hassle, expense, and risk of treatment. Answering these questions is difficult. You need to follow large groups of people for many years. There is no way to even begin to approach these questions by analyzing only the distribution of blood pressure values collected at one time. You also need to assess blood pressure in a standard manner so that the results are comparable. In this "study" inexperienced students measured each other's pressures. The measurements are likely to be far from the true arterial pressure (it takes experience to measure blood pressure), and the arterial pressure is likely to be far from its resting value (some will find it stressful to have a friend take their blood pressure in public).

- Our definition of normal and abnormal really should depend on other factors such as age and sex. What is abnormal for a 25 year old may be normal for an 80 year old.
- In many cases, we don't really want a crisp boundary between "normal" and "abnormal." It often makes sense to label some values as "borderline."
- You don't need to decide based on one measurement. Before labeling someone as having abnormally high blood pressure, you should accurately measure the pressure several times in a controlled environment.
- Even if the population is approximately Gaussian, it is unlikely to follow a Gaussian distribution exactly. The deviations from Gaussian are likely to be most apparent in the tails of the distribution. There is also no reason to think that the population distribution is symmetrical around the mean.
- Why do you want to define 2.5% of the population to be abnormally high and 2.5% of the population to be abnormally low? What's so special about 2.5%? Why should there be just as many abnormally high values as abnormally low?

The problem is that the word *normal* has at least three meanings:

- Mathematical statisticians use the term as a synonym for a Gaussian distribution. However, there is nothing unusual about variables distributed in some other way.
- Scientists usually use the term to denote values that are commonly observed.
- Clinicians sometimes use *normal* to mean usual, and sometimes use *normal* to mean values that are not associated with the presence of disease.

You should be a bit confused at this point. Determining a "normal range" is a complicated issue. The important point is to realize that it is usually far too simple to define the "normal range" as the mean plus or minus two SDs.

> Everybody believes in the normal approximation, the experimenters because they think it is a mathematical theorem, the mathemeticians because they think it is an experimental fact.
>
> G. Lippman (1845–1921)

SUMMARY

Many variables distribute in a Gaussian distribution. This is expected when many independent facctors account for the variability, with no one factor being most important. Areas under portions of a Gaussian distribution are tabulated as the "z" distribution. The range of values that are 95% certain to contain the next value sampled from the population is known as the *prediction interval*. The term *normal limits* can be defined in many ways. Using the Gaussian distribution to make rules about "normal" values is not always useful.

OBJECTIVES

1. You should be familiar with the following terms:
   - Gaussian distribution
   - Probability distribution
   - Normal
   - Prediction interval
2 Using tables, you should be able to determine the fraction of a Gaussian population that lies more (or less) than z SDs from the mean.
3 You should understand the complexities of "normal limits" and the problem of defining normal limits as being the mean plus or minus 2 SDs.

PROBLEMS

1. The level of an enzyme in blood was measured in 20 patients, and the results were reported as 79.6 ± 7.3 units/ml (mean ± SD). No other information is given about the distribution of the results. Estimate what the probability of distribution might have looked like.
2. The level of an enzyme in blood was measured in 20 patients and the results were reported as 9.6 ± 7.3 units/ml (mean ± SD). No other information is given about the distribution of the results. Estimate what the probability of distribution might have looked like.
3. The values of serum sodium in healthy adults approximates a Gaussian distribution with a mean of 141 mM and a SD of 3 mM. Assuming this distribution, what fraction of adults has sodium levels less than 137 mM? What fraction has sodium levels between 137 and 145 mM?
4. You measure plasma glaucoma levels in five patients and the values are 146, 157, 165, 131, 157 mg/dl. Calculate the mean and SD. What is the 95% prediction interval? What does it mean?
5. The Weschler IQ scale was created so that the mean is 100 and the standard deviation is 15. What fraction of the population has an IQ greater than 135?

5

The Confidence Interval of a Mean

INTERPRETING A CONFIDENCE INTERVAL OF A MEAN

You already learned how to interpret the CI of a proportion in Chapter 2. So interpreting the CI of a mean is easy. The 95% CI is a range of values, and you can be 95% sure that the CI includes the true population mean.

Example 5.1

This example is the same as the example followed in the last two chapters. You've measured the systolic blood pressure (BP) of all 100 students in the class. The sample mean is 123.4 mmHg, and the sample SD is 14.0 mmHg. You consider this class to be representative of other classes (different locations, different years), so you know that the sample mean may not equal the population mean exactly. But with such a large sample size, you expect the sample mean to be close to the population mean. The 95% CI for the population mean quantifies this. You'll learn how to calculate the CI later in the chapter. For now, accept that a computer does the calculations correctly and focus on the interpretation. For this example, the 95% CI ranges from 120.6 mmHg to 126.2 mmHg. You can be 95% sure that this range includes the overall population mean.

Example 5.2

This example continues the study of BPs. But now you randomly select a sample of five students, whose systolic BPs (rounded off to the nearest 5) are 120, 80, 90, 110, and 95 mmHg. The mean of this sample is 99.0 mmHg and the sample SD is 15.97 mmHg. But the sample mean is unlikely to be identical to the population mean. With such a large SD and such a small sample, it seems likely that the sample mean may be quite far from the population mean. The 95% CI of the population mean quantifies the uncertainty. In this example, the 95% CI ranges from 79.2 to 118.8 mmHg. You can be 95% sure that this wide range of values includes the true population mean.

What Does it Mean to be 95% Sure?

Note that there is no uncertainty about the sample mean. We are 100% sure that we have calculated the sample mean correctly. By definition, the CI is always centered on the sample mean. We are not sure about the value of the population mean. However, we can be 95% sure that the calculated interval contains it.

What exactly does it mean to be "95% sure"? If you calculate a 95% CI from many independent samples, the population mean will be included in the CI in 95% of the samples, but will be outside of the CI in the other 5% of the samples. When you only have measured one sample, you don't know the value of the population mean. The population mean either lies within the 95% CI or it doesn't. You don't know, and there is no way to find out. But you can be 95% certain that the 95% CI includes the population mean.

The correct syntax is to express the CI as 79.2 to 118.8 mmHg, or as [79.2, 118.8]. It is considered bad form to express the CI as 99.0 $\pm$ 19.8 or 79.2 − 118.8 (the latter form would be confusing with negative numbers).

ASSUMPTIONS THAT MUST BE TRUE TO INTERPRET THE 95% CI OF A MEAN

The CI depends on these assumptions:

- Your sample is randomly selected from the population.

    In Example 5.1, we measured the BP of every student in the class. If we are only interested in that class, there would be no need to calculate a CI. We know the class mean exactly. It only makes sense to calculate a CI to make inferences about the mean of a larger population. For this example, the population might include students in future years and in other cities.

    In Example 5.2, we randomly sampled five students from the class, so the assumption is valid. If we let students volunteer for the study, we'd be measuring BP in students who are especially interested in their BP (perhaps they've been told it was high in the past). Such a sample is not representative of the entire population, and any statistical inferences are likely to be misleading.

    In clinical studies, it is not feasible to randomly select patients from the entire population of similar patients. Instead, patients are selected for the study because they happened to be at the right clinic at the right time. This is called a *convenience sample* rather than a *random sample*. For statistical calculations to be meaningful, we must assume that the convenience sample adequately represents the population, and the results are similar to what would have been observed had we used a true random sample.

- The population is distributed in a Gaussian manner, at least approximately. This assumption is not too important if your sample is large. Since Example 5.1 has 100 subjects, the Gaussian assumption matters very little (unless the population distribution is very bizarre). Example 5.2 has only five students. For such a small sample, the CI can only be interpreted if you assume that the overall population is approximately Gaussian.

- All subjects come from the same population, and each has been selected independently of the others. In other words, selecting one subject should not alter the chances of selecting any other. Our BP example would be invalid if there were really less than 100 students, but some were measured twice. It would also be invalid if some of the students were siblings or twins (because BP is partly controlled by genetic factors, so two siblings are likely to have pressures more similar than two randomly selected subjects).

In many situations, these assumptions are not strictly true: The patients in your study may be more homogeneous than the entire population of patients. Measurements made in one lab will have a smaller SD than measurements made in other labs at other times. More generally, the population you really care about may be more diverse than the population that the data were sampled from. Furthermore, the population may not be Gaussian. If any assumption is violated, the CI will probably be too optimistic (too narrow). The true CI (taking into account any violation of the assumptions) is likely to be wider than the CI you calculate.

CALCULATING THE CONFIDENCE INTERVAL OF A MEAN

To calculate the CI, you need to combine the answers to four questions:

1. What is the sample mean? It is our best guess of the population mean.
2. How much variability? If the data are widely scattered (large SD), then the sample mean is likely to be further from the population mean than if the data are tight (small SD).
3. How many subjects? If the sample is large, you expect the sample mean to be close to the population mean and the CI will be very narrow. With tiny samples, the sample mean may be far from the population mean so the confidence interval will be wide.
4. How much confidence? Although CIs are typically calculated for 95% confidence, any value can be used. If you wish to have more confidence (i.e. 99% confidence) you must generate a wider interval. If you are willing to accept less confidence (i.e. 90% confidence), you can generate a narrower interval.

Equation 5.1 calculates the 95% CI of a mean from the sample mean (m), the sample standard deviation (s) and the sample size (N):

$$95\% \text{ CI: } \left(m - t^* \cdot \frac{s}{\sqrt{N}} \right) \text{ to } \left(m + t^* \cdot \frac{s}{\sqrt{N}} \right) \tag{5.1}$$

The CI is centered on the sample mean m. The width of the interval is proportional to the SD of the sample, s. If the sample is more scattered, the SD is higher and the CI will be wider. The width of the interval is inversely proportional to the square root of sample size, N. Everything else being equal, the CIs from larger samples will be narrower than CIs from small samples.

The width of the CI also depends on the value of a new variable, t*. As you expect, its value depends on how confident you want to be. If you want 99% confidence instead of 95% confidence, t* will have a larger value. If you only want 90% confidence, t* will have a smaller value. Additionally, the value of t* depends on the sample size. The value comes from the t distribution, which you'll learn more about later in the chapter. The value of t* is tabulated in Table 5.1, and more fully in Table A5.3 in the Appendix. Since this table has several uses, the rows do not directly show sample size,

Table 5.1. Critical Value of t for 95% Confidence

df	t*	df	t*
1	12.706	12	2.179
2	4.303	13	2.160
3	3.182	14	2.145
4	2.776	15	2.131
5	2.571	20	2.086
6	2.447	25	2.060
7	2.365	30	2.042
8	2.306	40	2.021
9	2.262	60	2.000
10	2.228	120	1.980
11	2.201	∞	1.960

but rather show degrees of freedom (df). When calculating the confidence interval of a mean, df = N−1 as explained previously on pages 27–28.

In Example 5.1, the sample mean is 123.4 mmHg and the SD is 14.0 mmHg. Because the sample has 100 subjects, there are 99 df. The table does not show t* for 99 df. Interpolating between the values shown for 60 df and 120 df, t* is about 1.99. The CI ranges from 120.6 to 126.2 mmHg.

Note that as the sample size increases, the value of t* approaches 1.96 (for 95% confidence). It is enough to remember that the value is approximately 2, so long as the sample has at least 20 or so subjects.

For Example 5.2, the mean of this sample is 99.0 mmHg with a SD of 15.97 mmHg. With only five subjects, there are only 4 df, and t* equals 2.776. The 95% CI of the population mean ranges from 79.2 to 118.8 mmHg. With a small sample, the confidence interval is wide.

To understand why Equation 5.1 works, you need to understand two things. First, you need to understand where the ratio $s/\sqrt{N}$ comes from. Next you need to understand where the value of t* comes from. Both of these topics are complicated, but I'll try to explain them briefly in the next two sections.

THE CENTRAL LIMIT THEOREM

Where does the ratio $s/\sqrt{N}$ in Equation 5.1 come from? To understand the answer, we first need to digress a bit.

If you collect many samples from one population, they will not all have the same mean. What will the distribution of means look like? Mathematicians approach this kind of question by manipulating equations and proving theorems. We'll answer the question by simulating data, an approach that is far more intuitive to many.

Figure 5.1 shows simulated data from throwing dice 2000 times. The X axis shows the average of the numbers that appear on top of the dice, and the Y axis shows the number of times each average appears. We are sampling from a "population" of numbers with a flat distribution—each die is equally likely to land with 1, 2, 3, 4, 5, or 6 on top. The left panel shows a histogram of 2000 throws of one die. The distribution

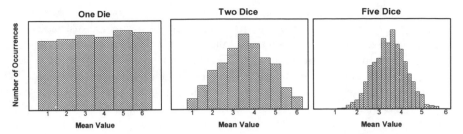

Figure 5.1. An illustration of the central limit theorem. Each panel shows the simulated results of throwing dice 2000 times. The X axis shows the mean of the values on top of the dice, and the Y axis shows how frequently each value appeared. If you only throw one die at a time (left panel) the distribution of values is flat. Each number is equally likely to appear. If you throw two dice at a time (middle panel), the distribution is not flat. You are more likely to get a mean of 3.5 (a sum of 7) than a mean of 1 or 6. The right panel shows the results with five dice. The histogram starts to look like a Gaussian distribution. The central limit theorem states this: No matter how the original population is distributed, the distribution of sample means will approximate a Gaussian distribution if the samples are large enough.

of this sample of 2000 observations closely resembles the overall "population," with some random variability. In the middle panel, two dice were thrown 2000 times, and we plot the distribution of the sample of 2000 mean values. This distribution differs substantially from the original population. If you've every played with dice, this comes as no surprise. There are six ways for a pair of dice to sum to seven (1&6, 2&5, 3&4, 4&3, 5&2, or 6&1), but there is only one way for the sum to equal 12 (6&6) or 2 (1&1). The figure shows average values, and shows that the average of 3.5 (two dice totaling seven) occurred about six times more often than did the average 1 or 6. The right panel shows the results when we threw five dice in each of 2000 trials. The distribution is approximately bell-shaped.

This example illustrates an important statistical theorem, known as the central limit theorem. The proof of this theorem is not accessible to nonmathematicians. But the theorem itself can be understood: No matter how the values in the population are distributed, the distribution of means from independently chosen samples will approximate a Gaussian distribution, so long as your samples are large enough.

The central limit theorem explains the computer generated data of Figure 5.1. Even though the "population" has a flat and chunky (only integers) distribution, the distribution of means approximates a bell-shaped distribution. As you increase the sample size, the distribution of sample means comes closer and closer to the ideal Gaussian distribution.

No matter how the population is distributed, the distribution of sample means will approximate a Gaussian distribution if the sample size is large enough. How large is that? Naturally, the answer is "it depends"! It depends on your definition of "approximately" and on the distribution of the population. Even if the population distribution is really weird, a sample of 100 values is large enough to invoke the central limit theorem. If the population distribution is approximately symmetrical and unimodal (it looks like a mountain and not like a mountain range), then you may invoke the central limit theorem even if you have only sampled a dozen or so values.

What is the standard deviation (SD) of the distribution of sample means? It is not the same as the SD of the population. Since sample means are distributed more compactly than the population, you expect the SD of the distribution of sample means to be smaller than the SD of the population. You also expect it to depend on the SD of the population and the sample size. If the population is very diverse (large SD), then the sample means will be spread out more than if the population is very compact (small SD). If you collect bigger samples, the sample means will be closer together so the SD of sample means will be smaller.

The central limit theorem proves that the SD of sample means equals the SD of the population divided by the square root of the sample size. This is the origin of the ratio $s/\sqrt{N}$ in Equation 5.1.

THE STANDARD ERROR OF THE MEAN

Because the phrase *standard deviation of sample means* is awkward, this value is given a shorter name, the *standard error of the mean,* abbreviated SEM. Often the SEM is referred to as the *standard error,* with the word *mean* missing but implied. The term is a bit misleading, as the standard error of the mean usually has nothing to do with standards or errors.

The central limit theorem calculates the SEM from the SD of the population and the sample size with Equation 5.2.

$$SEM = \frac{SD}{\sqrt{N}} \tag{5.2}$$

The SEM quantifies the precision of the sample mean. A small SEM indicates that the sample mean is likely to be quite close to the true population mean. A large SEM indicates that the sample mean is likely to be far from the true population mean.

Note that the SEM does *not* directly quantify scatter or variability in the population. Many people misunderstand this point. A small SEM can be due to a large sample size rather than due to tight data. With large sample sizes, the SEM is always tiny.

We can substitute Equation 5.2 into Equation 5.1 to calculate the 95% CI of a mean more simply.

$$95\% \text{ CI: } (m - t^* \cdot SEM) \text{ to } (m + t^* \cdot SEM) \tag{5.3}$$

Since the value of t^* is close to 2 for large sample sizes, you can remember that the 95% confidence interval of a mean extends approximately two standard errors on either side of the sample mean.

THE t DISTRIBUTION

Equations 5.1 and 5.3 include the variable t^*. To understand those equations, therefore, you need to understand the t distribution.

Imagine this experiment. Start with a Gaussian population of known mean and SD. From the population, collect many samples of size N. For each sample, calculate the sample mean and SD and then calculate the ratio defined by Equation 5.4.

$$t = \frac{\text{Sample mean} - \text{Population mean}}{\text{Sample SD}/\sqrt{N}} = \frac{m - \mu}{\text{SEM}} \qquad (5.4)$$

Since the sample mean is equally likely to be larger or smaller than the population mean, the value of t is equally likely to be positive or negative. So the t distribution is symmetrical around t = 0. With small samples, you are more likely to observe a large difference between the sample mean and the population mean, so the numerator of the t ratio is likely to be larger with small samples. But the denominator of the t ratio will also be larger with small samples, because the SEM is larger. So it seems reasonable that the distribution of the t ratio would be independent of sample size. But it turns out that that isn't true—including N in the equation does not entirely correct for differences in sample size, and the expected distribution of t varies depending on the size of the sample. Therefore, there is a family of t distributions for different sample sizes.

Figure 5.2 shows the distribution of t for samples of five subjects (df = 4). You'll see t distributions for other numbers of df in Chapter 23. The t distribution looks similar to the Gaussian distribution, but the t distribution is wider. As the sample size increases, the t distribution becomes more and more similar to the Gaussian distribution in accordance with the central limit theorem.

Since Figure 5.2 plots probability density, the area under the entire curve represents all values in the distribution. The tails of the distribution are shaded for all values of t greater than 2.776 or less than −2.776. Calculations prove that each of these tails represents 2.5% of the distribution. This means that the t ratio will be between −2.776 and 2.776 in 95% of samples (N = 5) randomly chosen from a Gaussian population. If the samples were of a different size, the cutoff would not be 2.776 but rather some other number as shown in Table 5.1.

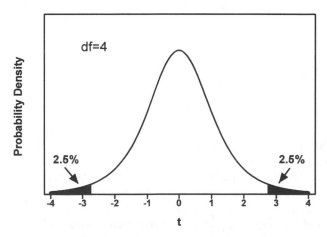

Figure 5.2. The t distribution for four degrees of freedom. Here are the steps you'd need to do to create this graph experimentally. Start with a population whose mean you know. Collected many samples of five subjects (df = 4) from this population. Calculate the t ratio for each using Equation 5.4. Of course, mathematicians can derive the distribution from first principles without the need for endless sampling. The largest and smallest 2.5% of the distribution are shaded. In 95% of the samples, t will be between −2.776 and 2.776.

So far, the logic has been in the wrong direction—we've started with a known population and are looking at the variability between samples. Data analysis goes the other way. When calculating CI, you don't know the population mean. All you know is the mean and SD of one particular sample. But we can use the known distribution of t to make inferences about the population mean. Let's compute the CI for example 5.2. The sample has five subjects, so you can be 95% sure that the t ratio will be between -2.776 and 2.776. Rearrange Equation 5.4 to solve for the population mean as a function of the sample mean (m), the sample standard deviation (s), the sample size (N), and t.

$$\mu = m - t \cdot \frac{s}{\sqrt{N}} = m - t \cdot \text{SEM} \qquad (5.5)$$

You know the sample mean (m = 99.0 mmHg), the sample SD (s = 15.97 mmHg), and the sample size (N = 5). You also know that you can be 95% sure that t will be between -2.776 and 2.776. To calculate the confidence limits, first set t = 2.776 and calculate that μ = 79.2 mmHg. Then set t = -2.78 and calculate that μ = 118.8 mmHg. You've calculated the lower and upper limits of the 95% confidence interval. The 95% CI ranges from 79.2 to 118.8 mmHg.

In deriving the t distribution, statisticians started with a hypothetical population with known population mean and SD, and calculated the distribution of t in many samples. We then use the distribution backwards. Instead of making inferences about the distribution of samples from a population, you can make inferences about the population from a single sample. The ability to use probability distributions backwards is mathematically simple but logically profound, and of immense practical importance as it makes statistics useful to experimenters.

THE GAUSSIAN ASSUMPTION AND STATISTICAL INFERENCE

As noted earlier, the interpretation of the 95% CI depends on the assumption that the data were sampled from a Gaussian distribution. And, as you'll see in future chapters, this assumption is common to many statistical tests. Is this assumption reasonable for the blood pressure example? A few notes:

- The Gaussian distribution, by definition, extends infinitely in each direction. It allows for a very small proportion of BPs less than 0 (physically impossible) and a very small proportion of BPs greater than 300 (biologically impossible). Therefore, BPs cannot follow a Gaussian distribution exactly.
- The assumption doesn't need to be completely true to be useful. Just like the mathematics of geometry is based on mathematical abstractions such as a perfect rectangle, the mathematics of statistics is based upon the abstract Gaussian distribution. Geometry tells us that the area of a perfect rectangle equals its length times its width. No room is a perfect rectangle, but you can use that rule to figure out how much wallpaper you need. The calculation, based on an ideal model, is useful even if the walls are a bit warped. Similarly, statistical theory has devised methods for analyzing data obtained from Gaussian populations. Those methods are useful even when the distributions are not exactly Gaussian.
  - You can look beyond our one sample. BP has been measured in lots of studies, and we can use information from these studies. While the distribution of BPs is often a

bit skewed (more high values than low) they tend to distribute according to a roughly Gaussian distribution. Since this has been observed in many studies with tens of thousands of subjects, it seems reasonable to assume that BP in our population is approximately Gaussian.

- You can think about the sources of scatter. The variability in BP is due to numerous genetic and environmental variables, as well as imprecise measurements. When scatter is due to the sum of numerous factors, you expect to observe a Gaussian distribution, at least approximately.*
- You can perform formal calculations to test whether a distribution of data is consistent with the Gaussian distribution. For more information, read about the Kolmogorov-Smirnov test in a more advanced book.

What should you do if the distribution of your data deviate substantially from a Gaussian distribution? There are three answers.

- You can mathematically transform the values to convert a nongaussian population into a Gaussian one. This is done by converting each value into its logarithm, reciprocal, or square root (or some other function). While this sounds a bit dubious, it can be a good thing to do. Often, there is a good biological or chemical justification for making the transformation. For example, it often makes sense both biologically and statistically to express acidity as pH rather than concentration of hydrogen ions, to express potency of a drug as $\log(EC_{50})$ rather than EC_{50}†, and to express kidney function as the reciprocal of plasma creatinine concentration rather than the plasma creatinine concentration itself.
- You can rely on the central limit theorem and analyze large samples using statistical methods based on the Gaussian distribution, even if the populations are not Gaussian. You can rely on the central limit theorem when both of the following are true: (1) You are making inferences about the population means, and not about the details of the distribution itself. (2) Either the samples are very large or the population is approximately Gaussian.
- You can use statistical methods that are not based on the Gaussian distribution. For example, it is possible to calculate the 95% CI of a median without making any assumption about the distribution of the population. We'll discuss some of these methods, termed nonparameteric methods, later in the book.

CONFIDENCE INTERVAL OF A PROPORTION REVISITED*

You previously saw the CI of the proportion defined in Equation 2.1. This can be rewritten as shown in Equation 5.6.

Approximate 95% CI of proportion:

$$\left(p - z^* \cdot \sqrt{\frac{p(1-p)}{N}}\right) \text{ to } \left(p + z^* \cdot \sqrt{\frac{p(1-p)}{N}}\right) \tag{5.6}$$

---

*You also need to assume that the various factors all have nearly equal weight.
†The EC_{50} is the concentration of drug required to get half the maximal effect.

The value of z* comes from the Gaussian distribution. It equals 1.96 for 95% CI, because 95% of observations in a Gaussian population lie within 1.96 SD of the mean. You may substitute other numbers from the Gaussian distribution if you want to change the degree of confidence. For example, set z* = 2.58 to generate a 99% CI. This is the correct value, because 99% of a Gaussian distribution lie within 2.58 SD of the mean. The value of z* does not depend on the size of your sample.

This approximation works because the binomial distribution approximates the Gaussian distribution with large samples. It is a reasonable approximation when the numerator of the proportion is at least five, and the denominator is at least five larger than the numerator. If you have small samples or want to calculate the CI of a proportion more exactly, you must use tables or programs that are based on the binomial distribution. You should *not* replace z* in Equation 5.6 with t*.

ERROR BARS

Graphs of data often include error bars, and the text of articles often gives values plus or minus an error, i.e. 10.34 ± 2.3. The error bar or the plus/minus value usually denotes the SD or the SEM. Occasionally the error bars denote the range of the data, or the 95% CI of the mean. You must look at the methods section or figure legend to figure out what the authors are trying to say. If there is no explanation as to how the error bars were calculated, the error bar is nearly meaningless.

Figure 5.3 shows the appearance of several error bars. Note that some are capped while others are not. The difference is a matter of aesthetics, not statistics. Also note that some error bars extend above and below the data point, while others extend only above or only below. Usually this is done to prevent error bars from different data sets from overlapping. This is an artistic decision by the investigator. The true uncertainty always extends in both directions, even if the author only shows the error bar in one direction.

In theory, the choice of showing the SD or SEM should be based on statistical principles. Show the SD when you are interested in showing the scatter of the data. Show the SEM when you want to show how well you know the population mean. Sometimes people display the SEM for another reason: The SEM is the smallest measure of ''error'' and thus looks nicest.

Because the SD and SEM are used so frequently, anyone reading the biomedical literature must be able to convert back and forth without looking up the equations in a book. There are only a few equations in statistics that must be committed to memory, and these are two of them:

$$SEM = SD / \sqrt{N} \qquad\qquad (5.7)$$

$$SD = SEM \cdot \sqrt{N}$$

The only role of the standard error . . . is to distort and conceal the data. The reader wants to know the actual span of the data; but the investigator displays an estimated zone for the mean.

A. R. Feinstein, *Clinical Biostatistics*

*This section is more advanced than the rest. You may skip it without loss of continuity.

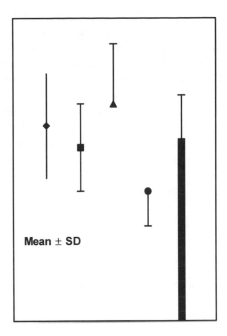

Figure 5.3. Error bars. Graphs show error bars in various ways. Note that the first bar is uncapped, and the others are capped. This is merely an artistic distinction that tells you nothing about the distribution of data. The first two error bars extend above and below the point, while the next two go only above or only below. This too is an artistic distinction that tells you nothing about the data. Note the legend telling us that these error bars represent the standard deviation. You can't interpret these error bars unless you know whether the size of the error bar shows you the standard deviation, the standard error of the mean, the extent of the 95% CI, or something else.

Remember that the scatter (however expressed) means different things in different contexts. Is the author showing the variability among replicates in a single experiment? Variability among experiments performed with genetically identical animals? Variability among cloned cells? Variability between patients? Your interpretation of error bars should depend heavily on such considerations.

SUMMARY

The sample mean you calculate from a list of values is unlikely to be exactly equal to the overall population mean (that you don't know). The likely discrepancy depends on the size of your sample and the variability of the values (expressed as the standard deviation). You can combine the sample mean, SD and sample size to calculate a

95% confidence interval of the mean. If your sample is representative of the population, you can be 95% sure that the true population mean lies somewhere within the 95% CI.

The *central limit theorem* states that the distribution of sample means will approximate a Gaussian distribution even if the population is not Gaussian. It explains why you can interpret a confidence interval from large samples even if the population is not Gaussian.

The *standard error of the mean* (SEM) is a measure of how close a sample mean is to the population mean. Although it is often presented in papers, it is easier to interpret confidence intervals, which are calculated from the SEM.

Papers often present the data as mean ± error, or show a graph with an *error bar.* These error values or error bars may reprsent the SD, the SEM or something else. You can not interpret them unless you know how they were defined.

OBJECTIVES

1. You must be familiar with the following terms:
   - Gaussian distribution
   - Central limit theorem
   - Standard error of the mean
   - Confidence interval of a mean
   - Confidence interval
   - Error bar
   - t distribution
2. Given a list of numbers, you should be able to calculate the 95% CI of the mean.
3. Given the mean, sample size, and SD or SEM, you should be able to calculate the 95% CI of the mean.
4. You should know the assumptions that must be true to interpret the CI.
5. You must be able to convert between SD and SEM without referring to a book.
6. You should be able to interpret error bars shown in graphs or tables of publications.

PROBLEMS

1. Figure 5.4 shows the probability distribution for the time that an ion channel stays open. Most channels stay open for a very short time, and some stay open longer. Imagine that you measure the open time of 10 channels in each experiment, and calculate the average time. You then repeat the experiment many times. What is the expected shape of the distribution of the mean open times?
2. An enzyme level was measured in cultured cells. The experiment was repeated on 3 days; on each day the measurement was made in triplicate. The experimental conditions were identical on each day; the only purpose of the repeated experiments was to determine the value more precisely. The results are shown as enzyme activity in units per minute per milligram of membrane protein.

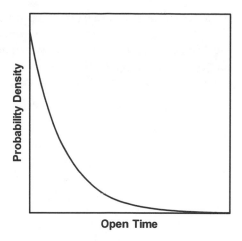

Figure 5.4.

	Replicate 1	Replicate 2	Replicate 3
Monday	234	220	229
Tuesday	269	967	275
Wednesday	254	249	246

Summarize these data as you would for publication. The readers have no interest in the individual results for each day; just one overall mean with an indication of the scatter. Give the results as mean, error value, and N. Justify your decisions.

3. Is the width of a 99% CI wider or narrower than the width of a 90% CI?

4. The serum levels of a hormone (the "Y factor") was measured to be 93 ± 1.5 (mean ± SEM) in 100 nonpregnant women and 110 ± 2.3 (mean ± SEM) in 100 women in the first trimester of pregnancy.

   A. Calculate the 95% CI for the mean value in each group.
   B. Assuming that the measurement of Y is easy, cheap, and accurate, is it a useful assay to diagnose pregnancy?

5. A paper reports that cell membranes have 1203 ± 64 (mean ± SEM) fmol of receptor per milligram of membrane protein. These data come from nine experiments.

   A. Calculate the 95% CI. Explain what it means in plain language.
   B. Calculate the 95% prediction interval. Explain what it means in plain language.
   C. Is it possible to sketch the distribution of the original data? If so, do it.
   D. Calculate the 90% CI.
   E. Calculate the coefficient of variation.

6. You measure BP in 10 randomly selected subjects and calculate that the mean is 125 and the SD is 7.5 mmHg. Calculate the SEM and 95% CI. Now you measure BP in 100 subjects randomly selected from the same population. What values do you expect to find for the SD and SEM?

7. Why doesn't it ever make sense to calculate a CI from a population SD (as opposed to a sample SD)?

8. Data were measured in 16 subjects, and the 95% CI was 97 to 132.

   A. Predict the distribution of the raw data.
   B. Calculate the 99% CI.

9. Which is larger, the SD or the SEM? Are there any exceptions?

6

Survival Curves

In the long run, we are all dead.
John Keynes

Note to basic scientists. This chapter describes a method commonly used in clinical research but rarely used in basic research. You may skip the entire chapter without loss in continuity.

We have discussed how to quantify uncertainty for outcomes expressed as proportions (Chapter 2) or as measurements (Chapters 3 to 5). However, in many clinical studies, the outcome is survival time. Analysis of survival data is trickier than you might first imagine.

What's wrong with calculating the mean survival time and its confidence interval (CI)? This approach is rarely useful. One problem is that you can't calculate the mean survival time until you know the survival time for each patient, which means you can't analyze the data until the last patient has died. Another problem is that survival times are unlikely to follow a Gaussian distribution. For these and other reasons, special methods must be used for analysis of survival data.

A SIMPLE SURVIVAL CURVE

Survival curves plot percent survival as a function of time. Figure 6.1 shows a simple survival curve. Fifteen patients were followed for 36 months. Nine patients died at known times, and six were still alive at the end of the study.

Time zero is not some specified calendar date; rather, it is the time that each patient entered the study. In many clinical studies, "time zero" spans several calendar years as patients are enrolled. At time zero, by definition, all patients are alive, so Y = 100%. Whenever a patient dies, the percent surviving decreases. If the study (and thus the X axis) were extended far enough, Y would eventually reach 0. This study ended at 36 months with 40% (6/15) of the patients still alive.

Each patient's death is clearly visible as a downward jump in the curve. When the first patient died, the percent survival dropped from 100.0% to 93.3% (14/15). When the next patient died, the percent survival dropped again to 86.7%. At 19 months, two patients died, so the downward step is larger.

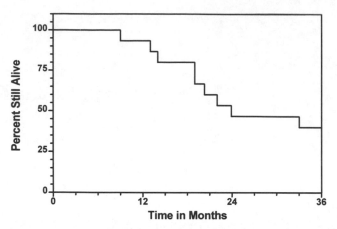

Figure 6.1. A simple survival curve. Fifteen subjects were followed for 36 months. Nine of the subjects died. You can see each death as a downward step in the curve. Two subjects died at 19 months, so the drop is twice as large. Note that time 0 does not have to be any particular day or year. Time 0 is the time that each subject was enrolled in the trial.

The term *survival* curve is a bit misleading, as "survival" curves can plot time to any well-defined end point, such as occlusion of a vascular graft, date of first metastasis, or rejection of a transplanted kidney. The event does not have to be dire. The event could be restoration of renal function, discharge from a hospital, or graduation. The event must be a one-time event. Recurring events should not be analyzed with survival curves.

CENSORED SURVIVAL DATA

In the previous example, we knew that all subjects either died before 36 months or survived longer than 36 months (the right end of our curve). Real data are rarely so simple. In most survival studies, some surviving subjects are not followed for the entire span of the curve. This can happen in two ways:

- Some subjects are still alive at the end of the study but were not followed for the entire span of the curve. Many studies enroll patients over a period of several years. The patients who enroll later are not followed for as many years as patients who enroll early. Imagine a study that enrolls patients between 1985 and 1989, and that ends in 1991. Patient A enrolled in 1989 and is still alive at the end of the study. Even though the study lasted 6 years, we only know that patient A survived at least 3 years.
- Some drop out of the study early. Perhaps they moved to a different city or got fed up with university hospitals. Patient B enrolled in 1986 but moved to another city (and stopped following the protocol) in 1988. We know that this subject survived at least 2 years on the protocol but can't evaluate survival after that.

In either case, you know that the subject survived up to a certain time but have no useful information about what happened after that. Information about these patients is said to be censored. Before the censored time, you know they were alive and following the experimental protocol, so these subjects contribute useful information. After they are censored, you can't use any information on the subjects. Either we don't have information beyond the censoring day (because the data weren't or can't be collected) or we have information but can't use it (because the patient no longer was following the experimental protocol). The word *censor* has a negative ring to it. It sounds like the subject has done something bad. Not so. It's the data that have been censored, not the subject!

CREATING A SURVIVAL CURVE

There are two slightly different methods to create a survival curve. With the *actuarial* method, the X axis is divided up into regular intervals, perhaps months or years, and survival is calculated for each interval. With the *Kaplan-Meier* method, survival is recalculated every time a patient dies. This method is preferred, unless the number of patients is huge. The term *life-table analysis* is used inconsistently, but usually includes both methods. You should recognize all three names.

The Kaplan-Meier method is logically simple but tedious. Since computer programs can do the calculations for you, the details will not be presented here. The idea is pretty simple. To calculate the fraction of patients who survived on a particular day, simply divide the number alive at the end of the day by the number alive at the beginning of the day (excluding any who were censored on that day from both the numerator and denominator).* This gives you the fraction of patients who were alive at the beginning of a particular day who were still alive at the beginning of the next day. To calculate the fraction of patients who survive from day 0 until a particular day, multiply the fraction of patients who survive day 1, times the fraction of those patients who survive day 2, times the fraction of those patients who survive day 3 . . . times the fraction who survive day k. This method automatically accounts for censored patients, as both the numerator and denominator are reduced on the day a patient is censored. Because we calculate the product of many survival fractions, this method is also called the *product-limit* method.

Figure 6.2 shows a survival curve with censored data. The study started with 15 patients. Nine died during the study (same as the previous example) and six were censored at various times during the study. On the left panel, each censored patient is denoted by upward blips in the survival curve. On the right panel, each censored patient is denoted by a symbol in the middle of a horizontal part of the survival curve. At the time a patient is censored, the survival curve does not dip down as no one has died. When the next patient dies, the step downward is larger because the denominator (the number of patients still being followed) has shrunk.

CONFIDENCE INTERVAL OF A SURVIVAL CURVE

In order to extrapolate from our knowledge of a sample to the overall population, a survival curve is far more informative when it includes a 95% CI. Calculating CIs is

---

*Note that *day* refers to day of the study, not a particular day on the calendar. Day 1 is the first day of the study for each subject.

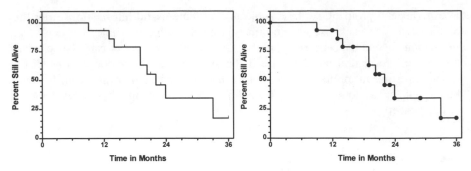

Figure 6.2. A survival curve with censored subjects. A subject is censored at a certain time for one of two reasons. (1) He stopped following the study protocol at that time. (2) The trial ended with the subject still alive. In the left panel, censored subjects are shown as upward blips. In the right panel, censored subjects are shown as solid circles in a horizontal portion of the curve. You'll see both kinds of graphs frequently.

not straightforward and is best left to computer programs. The interpretation of the 95% CI for a survival curve should be clear to you by now. We have measured survival exactly in a sample but don't know what the survival curve for the entire population looks like. We can be 95% sure that the true population survival curve lies within the 95% CI shown on our graph at all times.

Unfortunately, many published survival curves do not include CIs. Assuming that you can figure out how many patients are still alive at any given time, you can use Equation 6.1 to calculate an approximate 95% CI for the fraction surviving up to at any time t (p is fraction surviving up to time t, and N is the number of patients still alive and following the protocol at time t):*

$$\text{95\% CI of fractional survival (p) at time } t \approx p \pm 1.96 \cdot p\sqrt{(1 - p)/N}. \quad (6.1)$$

Let's use this equation to figure out approximate CIs for the example at 24 months. We started with 15 patients. Between 0 and 24 months, eight patients have died (just count the downward steps, remembering that the big step at 19 months represents two patients). Four patients have been censored before 24 months (count the ticks on the left panel of Figure 6.1 between 0 and 24 months). Thus three patients (15 − 8 − 4) are still alive and being followed at 24 months, so N = 3. Reading off the curve, p ≈ 0.35. Plugging p and N into the equation, the 95% CI is approximately 0.03 to 0.67.

Equation 6.1 is only an approximation and sometimes calculates values that are nonsense. It can calculate a lower confidence limit less than 0. In this case, set the lower limit to 0. It can also calculate an upper confidence limit greater than 100%. In this case set the limit to 100%. Figure 6.3 shows more exact CIs calculated by computer. If there are censored patients, the right side of a survival curve represents fewer patients than the left side, and the CIs become wider as time progresses (until survival converges on 0).

*Equation 6.1 is not well known. I got it from page 378 of D. G. Altman, *Practical Statistics for Medical Research,* Chapman & Hall, London, 1991.

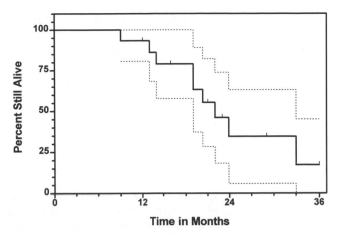

Figure 6.3. A survival curve with 95% CIs. The solid line shows the survival curve for the sample of 15 subjects. You can be 95% sure that the overall survival curve for the entire population lies within the dotted lines. The CIs are wide because the sample is so small.

MEDIAN SURVIVAL

It is easy to derive the median survival time from the survival curve. Simply draw a horizontal line at 50% survival and see where it crosses the curve. Then look down at the X axis to read off the median survival time. Figure 6.4 shows that the median survival in the example is about 22 months. Sometimes the survival curve is horizontal at 50% survival. In this case, the median encompasses a range of survival times. Most people define the median survival in this case as the average of the first and last time point at which the survival curve equals 50%.

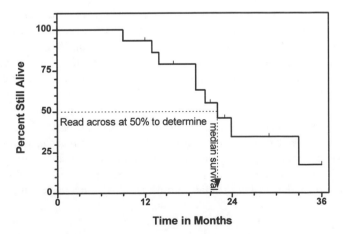

Figure 6.4. Median survival. The median is the 50th percentile. The median survival time is the time at which half the subjects have died and half are still alive. Read across at 50 percent to determine median survival. If fewer than half the subjects have died by the end of the study, you cannot determine median survival.

If the survival curve includes 95% CIs, you can determine the 95% CI of the median by seeing where the upper and lower CI crosses the horizontal line where survival equals 50%. From Figure 6.3 you can estimate that the 95% CI of median survival ranges from about 19 to 33 months. Obviously, you can't determine median survival if more than half the subjects are still alive when the study ends.

ASSUMPTIONS

The interpretation of survival curves (and their CIs) depends on these assumptions:

- *Random sample.* If your sample is not randomly selected from a population, then you must assume that your sample is representative of that population.
- *Independent observations.* Choosing any one subject in the population should not affect the chance of choosing any other particular subject.
- *Consistent entry criteria.* Patients are enrolled into studies over a period of months or years. In these studies it is important that the starting criteria don't change during the enrollment period. Imagine a cancer survival curve starting from the date that the first metastasis was detected. What would happen if improved diagnostic technology detected metastases earlier? Even with no change in therapy or in the natural history of the disease, survival time will apparently increase (patients die at the same age they otherwise would but are diagnosed at an earlier age and so live longer with the diagnosis).
- *Consistent criteria for defining "survival."* If the curve is plotting time to death, then the ending criterion is pretty clear. If the curve is plotting time to some other event, it is crucial that the event be assessed consistently throughout the study.
- *Time of censoring is unrelated to survival.* The survival of the censored patients must be identical (on average) to the survival of the remainder. If there are many censored subjects, this assumption is critical to the validity of the results. There is no reason to doubt that assumption for patients still alive at the end of the study. Patients who dropped out of the study are a different matter. You must ask why the patients left the study. If the reason is related to survival, then the survival curve would be misleading. A survival curve would be misleading, for example, if many patients quit the study because they were too sick to come to the clinic or because they felt too well to take medication. In well-run studies, a very small fraction of patients leave the study. When looking at papers reporting survival curves, check what fraction of the patients dropped out of the study and why.
- *Average survival does not change during the course of the study.* If patients are enrolled over a period of years, you must assume that overall survival is not changing over time. You can't interpret a survival curve if the patients enrolled in the study early are more (or less) likely to die than those who enroll in the study later.

PROBLEMS WITH SURVIVAL STUDIES

Since the survival curve plots time until death, you have to decide when to "start the clock." The starting point should be an objective date—perhaps the date of first diagnosis or first hospital admission. You may be tempted to use instead an earlier starting criteria, such as the time that a patient remembers first observing symptoms. Don't do it. Such data are invalid because a patient's recollection of early symptoms may be altered by later events.

If the curve is plotting deaths due to a particular form of cancer, you need to decide what to do with patients who die of another cause, say, an automobile accident. Some investigators count these as deaths, and others count them as censored subjects. Both approaches are sensible, but the approach should be decided before the study is started.

SUMMARY

You will frequently encounter survival curves in the medical literature. Survival curves can be used to plot time to any nonrecurrent event. The event does not have to be death, so the term *survival* can be misleading.

Creating a survival curves is a bit tricky, because you need to account for censored subjects. Subjects can be censored because they stop following the experimental protocol, or because they are still alive when the protocol ends. These subjects contribute data up until the time of censoring but contribute no data after that.

It is easiest to interpret a survival curve when you plot 95% confidence limits for survival at various times. You can be 95% sure that the survival curve for the overall population lies somewhere within those limits.

OBJECTIVES

1. You should be familiar with the following terms:
   - Survival curve
   - Censored subject
   - Kaplan-Meier curve
   - Product-limit method
   - Actuarial method
2. You should be able to recognize data that can appropriately be plotted on a survival curve.
3. You should be able to interpret a survival curve.
4. You should understand why it is difficult to construct a survival curve.
5. You should be able to read median survival from a survival curve.
6. You should know all assumptions that must be true for survival studies to be interpretable.
7. Referring to the book, you should be able to calculate an estimated 95% CI for a survival curve.

PROBLEMS

1. Why are the CIs in Figure 6.3 asymmetrical?
2. Why is it possible to calculate the median survival time when some of the subjects are still alive?

3. Why are survival curves drawn as staircase curves?
4. A survival curve includes many subjects who were censored because they dropped out of the study because they felt too ill to come to the clinic appointments. Is the survival curve likely to overestimate or underestimate survival of the population?
5. A study began on January 1, 1991 and ended on December 31, 1994. How will each of these subjects appear on the graph?

   A. Entered March 1, 1991. Died March 31, 1992.
   B. Entered April 1, 1991. Left study March 1, 1992.
   C. Entered January 1, 1991. Still alive when study ended.
   D. Entered January 1, 1992. Still alive when study ended.
   E. Entered January 1, 1993. Still alive when study ended.
   F. Entered January 1, 1994. Died December 30, 1994.
   G. Entered July 1, 1992. Died of car crash March 1, 1993.

6. A survival study started with 100 subjects. Before the end of the fifth year, no patients had been censored and 25 had died. Five patients were censored in the sixth year, and five died in the seventh year. What is the percent survival at the end of the seventh year? What is the 95% CI for that fraction?

II

COMPARING GROUPS WITH CONFIDENCE INTERVALS

Thus far we have only analyzed data from one group. When you analyze data, you usually want to compare two (or more) groups. This section shows you how to do so by calculating confidence intervals. Later you'll learn how to compare groups by calculating P values and determining statistical significance.

7

Confidence Interval of a Difference Between Means

INTERPRETING THE 95% CI FOR THE DIFFERENCE BETWEEN TWO MEANS

Example 7.1

Diarrhea is a major health problem for babies, especially in underdeveloped countries. Oral rehydration with a mixture of salt and sugar effectively replaces lost fluids and probably saves 1 million lives a year. But this treatment does not reduce the amount of diarrhea. Bismuth salicylate (sold as Pepto Bismol) reduces diarrhea in adults, and Figueroa-Quintanilla et al. wanted to know whether it would be beneficial to babies with diarrhea.* They studied several hundred infant boys with diarrhea in a clinic in Peru. All were given standard oral rehydration. One third were given a low dose of bismuth salicylate, one third were given a higher dose, and one third were given placebo. They analyzed the data in many ways and recorded the clinical outcome of the infants in several ways. We will look at only one end point, the total output of stool, and only compare control subjects with those given low dose bismuth salicylate. To adjust for body size, the authors divided stool output in milliliters by body weight in kilograms and compared stool output per kilogram between the two groups.

The authors report that the mean stool output was 260 ml/kg for the 84 placebo-treated babies with a standard deviation (SD) of 254 ml/kg. The mean stool output for the 85 treated babies was 182 ml/kg with a SD of 197 ml/kg.

The authors do not show us individual values or even the range of the data. Is it reasonable to assume that stool output follows a Gaussian distribution? From the data we do have (mean and SD), the answer is clearly NO. If stool output was distributed according to a Gaussian distribution, we'd expect that 95% of the values would be within about 2 SDs of the mean, or -248 to 768 ml/kg. If the data followed a Gaussian distribution, a reasonable fraction of the values would be negative. Since a negative stool output is impossible, the data cannot be sampled from a Gaussian distribution.

*D. Figueroa-Quintanilla et al. A controlled trial of bismuth subsalicylate in infants with acute watery diarrheal disease. *N Engl J Med* 328:1653–1658, 1993.

63

Because the data must be non-Gaussian, we *cannot* use the SD to calculate a prediction interval or "normal range." Since the samples are so large, we *can* invoke the central limit theorem and make inferences about the average stool output in the population. You already know how to calculate the standard error of the mean (SEM) and the 95% confidence interval (CI) for each group. For the control infants, the SEM is $254/\sqrt{84} = 27.7$ ml/kg. The 95% CI for the mean is $260 \pm 2.00 \times 27.7$, which equals 205 to 315 ml/kg. Assuming that the infants studied are representative of a larger population of infants with diarrhea, we can be 95% sure that the mean stool output lies within that range. For the treated infants, the SEM is $197/\sqrt{85} = 21.4$ ml/kg and the 95% CI is 139 to 225 ml/kg. The two CIs barely overlap.

On average the stool production of treated infants was $260 - 182$ or 78 ml/kg less than the sample control infants. We'd like to calculate a 95% CI for the mean decrease in the population. The next section will explain the calculations. The answer is that the 95% CI for the difference between the two means is 19 to 137 ml/kg.

We defined the difference to be the mean value in control minus the mean value in the treated infants. A positive value indicates that the treatment worked to reduce stool ouput. A negative value would indicate that the treatment increased stool output.

The interpretation of the 95% CI should be familiar to you. You can be 95% sure that the true difference between population means lies somewhere within the CI. If you were to perform this kind of experiment many times, each one would calculate a different CI for the difference between means. In 95% of the experiments, the true difference between population means would lie within the CI. In the other 5% of the samples, the true difference between means would lie outside the CI. When analyzing data you have only one pair of samples, and you can be 95% sure that the CI of the difference contains the true difference between population means.

The interpretation of the CI of the difference between population means rests on the same assumptions as the CI of the mean:

- Your samples are randomly selected from the populations, or at least are representative of the populations. In our example, the infants are not randomly selected from the population of infants with diarrhea, but there is no reason to think that they are not representative. The authors selected subjects from one population—children with diarrhea. They then randomly assigned two different treatments. In interpreting the results, you should think of two different populations—the population of children with diarrhea given oral rehydration only and another population of children with diarrhea given both oral rehydration and low-dose bismuth salicylate.
- The populations are distributed in a Gaussian manner. The larger the samples, the less this assumption matters. In our example, we know that the distribution cannot be Gaussian because the SD is so large compared with the mean. But with such large samples, we can ignore that assumption as long as the distribution was not really bizarre.
- The two populations have identical SDs.
- All subjects in each group come from the same population, and each has been selected independently of the others. There are many causes of diarrhea, and you want to get a random sampling of the various causes in both groups. You wouldn't want to include siblings or a group of children from only one orphanage.

Later in this chapter we will discuss the common situation when subjects in the two groups are not independent, but rather are paired or matched.

Since these assumptions seem reasonable, we can interpret the CI. We are 95% sure the true difference between means lies within the interval. The interval does not span 0 (does not include both negative and positive numbers) so we can be at least 95% sure that on average the drug reduces mean stool output in the population.

In assessing the clinical significance of this study, we (like the authors) need to look at more than just that one variable. The authors showed the effectiveness of the drug in several ways. Not only did the drug reduce stool volume, it also reduced the average duration of diarrhea and the average length of hospitalization. Moreover, there were no adverse effects of the drug. You can conclude that the drug worked.

CALCULATING THE 95% CI FOR THE DIFFERENCE BETWEEN MEANS OF UNPAIRED GROUPS*

To calculate the 95% CI of the difference between means, you need to first calculate the SE of the difference, calculated from the standard errors (SEs) of the two groups, SEM_a and SEM_b. If the two groups have equal numbers of data points, Equation 7.1 calculates the SEM of the difference between the means of the two groups:

$$\text{SE of the difference} = \sqrt{SEM_a^2 + SEM_b^2}. \tag{7.1}$$

If the two groups have unequal numbers of data points, a more complicated equation is needed (7.2), as the SEM of the larger group has to be weighted more heavily than the SEM of the smaller group. The equation is expressed in terms of the two sample SDs (SD_a and SD_b) and the two sample sizes (n_a and n_b):

$$\text{Pooled SD} = \sqrt{\frac{(n_a - 1) \cdot SD_a^2 + (n_b - 1) \cdot SD_b^2}{n_a + n_b - 2}}. \tag{7.2}$$

$$\text{SE of difference} = \text{Pooled SD} \sqrt{\frac{1}{n_a} + \frac{1}{n_b}}.$$

The 95% CI of the difference between means (Δ) is calculated using Equation 7.3:

$$95\% \text{ CI of mean difference} = (\Delta - t^* \cdot \text{SE of difference}) \text{ to } (\Delta + t^* \cdot \text{SE of difference}).$$
$$df = n_a + n_b - 2. \tag{7.3}$$

Applying the first equation to the sample data, the SE of the difference between the two means is 30.0 ml/kg. This is a bit higher than the SEM of either group. This makes sense, as the uncertainty with which we know the difference between the two means is greater than the uncertainty with which we know each individual mean. In other words, the errors propagate so that the difference has a larger SE than its components.

---

*This section contains the equations you need to calculate statistics yourself. You may skip it without loss of continuity.

The total number of degrees of freedom (df) is $84 + 85 - 2$, which equals 167. The critical value of t* for so many df is 1.96. The 95% CI of the difference is $78 \pm 30.0 * 1.96$, or 19 to 137 ml/kg. We can be 95% certain that the mean reduction in stool output in the population of treated infants (compared with the mean of control infants) lies within this range.

WHY ARE PAIRED SUBJECTS ANALYZED DIFFERENTLY?

Often experiments are designed so that the same measurement is made in each subject before and after an intervention. In such studies, it would be a mistake to calculate the 95% CI of the difference between two means using the equations presented in the previous section. Those equations lump together systematic variability between groups with variability between subjects. With unpaired data, there is no choice but to combine these two sources of variability. With paired data, however, you can (and should) separate the two sources of variability.

Paired analyses are appropriate in several kinds of experiments:

• When measuring a variable in each subject before and after an intervention.
• When recruiting subjects as pairs, matched for variables such as age, neighborhood, or diagnosis. One of the pair receives an intervention; the other receives an alternative treatment (or placebo).
• When measuring a variable in sibling or child/parent pairs.
• When running a laboratory experiment several times, each time with a control and treated preparation handled in parallel.

More generally, you should use methods for paired data whenever the value of one subject in the first group is expected to be closer to a particular subject in the second group than with a random subject in the second group. Ideally, you should decide whether to treat the data as pairs when you design the experiment, before collecting any data. Certainly, you must define the pairs using only information you knew before the experiment was performed. Generally, if the pairing was effective the CIs will be narrower if the correct methods are used (taking into account pairing).

Example 7.2

Ye and Grantham* were interested in the mechanism by which renal cysts accumulate fluid. They investigated fluid absorption of cysts surgically excised from patients with polycystic kidney disease. Before experimenting with various drugs, they simply incubated the cysts in cell culture medium and measured weight change (as an indirect measure of volume change). Before incubating, the average weight of nine cysts was 6.51 grams with a SEM of 2.26 grams. After incubating for 24 hours, the average weight was 7.02 grams with a SEM of 2.40 grams. Using Equation 7.3, the 95% CI

*M Ye, JJ Grantham. The secretion of fluid by renal cysts from patients with autosomal dominant polycystic kidney disease. *N Engl J Med* 329:310–313, 1993.

of the difference between means is −6.48 to 7.50. The lower limit is negative, indicating a loss in weight. The upper limit is positive, indicating a gain in weight. Looked at this way, the data appear to provide no evidence that renal cysts can take up fluid in culture. But this method is not appropriate for these data. You are not comparing nine cysts in one group with nine in another. Each cyst was weighed before and after the incubation. The data are paired.

It is easy to take into account the pairing. Calculate the weight gain (or loss) of each of the nine cysts. Then calculate the mean and the SEM of the mean change. You need the raw data to do this. The authors tell us that the mean change was 0.50 grams with a SE of .23 grams. Now you can use Equation 5.3 (repeated as Equation 7.4) to calculate that the 95% CI for the mean change. It is between −0.03 to 1.04 grams. You can see that accounting for the pairing narrowed the CI.

The interpretation of the CI of the mean difference of paired measurements depends on these assumptions:

• Your pairs of subjects are randomly selected from the population of pairs, or at least are representative of the populations. In the example the cysts were not randomly chosen from a large population of renal cysts, but it is reasonable to think that they are representative.
• In the overall population of pairs, the differences are distributed in a Gaussian manner. The larger the samples, the less this assumption matters. This assumption seems reasonable, although we don't have enough data to be very sure.
• The two measurements are before/after measurements on one subject or are measurements on two subjects matched before the data were collected. The example data are before/after measurements.
• All subjects come from the same population, and each subject (if before/after) or each pair of matched subjects has been selected independently of the others. This assumption was violated in the example. The nine cysts were excised from only three patients. Our data are not a sample of nine independent cysts from a larger population. It is quite likely that cysts removed from a particular patient will vary less than cysts removed from different patients.

HOW TO CALCULATE THE 95% CI OF THE MEAN DIFFERENCE OF PAIRED SUBJECTS*

Equation 7.4 calculates the 95% CI of the mean difference of paired measurements. First calculate the difference for each pair, keeping track of the sign. For before and after studies, subtract the *before* measurement from the *after* measurement for each subject. An increase is positive; a decrease is negative. For matched studies, calculate the difference for each matched pair, keeping track of the sign. Calculate the mean and SEM of that single list of differences. Define Δ as the mean of the paired differences.

---

*This section contains the equations you need to calculate statistics yourself. You may skip it without loss of continuity.

$$95\% \text{ CI of mean difference } =$$
$$(\Delta - t^* \cdot \text{SE of paired differences}) \text{ to } (\Delta + t^* \cdot \text{SE of paired differences}).$$

$$df = N_{pairs} - 1 \tag{7.4}$$

With some experimental measurements, it may make more biological sense to calculate the ratio, rather than the difference, of the paired measurements. Chapter 25 explains how to analyze these kind of data.

SUMMARY

When comparing a measured variable in two groups, you can calculate the 95% CI of the difference between the population means. You can be 95% certain that the true mean difference lies somewhere within the 95% CI. The best way to perform such a study is to pair subjects or to measure the same subject before and after an intervention. The 95% CI of the mean difference is calculated differently in a paired study. In most cases, such pairing reduces the influence of intrasubject variability and makes the 95% CI narrower.

OBJECTIVES

1. Using a text and a calculator, you should be able to calculate the 95% CI for the difference between two means for either paired or unpaired data.
2. You should be able to interpret the 95% CI of a difference and be able to state all assumptions.
3. You should be able to identify experimental protocols that should be analyzed using paired differences.

PROBLEMS

1. (Same data as in Problem 4 in Chapter 5). The serum levels of a hormone (the Y factor) was measured to be 93 ± 1.5 (mean ± SEM) in 100 nonpregnant women and 110 ± 2.3 (mean ± SEM) in 100 women in the first trimester of pregnancy.

   a. What is the 95% CI for the difference between mean levels of the Y factor?
   b. What assumptions do you have to make to answer that question?

2. Pullan et al. investigated the use of transdermal nicotine for treating ulcerative colitis.* The plasma nicotine level at baseline was 0.5 ± 1.1 ng/ml (mean ± SD; N = 35). After 6 weeks of treatment, the plasma level was 8.2 ± 7.1 ng/ml (N = 30).

   Calculate a 95% CI for the increase in plasma nicotine level.

*RD Pullan, J Rhodes, S Ganesh, et al. Transdermal nicotine for active ulcerative colitis. *N Engl J Med* 330:811–815, 1994.

What assumptions are you making? Are these valid?

If you had access to all the data, how might you analyze the data?

3. You measure receptor number in cultured cells in the presence and absence of a hormone. Each experiment has its own control. Experiments were performed several months apart, so that the cells were of different passage numbers and were grown in different lots of serum. The results are given in the following table.

	Control	Hormone
Experiment 1	123	209
Experiment 2	64	103
Experiment 3	189	343
Experiment 4	265	485

How would you summarize and graph these results? What CI would you calculate?

4. Why is the SE of a difference between two means larger than the SE of either mean?

Confidence Interval of the Difference or Ratio of Two Proportions: Prospective Studies

CROSS-SECTIONAL, PROSPECTIVE, AND RETROSPECTIVE STUDIES

In the last chapter, you learned how to compare two groups when the outcome variable was a measurement by calculating the 95% CI for the difference between the two means. In many studies that compare two groups, however, the outcome variable is a proportion. In this chapter and the next, you'll learn how to interpret and calculate 95% CI for the difference or ratio of two proportions.

The methods used to analyze the data depend on how the study was conducted. There are four kinds of studies whose results are expressed as two proportions:

- In a *retrospective study* (also called a *case-control study*), the investigators start with the outcome and look back to see the cause. They select two groups of subjects. One group has the disease or condition being studied. These are the cases. The other group is selected to be similar in many ways, but not to have the condition. These are the controls. The investigators then look back in time to compare the exposure of the two groups to a possible risk factor. The next chapter (Chapter 9) explains how to interpret data from case-control studies.
- In a *prospective study,* the investigators start with the exposure and look to see if it causes disease. They select two groups of subjects. One group has been exposed to a possible risk factor. The other group hasn't. The investigator then waits while the natural history of the disease progresses and compares the incidence rates in the two groups.
- In a *cross-sectional study,* the investigator selects a single sample of subjects, without regard to either the disease or the risk factor. The subjects are then divided into two groups based on previous exposure to the risk factor. The investigators then compare the prevalence* of the disease in the two groups.

Prevalence is the proportion of the group that now has the disease. *Incidence* is the proportion of the group that develops the disease within a defined period of time.

- In an *experimental study,* the investigator selects a sample of subjects, which are randomly divided into two groups. Each of the groups gets a different treatment (or no treatment) and the investigators compare the incidence of the disease.

The data from prospective, cross-sectional, and experimental studies are analyzed similarly, as discussed in the rest of this chapter. The next chapter explains how to interpret and analyze data from retrospective studies.

AN EXAMPLE OF A CLINICAL TRIAL

Example 8.1

Cooper et al. studied the effectiveness of zidovudine (also known as AZT) in treating asymptomatic people infected with the human immunodeficiency virus (HIV)*. It is well established that AZT benefits patients with the acquired immunodeficiency syndrome (AIDS) or asymptomatic patients infected with the HIV who have low numbers of T-helper (CD4+) cells. Patients can be infected with the HIV for many years before they develop any symptoms of AIDS and before their CD4+ cells counts drop. Does AZT help these patients? The investigators selected adults who were infected with the HIV but had no symptoms and randomly assigned them to receive AZT or placebo. The subjects were followed for 3 years. The authors analyzed the data in several ways and looked at several outcomes. We'll just look at one outcome—whether the disease progressed in 3 years. The authors defined disease progression to be when a patient developed symptoms of AIDS or when the number of CD4+ cells dropped substantially. They asked whether treatment with AZT reduces progression of the disease.

This study is called a randomized, double-blind prospective study. Let's look at all these terms one by one. See Chapter 20 for more information on clinical trials.

- It is a randomized study because the assignment of subjects to receive AZT or placebo was determined randomly. Patients or physicians could not request one treatment or the other.
- It is double blind because neither patient nor investigator knew who was getting AZT and who was getting placebo. Until the study was complete, the information about which patient got which drug was coded, and the code was not available to any of the participating subjects or investigators (except in a medical emergency).
- It is prospective because the subjects were followed forward over time. In the next chapter you'll lean about retrospective case-control studies where this is not true.

The results are shown in Table 8.1.

The treatment seemed to work. The disease progressed in 28% of the patients receiving placebo (129/461) and in only 16% of the patients receiving AZT (76/475).

---

*DA Cooper et al. Zidovudine in persons with asymptomatic HIV infection and CD4+ cell counts greater than 400 per cubic millimeter. *N Engl J Med* 329:297–303, 1993.

Table 8.1. Results of Example 8.1

Treatment	Disease Progressed	No Progression	Total
AZT	76	399	475
Placebo	129	332	461
Total	205	731	936

As always, we want to use data from this sample to make generalizations about the general population of patients infected with HIV. You already know one way to make inferences from the data by calculating the 95% confidence interval (CI) for each of the two proportions using the methods of Chapter 2. Disease progressed in 16% of the patients receiving AZT, and the 95% CI ranges from 13% to 20%. Disease progressed in 28% of the subjects receiving placebo, and we can be 95% sure that the true value in the population lies between 24% and 32%. However, rather than analyze each group separately, it is better to analyze the difference or ratio of the proportions.

DIFFERENCE BETWEEN TWO PROPORTIONS

One way to summarize the data from Example 8.1 is to calculate the difference between the two proportions. Disease progressed in 28% of placebo-treated subjects and in 16% of AZT-treated subjects. In our sample, the difference is 28% − 16% or 12%.

More generally, the difference between two proportions can be calculated with Equation 8.1 using the variables A to D defined in Table 8.2. This kind of table is called a *contingency table*.

$$\text{Difference between proportions} = \Delta = P_1 - P_2 = \frac{A}{A + B} - \frac{C}{C + D}. \quad (8.1)$$

The equation to calculate the 95% CI for the difference is given at the end of the chapter. For this example, the 95% CI of the difference is 6.7% to 17.3%. If we assume our subjects are representative of the larger population of adults infected with the HIV but not yet symptomatic, we are 95% sure that treatment with AZT will reduce the incidence of disease progression by somewhere between 6.7% and 17.3%. Note that these calculations deal with the actual difference in incidence rates, not the relative

Table 8.2. Sample Contingency Table

	Disease	No Disease	Total
Exposed or treated	A	B	A + B
Not exposed or placebo	C	D	C + D
Total	A + C	B + D	A + B + C + D

change. When investigating a risk factor that might increase the risk of disease, the difference between two incidence rates is called *attributable risk**.

RELATIVE RISK

It is often more intuitive to think of the ratio of two proportions rather than the difference. This ratio is termed the *relative risk*. The relative risk is the ratio of incidence rates (Equation 8.2).

$$\text{Relative risk} = \frac{p_1}{p_2} = \frac{\dfrac{A}{A + B}}{\dfrac{C}{C + D}}. \tag{8.2}$$

Disease progressed in 28% of placebo-treated subjects and in 16% of AZT-treated subjects. The ratio is 16%/28% or 0.57. In other words, subjects treated with AZT were 57% as likely as placebo-treated subjects to have disease progression. A relative risk between 0.0 and 1.0 means that the risk decreases with treatment (or exposure to risk factor). A relative risk greater than 1.0 means that the risk increases. A relative risk = 1.0 means that the risk is identical whether or not the subject was given the treatment (or was exposed to the risk factor).

We could also have calculated the ratio the other way, as 28%/16%, which is 1.75. This means that subjects receiving the placebo were 1.75 times more likely to have disease progression than subjects receiving AZT. When interpreting relative risks, make sure you know which group has the higher risk.

In this example, it is pretty clear that the *risk* refers to disease progression (rather than lack of disease progression). In other contexts, one alternative outcome may not be worse than the other, and the *relative risk* is more appropriately termed the *relative probability* or *relative rate*. Because of this potential ambiguity, it is important to state clearly what the relative risk refers to. Don't just blindly plug numbers into Equation 8.2. First make sure that the table is arranged in such a way that the equation makes sense.

As always, a CI helps you interpret the data. The equation needed to calculate the CI of relative risk is given later in the chapter, but the calculations are usually done by computer. In the AZT example, the 95% CI of the relative risk is 0.44 to 0.74. Assuming that our sample is representative of the entire population, we can be 95% sure that the population relative risk lies within this range.

ASSUMPTIONS

To interpret the results of a prospective or experimental study you need to make the following assumptions:

---

*The term *attributable risk* has been defined in four different ways. When you see the term in publications, make sure you know what the authors are referring to. For more information on attributable risk, see Chapter 4 of HA Kahn, CT Sempos. *Statistical Methods in Epidemiology*. New York, Oxford University Press, 1989.

- The subjects are randomly selected from a population, or at least are representative of that population. The patients in the example were certainly not randomly selected, but it is reasonable to think that they are representative of adult asymptomatic people infected with the human immunodeficiency virus (HIV).*
- Each subject was selected independently of the rest. Picking one subject should not influence the chance of picking anyone else. You want to pick each subject individually. You don't want to include several people from one family or clusters of individuals who are likely to have the same strain of HIV.
- The only difference between the two groups is exposure to the risk factor (prospective study) or exposure to the treatment (experimental study). In this study subjects were randomly assigned to receive drug or placebo, so there is no reason to think that the two groups differ. But it is possible that, just by chance, the two groups differ in important ways. In the paper the authors presented data showing that the two groups were indistinguishable in terms of age, T-cell counts, sex, and HIV risk factors.

Of course, you also have to assume that the data were collected accurately and that the outcome was defined reasonably. You also need to be aware of extrapolating the conclusions too far. This study can tell you about disease progression as defined by symptoms and CD4+ cell counts. It can not tell you about the variable you really care about—survival. It's easy to forget this point and to generalize results too far. In fact, some longer studies have shown that AZT therapy of asymptomatic patients may not increase life span.

Statistical calculations usually analyze one outcome at a time. In this example, the outcome was disease progression defined in a particular way. The investigators measured other outcomes and analyzed these results too. They also looked at drug-induced side effects. Before you reach an overall conclusion from this study, you need to look at all the data. Even though statistical tests focus on one result at a time, you need to integrate various results before reaching an overall conclusion.

HOW THE RELATIVE RISK CAN BE MISLEADING

The relative risk is useful because it summarizes results in one number. But this simplification can also be misleading. One number is not really sufficient to summarize the data.

Here is an example in which the relative risk is insufficient. What is the public health importance of a vaccine that halves the risk of a particular infection? In other words, the vaccinated subjects have a relative risk of 0.5 of getting that infection compared with unvaccinated subjects. The answer depends on the prevalence of the disease that the vaccine prevents. If the risk in unexposed people is 2 in 10 million, then halving the risk to 1 in 10 million isn't so important. If the risk in unexposed people is 20%, then halving the risk to 10% would have immense public

---

*Don't confuse the two uses of the term *random*. Each subject was randomly assigned to receive drug or placebo. They were not randomly selected from the population of all people with asymptomatic HIV infection.

health consequences. The relative risk alone does not differentiate between the two cases.

Expressing the data as the difference in risks (rather than the ratio) is more helpful in this example. In the first example, the difference is 0.0000001; in the second case it is 0.1. Since many people have trouble thinking intuitively about small fractions, Laupacis and colleagues* have suggested taking the reciprocal of the difference and term the reciprocal the *number needed to treat* or NNT. In our example, NNT = 10,000,000 for the first example and 10 in the second. In other words, to prevent one case of disease you have to vaccinate 10 million people in the first case, and only 10 people in the second.

PROBABILITIES VERSUS ODDS

So far we have summarized the data in two ways (difference in risks and relative risk). There is still another way to summarize the data, the odds ratio. Before you can learn about odds ratios, you first need to learn about odds.

Likelihood can be expressed either as a probability or as odds.

- The *probability* that an event will occur is the fraction of times you expect to see that event in many trials.
- The *odds* are defined as the probability that the event will occur divided by the probability that the event will not occur.

Probabilities always range between 0 and 1. Odds may be any positive number (or zero). A probability of 0 is the same as odds of 0. A probability of 0.5 is the same as odds of 1.0. The probability of flipping a coin to heads is 50%. The odds are "fifty:fifty," which equals 1.0. As the probability goes from 0.5 to 1.0, the odds increase from 1.0 to approach infinity. For example, if the probability is 0.75, then the odds are 75:25, three to one, or 3.0.

Probabilities and odds are two ways of expressing the same concept. Probabilities can be converted to odds, and odds can be converted to probabilities. Convert between probability and odds using Equations 8.3:

$$\text{Odds} = \frac{\text{probability}}{1 - \text{probability}}.$$

$$\text{Probability} = \frac{\text{odds}}{1 + \text{odds}}. \tag{8.3}$$

In most contexts, there is no particular advantage to thinking about odds rather than probabilities. Most people are more comfortable thinking about probabilities. But, as you will see, there are a few situations where there is an advantage to using odds.

*A Laupacis, DL Saenett, RS Roberts. An assessment of clinically useful measures of the consequences of treatment. *N Engl J Med* 318:1728–33, 1988.

THE ODDS RATIO

For the AZT example, the odds of disease progression in the AZT-treated subjects is 0.19, and the odds of disease progression in the control patients is 0.39. The data can be summarized by taking the ratio of these values, called the *odds ratio*. Compared to control patients, the odds of disease progression in AZT-treated subjects is 0.19/ 0.39, which equals 0.49. In other words, the odds of disease progression in AZT-treated subjects is about half that of control patients.

More generally, the odds of the exposed (or treated) patients having the disease is (Equation 8.4)

$$\text{Odds of disease in exposed} = \frac{A/(A + B)}{B/(A + B)} = A/B.$$

$$\text{Odds of disease in unexposed} = \frac{C/(C + D)}{D/(C + D)} = C/D. \tag{8.4}$$

$$\text{Odds ratio} = \frac{\text{Odds in exposed}}{\text{Odds in unexposed}} = \frac{A/B}{C/D.} \tag{8.5}$$

If any of the values A through D are 0, Equation 8.5 cannot be calculated. In such cases some investigators add 0.5 to each of the four values A through D before calculating the odds ratio and its CI.

The equation needed to calculate the CI of the odds ratio is given later in this chapter, but the calculations are usually performed by computer. The CI of the odds ratio of our example is approximately 0.36 to 0.67. The interpretation should be familiar to you. If our sample is representative of the entire population, we can be 95% sure that the population odds ratio lies within that interval.

While most people find the idea of a relative risk pretty easy to grasp, they find odds ratios to be a bit strange. When analyzing prospective studies, cross-sectional studies, or experimental studies, there is no particular advantage to calculating an odds ratio. As you'll see in the next chapter, odds ratios are essential when analyzing retrospective case-control studies.

RELATIVE RISKS FROM SURVIVAL STUDIES

In Chapter 6, you learned how to plot survival curves for one group. How can you compare survival in two groups? One way is to calculate the overall relative risk. If the relative risk is 2.1, this means that subjects in one group died (on average) at a rate 2.1 times that the other group.

Calculating a relative risk from survival curves is difficult. You can't use the equations presented in this chapter because the subjects were not all followed for the same length of time. Special methods are used, and the details are not presented in this book. Even though the calculations are tricky, it is easy to interpret the relative risk (and its 95% CI).

When calculated from survival data, the relative risk is often referred to as the *hazard ratio*. See Chapter 33 for a more formal definition of hazard and hazard ratios.

By summarizing two entire survival curves with a single number, you are losing a lot of information. The relative risk only gives the overall difference between the groups. Look at the survival curves to see how the curves differ over time.

Rather than do complicated survival curve calculations, some investigators calculate risk per person-year rather than per person. For example, if 5 people are followed for 1 year, and 3 people are followed for 2 years, together you have 11 person-years (5 × 1) + (3 × 2). It is only fair to calculate data this way when the risk is consistent over time, so a subject followed for 2 years has twice the risk of a subject followed for 1 year. See an epidemiology text for more details.

WHAT IS A CONTINGENCY TABLE?

The data from several kinds of studies can be presented on a contingency table. The tables shown earlier in this chapter (Tables 8.1 and 8.2) are contingency tables. The next chapter continues with analyses of contingency tables.

The rows and columns of contingency tables can have many different meanings, depending on the experimental design. The rows usually represent exposure (or lack of exposure) to treatments or putative risk factors. Columns usually denote alternative outcomes. No matter what experimental design was used, the two rows must reprseent mutually exclusive categories and the two columns must also be mutually exclusive. Each "cell" in the table contains the number of subjects that are classified as part of one particular column and row.

Not all tables with two rows and two columns are contingency tables. Contingency tables show the number of subjects in various categories. Thus each number must be an integer. Tables of fractions, proportions, percentages, averages, changes, or durations are not contingency tables, and it would not make sense to calculate the relative risk or odds ratio. This is an important point. Applying methods appropriate for contingency tables to other kinds of data will generate meaningless results.

CALCULATING CONFIDENCE INTERVALS*

The equations for calculating CIs are a bit tedious and are best left to a computer. It is very difficult to calculate the exact CI, even with a computer, so it is common to approximate the intervals. If you want to calculate the approximate CI yourself, here are the equations.

CI of the Difference of Two Proportions

The equation for calculating the 95% CI for a difference between two proportions is a straightforward extension of the one used for calculating the 95% CI of one proportion.

---

*This section contains the equations you need to calculate statistics yourself. You may skip it without loss of continuity.

To calculate the difference in incidence rates from a prospective or cross-sectional study, use Equation 8.6:

$$P_1 = \frac{A}{A + B} \quad P_2 = \frac{C}{C + D}$$

$$D = P_1 - P_2$$

$$SE\ of\ difference = \sqrt{\frac{P_1(1 - P_1)}{N_1} + \frac{P_2(1 - P_2)}{N_2}}$$

$$95\%\ CI\ of\ difference =$$
$$(D - 1.96 \times SE\ of\ difference)\ to\ (D + 1.96 \times SE\ of\ difference). \quad (8.6)$$

The equation is not valid when any of the values A through D is less than 5.

CI of Relative Risk

Calculating the CI of a relative risk is not straightforward, and several methods have been developed to calculate an approximate CI. The CI is not symmetrical around the relative risk. This makes sense, as the relative risk can never be less than 0 but can get very high. However, the CI of the logarithm of the relative risk is approximately symmetrical. Katz's method takes advantage of this symmetry (Equation 8.7):

$$95\%\ CI\ of\ \ln(RR) = \ln(RR) \pm 1.96 \sqrt{\frac{B/A}{A + B} + \frac{D/C}{C + D}}. \quad (8.7)$$

This equation calculates the two 95% confidence limits of the natural logarithm of the RR. Take the antilogarithm (e^x) of each limit to determine the 95% CI of the RR.

Odds Ratio

Like the relative risk, the CI of the odds ratio is not symmetrical. This makes sense, as the odds ratio cannot be negative but can be any positive number. The asymmetry is especially noticeable when the odds ratio is low.

Several methods can be used to calculate an approximate the CI of the odds ratio. Woolf's method is shown in Equation 8.8:

$$95\%\ CI\ of\ \ln(OR) = \ln(OR) \pm 1.96 \sqrt{\frac{1}{A} + \frac{1}{B} + \frac{1}{C} + \frac{1}{D}}. \quad (8.8)$$

This method calculates the symmetrical 95% CI of the natural logarithm of the OR. Take the antilogarithm of both values to obtain the 95% CI of the OR.

SUMMARY

The results of cross-sectional, prospective, and some experimental studies can be displayed on contingency tables with two rows and two columns. The data can be summarized as the difference of two incidence rates (prevalence rates for cross-sectional studies), or as the ratio of the incidence (or prevalence) rates. This ratio is called the relative risk. You can calculate 95% confidence intervals for either the difference or the ratio. The results can also be expressed as the odds ratio. However, there is little advantage to presenting odds ratios from prospective, cross-sectional, or experimental studies.

OBJECTIVES

1. You should be familiar with the following terms:
   * Prospective study
   * Randomized double-blind study
   * Contingency table
   * Relative risk
   * Odds ratio
   * Cross-sectional study
   * Odds
2. You should be able to create a 2 × 2 contingency table from clinical data.
3. You should be able to recognize when a 2 × 2 table is a contingency table and when it is not.
4. Without using a book, you should be able to calculate the relative risk.
5. You must be able to recognize data for which the relative risk cannot be meaningfully calculated.
6. Using a book and calculator, you should be able to calculate 95% CI for the relative risk or the difference between two proportions.

PROBLEMS

1. The relative risk of death from lung cancers in smokers (compared with nonsmokers) is about 10. The relative risk of death from coronary artery disease in smokers is about 1.7. Does smoking cause more deaths from lung cancer or from coronary artery disease in a population where the mortality rate in nonsmokers is 5/100,000 for lung cancer and 170/100,000 for coronary artery disease?
2. Goran-Larsson et al. wondered whether hypermobile joints caused symptoms in musicians.* They sent questionnaires to many musicians and asked about hypermobility of the joints and about symptoms of pain and stiffness. They asked about all joints, but this problem concerns only the data they collected about the wrists. Of 96 musicians with hypermobile wrists, 5% had pain and stiffness of the wrists. In contrast, 18% of 564 musicians without hypermobility had such symptoms.

   A. Is this a prospective, retrospective, or cross-sectional study?
   B. Analyze these data as fully as possible.

*L Goran-Larsson, J Baum, GS Mudholkar, GD Lokkia. Benefits and disadvantages of joint hypermobility among musicians. *N Engl J Med* 329:1079–1082, 1993.

3. The same number of cells (100,000 per ml) were placed into four flasks. Two cell lines were used. Some flasks were treated with drug, while the others were treated only with vehicle (control). The data in the following table are mean cell counts (thousands per milliliter) after 24 hours. Analyze these data as fully as possible. If you had access to all the original data, how else might you wish to summarize the findings?

	No drug	Drug
Cell line 1	145	198
Cell line 2	256	356

4. Cohen et al. investigated the use of active cardiopulmonary resuscitation (CPR).* In standard CPR the resuscitator compresses the victim's chest to force the heart to pump blood to the brain (and elsewhere) and then lets go to let the chest expand. Active CPR is done with a suction device. This enables the resuscitator to pull up on the chest to expand it as well as pressing down to compress it. These investigators randomly assigned cardiac arrest patients to receive either standard or active CPR. Eighteen of 29 patients treated with active CPR were resuscitated. In contrast, 10 of 33 patients treated with standard CPR were resuscitated.

   A. Is this a cross-sectional, prospective, retrospective, or experimental study?
   B. Analyze these data as you see fit.

5. In a letter in *Nature* (356:992, 1992) it was suggested that artists had more sons than daughters. Searching *Who's Who in Art,* the author found that arists had 1834 sons and 1640 daughters, a ratio of 1.118. As a comparison group, the investigator looked at the first 4002 children of nonartists listed in *Who's Who* and found that the ratio of sons/daughters was 1.0460. Do you think the excess of sons among the offspring of artists is a coincidence?

---

*TJ Cohen, BG Goldner, PC Maccaro, AP Ardito, S Trazzera, MB Cohen, SR Dibs. A comparison of active compression-decompression cardiopulmonary resuscitation with standard cardiopulmonary resuscitation for cardiac arrests occurring in the hospital. *N Engl J Med* 329:1918–1921, 1993.

Confidence Interval of the Ratio
of Two Proportions:
Case-Control Studies

WHAT IS A CASE-CONTROL STUDY?

Example 9.1

Cat scratch disease causes swollen lymph nodes. As the name suggests, it usually follows a scratch from a cat. Although it is usually a mild condition, thousands of people are hospitalized for it each year. Zangwill et al. studied risk factors for the disease.* One of their goals was to find out whether cat scratch disease is more likely in owners of cats with fleas. If so, it would make sense to do further studies to find out if the organism causing the disease is spread by fleas rather than cats. To perform a prospective study, they would need to follow a large group of cat owners and see who gets cat scratch disease. Since only a small proportion of cat owners get the disease, they would need to follow a large group of people for a long time. Rather than do that, the authors used an alternative approach. They performed a case-control study. They sent a letter to all primary care physicians in the state asking them to report all cases of cat scratch disease they had seen in the last year. They then called random phone numbers to find cat owners who did not have cat scratch disease. They asked both groups whether their cats had fleas.

This kind of study is called a *case-control study,* because the investigators pick cases and controls to study. It is also called a *retrospective study,* because the investigators start with the disease and look back in time to try to learn about the cause.

The results are summarized in Table 9.1.

This looks very similar to the contingency table from the previous example (AZT). But there is a big difference. In the AZT example, the investigators set the row totals by choosing how many subjects got each treatment. In this example, the investigators set the column totals by choosing how many cases and controls to study.

---

*KM Zangwill, DH Hamilton, BA Perkins, et al. Cat scratch disease in Connecticut. *N Engl J Med* 329:8–13, 1993.

Table 9.1. Results of Example 9.1

	Cases (Cat Scratch Disease)	Controls (Healthy)
Cat with fleas?		
Yes	32	4
No	24	52
Total	56	56

WHY CAN'T YOU CALCULATE THE RELATIVE RISK FROM CASE-CONTROL DATA?

You shouldn't try to calculate the relative risk from a case-control study. Here's why. In a case-control study, the investigator selects subjects known to have the disease or known to be disease free. Therefore the study gives no information at all about the risk or incidence of getting the disease. If you want to know about risk, you have to select subjects based on exposure to a treatment or toxin, and then find out whether they get the disease. In a case-control study, you select patients because you know they have the disease, so you can't use the data to determine the risk or relative risk.

    If you were to try to plug these numbers into Equation 8.2 to calculate a relative risk, your first step would be to calculate the risk that someone whose cat had fleas would get the disease as 32/(32 + 4). But this is not a helpful calculation. The authors happened to choose to study an equal number of cases and controls. They could just as easily have chosen to study twice as many controls. If they had done so, they would probably have found that about 8 of the 112 subjects recalled fleas. Now the risk would be calculated as 32(32 + 8), and the relative risk would have been different. Calculating the relative risk from case-control studies does not lead to a meaningful result.

HOW TO INTERPRET THE ODDS RATIO

How then can you summarize data from a case-control study? The answer is surprising. Although the relative risk equation does not work with case-control data, you can meaningfully calculate the odds ratio from retrospective studies. Moreover, if the disease is fairly rare (affects less than a few percent of the population studied), then the odds ratio calculated from a case-control study will be approximately equal to the true relative risk. This is a surprising conclusion, and I'll prove it in the next section with algebra. It's such an important idea that it is worth repeating and emphasizing: *If the disease is fairly rare, then the odds ratio calculated from a case-control study will approximately equal the true relative risk.*

    For this example, the odds ratio is 17.3. If we assume that cat scratch disease is rare among cat owners,* we can conclude that owners of cats with fleas are approxi-

---

*Other data tells us that this is a reasonable assumption. The case-control study tells us nothing about whether cat scratch disease is common or rare among cat owners. Since we are only studying the population of cat owners, the incidence of the disease in other populations (or the general population) is irrelevant.

mately 17.3 times more likely to get cat scratch disease than owners of cats without fleas. An equation for calculating the 95% confidence interval (CI) is shown at the end of the chapter. For this example, the 95% CI of the odds ratio ranges from 5.5 to 54.6. In the overall population, we can be 95% sure the true odds ratio is somewhere between 5.5 and 54.6. Because this range is so wide, some might say that the study is quite inconclusive. Others would point out that even if the true odds ratio is near the lower end of the CI, it is still substantial (fivefold increase in risk). Even though the study is too small to determine the true odds ratio very exactly, it shows clearly that the true odds ratio is almost certainly far from 1.0.

WHY THE ODDS RATIO FROM A CASE-CONTROL STUDY APPROXIMATES THE REAL RELATIVE RISK*

In the last section, I stated (but did not prove) an important point: If the disease is fairly rare, then the odds ratio calculated from a case-control study will approximately equal the true relative risk. Here I'll prove the point. Imagine what would happen if we collected data from all cat owners in the state to find out whether they get cat scratch disease. The data would look like Table 9.2.

This is called a cross sectional study. We don't pick subjects based on either row variable (fleas vs. no fleas) or column variable (disease or not). We just pick subjects from the population of interest (cat owners) and determine all the numbers.

If we had all this data, we could easily calculate the relative risk of cat scratch disease in the kitten owners as compared to the cat owners as follows:

$$\text{Relative risk} = \frac{\dfrac{A}{A + B}}{\dfrac{C}{C + D}}. \tag{9.1}$$

If we assume that cat scratch disease is rare among the population of cat owners, then A must be much smaller than B, and C must be much smaller than D. In this case, Equation 9.1 can be simplified as shown in Equation 9.2. This equation shows that the relative risk and the odds ratio are approximately equal when calculated from cross-sectional studies of rare diseases:

$$\text{Relative risk} = \frac{\dfrac{A}{A + B}}{\dfrac{C}{C + D}} \approx \frac{\dfrac{A}{B}}{\dfrac{C}{D}} \approx \text{odds ratio}. \tag{9.2}$$

When performing the case-control study, the investigators choose the relative number of cases and controls. In other words, the investigators choose the column totals. Of all the cases, they chose some fraction (we'll call it M) of them to study. In this study, M is probably a pretty big fraction because the investigators attempted to study all cases in the state. Of all the healthy cat owners in the state, the investigators

---

*This section is more advanced than the rest. You may skip it without loss of continuity.

Table 9.2. Hypothetical Cross-Sectional Data

Cross-sectional	Cat Scratch Disease	Healthy	Total
Fleas	A	B	A + B
No fleas	C	D	C + D
Total	A + C	B + D	N = A + B + C + D

studied some fraction (we'll call it K) of them. For this study, K is a tiny fraction. The data are given in Table 9.3.

This table has six variables (A, B, C, D, M, K) but we don't know any of them. The only way we could know A through D would be to a full cross-sectional study. We know the total number of cases and controls we studied (the column totals), but we don't know what fraction of all cases and controls in the state these represent. So we don't know K and M. We do know that A * M = 32, that B * K = 4, that C * M = 24, and that D * K = 52.

Equation 9.3 calculates the odds ratio from our cross-sectional data:

$$\text{Odds ratio} = \frac{32 \, / \, 4}{24 \, / \, 52} = 17.3. \tag{9.3}$$

$$\text{Odds ratio} = \frac{AM \, / \, BK}{CM \, / \, DK} = \frac{A \, / \, B}{C \, / \, D}.$$

The values of K and M (which you don't know but were set by your experimental design) drop out of the calculations. The odds ratio calculated from the case-control study is identical to the odds ratio you would have calculated from a cross-sectional study. We have already seen that the odds ratio from a cross-sectional study has a value very close to the relative risk when the disease is rare. So now we have shown that the odds ratio calculated from a case-control study has a value very close to the relative risk you would have calculated from a cross-sectional study.*

Using the same nomenclature, let's review why direct calculations of relative risk are not useful with data from a case-control study. Equation 9.4 shows that the values of K and M do not drop out of the relative risk calculations. Therefore the value you come up with depends in part on the ratio of cases to controls you chose to use. The relative risk value is not meaningful:

$$\text{Relative risk} = \frac{\dfrac{AM}{AM + BK}}{\dfrac{CM}{CM + DK}} \neq \frac{\dfrac{A}{A + B}}{\dfrac{C}{C + D}}. \tag{9.4}$$

---

*This analysis neglects the contribution of sampling variation. Any particular odds ratio from a case-control study may be higher or lower than the true relative risk. On average, the odds ratio from a case-control study will equal the true relative risk.

Table 9.3. Hypothetical Retrospective Data

Retrospective	Cases	Controls
Exposed	A * M	B * K
Not exposed	C * M	D * K
Total	(A + C) * M	(B + D) * K

ADVANTAGES AND DISADVANTAGES OF CASE-CONTROL STUDIES

The advantages of case-control studies are clear:

- They require fewer subjects than prospective studies. To perform a prospective study of the cat scratch example, you'd need to start with many more subjects because only a small fraction of cat owners get cat scratch disease.
- They can be performed more quickly. You don't have to wait for the natural history of the disease to unfold. You start with the disease and look back at risk factors. The time required to complete the study depends on the number of subjects and on the energy and resources of the investigators.
- You can perform some case-control studies by examining existing records. You can identify appropriate cases and controls through hospital records and often find the information about exposure or risk by reading their medical charts.

The problem with case-control studies is that it is hard to pick the right controls. You want to control for extraneous factors that might confuse the results but not control away the effects you are looking for.

In the cat scratch example, the authors picked controls by dialing random phone numbers. If the person answering owned a cat (and was of the correct age), he or she was invited to become part of the study by answering some questions. There are some problems with this approach:

- The controls were picked because they owned a cat or kitten. Therefore, the study could not determine anything about the association of cats with the disease.
- The controls were picked to have the same age as the subjects. Therefore, the study could not determine anything about whether age is a risk factor.
- The subjects obviously know whether or not they had suffered from cat scratch disease. The cases (those who had the disease) may recall their cats more vividly than do the controls.
- The interviewers knew whether they were talking to a control subject or to a case (someone who had cat scratch disease). Although they tried to ask the questions consistently, they may have inadvertently used different emphasis with the cases.
- The subjects had suffered from cat scratch disease and may be motivated to help the researchers learn more about the disease. The controls just happened to answer the phone at the wrong time. They probably never heard of cat scratch disease, and their main motive may be to end the interview as quickly as possible. Thus cases and controls may not give equally detailed or accurate information.

- The only way patients became part of the study was through physician referral. The study did not include people who had mild cases of the disease and did not seek medical attention. Not everyone who is sick seeks medical attention, and this study selects for people who go to doctors when mildly ill. This selection criteria was not applied to the controls.
- The only way that controls became part of the study was to be home when the investigators randomly dialed their phone number. This method selects for people who stay home a lot and for people who have more than one phone line. It selects against people who don't have phones, who travel a lot, or who screen their calls with answering machines. None of these selection criteria were applied to the cases.

It would be easy to design a study to circumvent some of the problems listed above, but you'd probably introduce new problems in the process. And you don't want to match cases and controls too carefully, or you may match away the variable you care about (if they had matched for whether cats had fleas, they couldn't have asked about the association of fleas and the disease). As a result, it is usually possible to find an alternative explanation for case-control data, and often these alternative explanations are plausible. Here are five alternative explanations for the data in the cat scratch study:

- Cat scratch disease is caused by an organism spread by fleas. This is the hypothesis the investigators believe.
- The patients knew that they had gotten sick from their cat and thus have more vivid memories of their cats. Because they have a personal interest in cat scratch disease, they are likely to think more carefully about their answers to the investigators' questions. The two groups of cats may have had equal numbers of fleas, but the patients may be more likely to remember them than the controls.
- Fleas are more common in rural areas of the state, and cats in rural areas are more apt to be bitten by ticks. Perhaps ticks are the real vector of the disease.
- The controls are more apt to stay home (and to have been home when the investigators called, see above) and thus interact with their cats more. The patients travel more and were more apt to get scratched when they came home.
- The patients are more apt to go to doctors for mild illnesses. Perhaps they are also more likely to take medications when they have a cold or flu. Perhaps their cats hate the smell of one of these medications and thus are apt to scratch their owner.

The first two explanations seem plausible. The next three are a bit far fetched, but not impossible. The point is that case-control data are rarely definitive. Case-control studies are a terrific way to test new hypotheses, but prospective studies are usually needed for definitive results.

ASSUMPTIONS IN CASE-CONTROL STUDIES

To interpret the results of a case-control study you need to make the following assumptions:

- The cases and controls are randomly selected from the populations, or at least are representative of those populations. In our example, the controls were picked by dialing random phone numbers and so are a fairly random sample. The cases were not randomly selected. But it is reasonable to think that cat owners in Connecticut are representative of cat owners elsewhere.

• Each subject was selected independently of the rest. Picking one subject should not influence the chance of picking anyone else. You wouldn't want to include several subjects from one family, because they are all exposed to the same cat.

• When selecting controls, you have not selected subjects that differ systematically from the cases in any way except for the absence of disease. As we have discussed, it is very difficult to be sure a case-control study has complied with this assumption.

MATCHED PAIRS

In a standard case-control study, the investigator compares a group of controls with a group of cases. As a group, the controls are supposed to be similar to the cases (except for the absence of disease). Another way to perform a case-control study is to match individual cases with individual controls based on age, gender, occupation, location, and other relevant variables.

Displaying and analyzing data from matched case-control studies obscures the fact that the cases and controls were matched. Matching makes the experiment stronger, so the analysis ought to take it into account.

An alternative way of tabulating the data to emphasize the matching is shown in Table 9.4.

In this graphic example, the investigators studied 134 cases and 134 matched controls. Each entry in the table represents one pair (a case and a control). This is not a contingency table, so the usual equations are not valid. It turns out that the odds ratio can be calculated very simply. The 13 pairs in which both cases and controls were exposed to the risk factor provide no information about the association between risk factor and disease. Similarly, the 92 pairs in which both cases and controls were not exposed to the risk factor provide no information. The odds ratio is calculated as the ratio of only the other two values: pairs in which the control has the risk factor but the case doesn't and pairs in which the case has the risk factor but the control doesn't. In this example the odds ratio for the risk factor being associated with disease is $25/4 = 6.25$. The 95% CI of the odds ratio is 2.18 to 17.81 (the equations are given below).

You can interpret the odds ratio from a matched case-control study just as you would interpret the odds ratio from an ordinary case-control study. If we assume that the disease is fairly rare, then we can conclude that exposure to the risk factor increases

Table 9.4. A Matched Pairs Case-Control Study

	Cases	
	Risk Factor +	Risk Factor −
Controls		
Risk factor +	13	4
Risk factor −	25	92

one's risk 6.25-fold. We can be 95% sure that the true odds ratio lies between 2.18 and 17.81.

CALCULATING THE 95% CI OF AN ODDS RATIO

Unpaired Studies

Several methods can be used to calculate an approximate CI interval of the odds ratio. Woolf's method is shown in Equation 9.5:

95% CI of ln(OR):

$$\left(\ln(OR) - 1.96 \sqrt{\frac{1}{A} + \frac{1}{B} + \frac{1}{C} + \frac{1}{D}} \right) \text{ to} \qquad (9.5)$$

$$\left(\ln(OR) + 1.96 \sqrt{\frac{1}{A} + \frac{1}{B} + \frac{1}{C} + \frac{1}{D}} \right).$$

This method calculates the symmetrical 95% CI of the natural logarithm of the OR. Take the antilogarithm of both values to obtain the asymmetrical 95% CI of the OR. *Asymmetrical* means that the interval is not centered on the OR.

Paired Studies

The odds ratio is calculated from the number of discordant pairs (one subject exposed to the risk factor, the other not). Call these two numbers R and S. The odds ratio is R/S. To calculate the 95% CI, calculate the symmetrical approximate CI of the log odds ratio using Equation 9.6 and then convert back:

$$95\% \text{ CI of ln(OR): } \left(\ln(OR) - 1.96 \sqrt{\frac{1}{R} + \frac{1}{S}} \right) \text{ to } \left(\ln(OR) + 1.96 \sqrt{\frac{1}{R} + \frac{1}{S}} \right).$$

$$(9.6)$$

SUMMARY

In a *case-control study* (also called *retrospective* studies) the investigators work backwards and start with the outcome. They compare two groups of subjects. One group has the disease or condition being studied. These are the cases. The other group is selected to be similar in many ways, but not to have the condition. These are the controls. The investigators then look back in time to compare the exposure of the two groups to a possible risk factor. The results are best summarized as an odds ratio, with its CI. While it is meaningless to calculate a relative risk from a case-control study, the odds ratio can be interpreted just like a relative risk so long as the incidence of the disease is low. In some studies, individual cases and controls are matched. When this is done, special equations should be used to compute the odds ratio.

OBJECTIVES

1. You should be familiar with the following terms:
   - Contingency table
   - Odds ratio
   - Prospective
   - Retrospective
2. Without using a book, you should be able to convert between odds and probability.
3. You should be able to create 2 × 2 tables from clinical data.
4. You should be able to recognize when a 2 × 2 table is a contingency table and when it is not.
5. Without using a book, you should be able to calculate the odds ratio.
6. You must be able to recognize data for which the relative risk cannot be meaningfully calculated.
7. Using a book and calculator, you should be able to calculate 95% CI for the odds ratio.
8. You should be able to recognize data for which a paired analysis is appropriate.

PROBLEMS

1. Can an odds ratio be greater than one? Negative? Zero?
2. Can the logarithm of the odds ratio be greater than one? Negative? Zero?
3. Gessner and colleagues investigated an outbreak of illness in an Alaskan community.* They suspected that one of the town's two water supplies delivered too much fluoride, leading to fluoride poisoning. They compared 38 cases with 50 controls. Thirty-three of the cases recalled drinking water from water system 1, while only four of the controls had drunk water from that system.

   Analyze these data as you think is appropriate. State your assumptions.

   How could these investigators have conducted a prospective study to test their hypothesis?

---

*BD Gessner, M Beller, JP Middaugh, GM Whitford. Acute fluoride poisoning from a public water system. *N Engl J Med* 330:95–99, 1994.

III

INTRODUCTION
TO P VALUES

I've put it off for ninc chapters, but I can't delay any longer. It's time to confront P values. If you've had any exposure to statistics before, you've probably already heard about *P values* and *statistical significance*. It's time to learn what these phrases really mean. These chapters explain P values generally, without explaining any particular statistical tests in any detail. You'll learn more about specific tests in Part VII.

What Is a P Value?

INTRODUCTION TO P VALUES

When using statistics to compare two groups, there are two approaches you can use:

- You've already learned about one approach—calculating the confidence interval (CI) for the difference between means or the difference (or ratio) of two proportions. With this approach, you are focusing on the question: "How large is the difference in the overall population?" You start with the difference (or ratio) you know in the sample, and calculate a zone of uncertainty (the 95% CI) for the overall population.
- This chapter introduces you to the second approach, calculating P values. This approach focuses on a different question: "How sure are we that there is, in fact, a difference between the populations?" You observed a difference in your samples, but it may be that the difference is due to a coincidence of random sampling rather than due to a real difference between the populations. Statistical calculations cannot tell you whether that kind of coincidence has occurred but can tell you how rare such a coincidence would be.

Calculations of P values and CIs are based on the same statistical principles and the same assumptions. The two approaches are complementary. In this book, I separated the two approaches to aid learning. When analyzing data, the two approaches should be used together. The easiest way to understand P values is to follow an example.

A SIMPLE EXAMPLE: BLOOD PRESSURE IN MEDICAL STUDENTS*

You want to test whether systolic blood pressure differs between first- and second-year medical students (MS1 and MS2, respectively). Perhaps the stress of medical school increases blood pressure. Measuring the blood pressure in the entire class seems like a lot of work for a preliminary study, so instead you randomly selected five students from each class and measured his or her systolic blood pressure rounded to the nearest 5 mmHg:

MS1: 120, 80, 90, 110, 95

MS2: 105, 130, 145, 125, 115

*You've already encountered these fake data in Example 5.2, and you'll see them again late in the book.

First you should look at the data. It helps to see a graph. Figure 10.1 shows blood pressures for each individual. Clearly the blood pressure tends to be lower in the first-year students than in the second-year students, although the two overlap. The mean values are 99 for MS1 and 124 for MS2. The difference between the mean values is 25 mmHg.

Figure 10.2 shows only the mean and standard error of the mean (SEM). Note how easy it is to be misled by the SEM error bars. It is easy to forget that the SEM error bars do not directly show you the scatter of the data. Biomedical research papers often show data in the format of Figure 10.2, even though the format of 10.1 is more informative. To make any sense at all of Figure 10.2, you would need to read the figure legend or the methods section to find out the sample size and whether the error bars represented standard deviation (SD) or SEM (or something else). The right half of Figure 10.2 also shows mean and SEM, but the Y axis doesn't begin at 0. This appears to magnify the difference between the two classes. If you don't notice the scale of the axis, you could be deceived by the right half of Figure 10.2.

Next, you should think about the biological or clinical implications of the data. A change of 25 mmHg in blood pressure is substantial and would have important clinical ramifications if the difference were consistent. This interpretation does not come from any statistical calculation; it comes from knowing about blood pressure. In contrast, a change of 25 units in some other variable might be trivial. Statistical calculations can't determine whether differences are clinically or scientifically important.

Before continuing with data analysis, you need to think about the design of the study. Were the data collected in a way that any further analysis is useful? In this

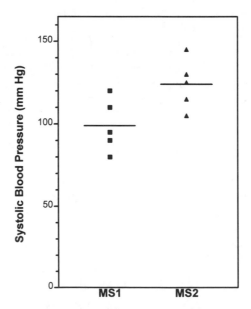

Figure 10.1. Sample data shown on a column scatter graph. Each square shows the systolic blood pressure of one first-year student (MS1). Each triangle shows the systolic blood pressure of one second-year student (MS2).

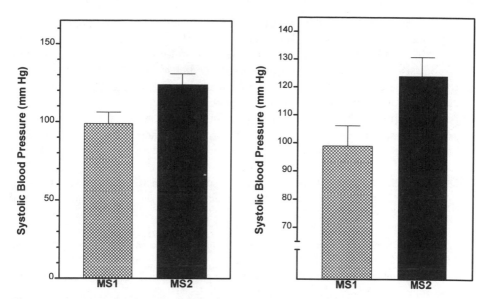

Figure 10.2. Sample data shown on a bar graph. This graph shows the mean and standard error of the blood pressures for the two samples. In the left panel, the Y axis begins at 0. In the right panel, the difference between the two means appears to be amplified because the axis doesn't begin at 0. You can be misled by this kind of graph unless you notice where the axis begins. You can't interpret these graphs unless you know that the error bar represents the standard error.

study, you would need to know how the students were selected and how their blood pressure was measured. We've already discussed this problem in Chapter 3. Let's assume that the experimental design was impeccable.

Two possible explanations remain for the difference between the two samples. The first possibility is that the overall populations of MS1s and MS2s have identical distributions of blood pressures, and that the difference observed in this experiment was just a coincidence—we just happened to select MS2s with higher blood pressure. The second possibility is that the mean systolic blood pressure of MS2s is really higher than that of MS1s. Before believing that these data represent a real difference, we would like to ask "what is the probability that the difference is due to chance?" Unfortunately, statistical calculations cannot answer that question.

Instead, statistical calculations can answer a related question: If one assumes that the distribution of blood pressure is identical in the two populations, what is the probability that the difference between the means of randomly selected subjects will be as large or larger than actually observed? The answer to this question is a probability, the P value.

There are several methods to calculate the P value, and these will be discussed later in the book. The general approach can be summarized as follows:

1. Assume that the individuals measured (the samples) are randomly selected from a larger group (the populations). If the subjects weren't randomly selected, at least assume that the subjects are representative of a larger population. Furthermore, assume that the experimental design is without biases or flaws.

2. Tentatively hypothesize that the distribution of values in the two populations are the same. This hypothesis is called the *null hypothesis,* sometimes abbreviated H_o. Most likely, the investigator did not believe (or did not want to believe) the null hypothesis. Rather, the investigator's experimental hypothesis (the whole reason for doing the experiment) was that the populations are different. This is sometimes called the *experimental* or *alternative hypothesis.*
3. Assuming the null hypothesis is true, calculate the likelihood of observing various possible results. You can use several methods (discussed later in this book) depending on the nature of the data and on which additional assumptions you are willing to make.
4. Determine the fraction of those possible results in which the difference between means is as large or larger than you observed. The answer is given in the form of a probability, called the *P value.*

Thinking about P values seems quite counterintuitive at first, as you must use backwards, awkward logic. Unless you are a lawyer or a Talmudic scholar used to this sort of argument by contradiction, you will probably find this sort of reasoning a bit uncomfortable. Four aspects are awkward:

- The hypothesis you are testing (the null hypothesis) is opposite to the hypothesis the experimenter expects or hopes to be true.
- Although mathematicians are comfortable with the idea of probability distributions, clinicians and scientists find it strange to calculate the theoretical probability distribution of the results of experiments that will never be performed.
- The derivation of the theoretical probability distributions depends on mathematics beyond the ready reach of most scientists.
- The logic goes in a direction that seems intuitively backwards. You observed a sample and want to make inferences about the population. Calculations of the P value start with an assumption about the population (the null hypothesis) and determine the probability of randomly selecting samples with as large a difference as you observed.

This book will present three methods for calculating a P value to compare a measured variable in two groups (t test, randomization test, and Mann-Whitney test). The t test is used most commonly. By plugging the data from our example into a computer program, we would learn that the P value is 0.034 (two tailed, see below).

Interpreting the P value is straightforward. If this null hypothesis were true, then 3.4% of all possible experiments of this size would result in a difference between mean blood pressures as large as (or larger) than we observed. In other words, if the null hypothesis were true, there is only a 3.4% chance of randomly selecting samples whose means are as far apart (or further) than we observed.

What conclusion should you reach? That's up to you. Statistical calculations provides the P value. You have to interpret it. Assuming that the study design was perfect, there are two possibilities: One possibility is that the two populations have different mean blood pressures. The other possibility is that the two populations are identical, and the observed difference is a coincidence of sampling. Statistical calculations determine how rare that coincidence would be. In this example we can say that such a coincidence will occur 3.4% of the time *if* there really is no difference between populations.

Often, the P value is used to make a statement about statistical significance. This is explained in the next chapter.

OTHER NULL HYPOTHESES

When we compare the means of two groups, the null hypothesis is that the two populations have identical means. When you make other kinds of comparisons, the null hypotheses are logical:

- When you compare two proportions, the null hypothesis is that the two proportions are identical in the population.
- When you compare two survival curves, the null hypothesis is that the two survival curves are identical in the population.

COMMON MISINTERPRETATIONS OF P VALUES

P values are easy to misinterpret. The best way to avoid misinterpreting the P value is to keep firmly in mind what it does tell you: The P value is the probability of getting a difference as big (or bigger) than you got if the null hypothesis is really correct. Thus, a P value of 0.03 means that even if the two populations have identical means, 3% of experiments like the one you conducted would yield a difference at least as large as you found.

It is very tempting to jump from this and say, "Oh, well, if there is only a 3% probability that my difference would have been caused by random chance, then there must be a 97% probability that it was caused by a real difference." Wrong! What you can say is that if the null hypothesis were true, then 97% of experiments would lead to a difference smaller than the one you observed, and 3% of experiments would lead to a difference as large or larger than the one you observed.

Calculation of a P value is predicated on the assumption that the null hypothesis is correct. P values cannot tell you whether this assumption is correct. P value tells you how rarely you would observe a difference as larger or larger than the one you observed if the null hypothesis were true. The question that the scientist must answer is whether the result is so unlikely that the null hypothesis should be discarded.

ONE-TAILED VERSUS TWO-TAILED P VALUES

A two-tailed P value is the probability (assuming the null hypothesis) that random sampling would lead to a difference as large as or larger than the observed difference with either group having the larger mean. A one-tailed P value, in contrast, is the probability (assuming the null hypothesis) that random sampling would lead to a difference as large as or larger than the observed difference, and that the group specified in advance by the experimental hypothesis has the larger mean.

If the observed difference went in the direction predicted by the experimental hypothesis, the one-tailed P value is half the two-tailed P value.* The terms *one-sided* and *two-sided* P values are equivalent to one and two tails.

In the blood pressure example, I chose to use a two-tailed P value. Thus the P value answers this question: If the null hypothesis is true, what is the chance that the difference between two randomly selected samples of five subjects would have been 25 mmHg or greater with either group having the higher mean BP. A one-tailed P value is the probability (under the null hypothesis) of observing a difference of 25 mmHg or greater, with the second-year class having the larger mean value.

A one-tailed test is appropriate when previous data, physical limitations, or common sense tells you that the difference, if any, can only go in one direction. Here is an example in which you might appropriately choose a one-tailed P value: You are testing whether a new antibiotic impairs renal function, as measured by serum creatinine. Many antibiotics poison kidney cells, resulting in reduced glomerular filtration and increased serum creatinine. As far as I know, no antibiotic is known to decrease serum creatinine, and it is hard to imagine a mechanism by which an antibiotic would increase the glomerular filtration rate. Before collecting any data, you can state that there are two possibilities: Either the drug will not change the mean serum creatinine of the population, or it will increase the mean serum creatinine in the population. You consider it impossible that the drug will truly decrease mean serum creatinine of the population and plan to attribute any observed decrease to random sampling. Accordingly, it makes sense to calculate a one-tailed P value. In this example, a two-tailed P value tests the null hypothesis that the drug does not alter the creatinine level; a one-tailed P value tests the null hypothesis that the drug does not increase the creatinine level.

Statisticians disagree about when to use one- versus two-tailed P values. One extreme position is that it is almost never acceptable to report a one-tailed P value—that a one-tailed P value should be reported only when it is physically impossible for a difference to go in a certain direction. Thus a two-tailed P value ought to be used in the preceding example, because a drug-induced decrease in serum creatinine is not impossible (although it is unprecedented and unexplainable).

The other extreme position is that one-tailed P values are almost always appropriate. The argument is that formal studies (requiring formal statistical analysis) should only be performed after the investigator has proposed an experimental hypothesis based on theory and previous data. Such a hypothesis should specify which group should have the larger mean (or larger proportion, or longer survival, etc.). Indeed one could question whether taxpayer's money should be spent on research so vaguely conceived that the direction of the expected difference can't be specified in advance.

The issue here in deciding between one- and two-tailed tests is not whether or not you expect a difference to exist. If you already knew whether or not there was a difference, there is no reason to collect the data. Rather, the issue is whether the direction of a difference (if there is one) can only go one way. You should only use a one-tailed P value when you can state with certainty (and before collecting any data) that in the overall populations there either is no difference or there is a difference in a specified direction. If your data end up showing a difference in the "wrong" direction, you should be willing to attribute that difference to random sampling without even

*There are exceptions, such as Fisher's exact test.

considering the notion that the measured difference might reflect a true difference in the overall populations. If a difference in the "wrong" direction would intrigue you (even a little), you should calculate a two-tailed P value.

Two-tailed P values are used more frequently than one-tailed P values. I have chosen to use only two-tailed P values for this book for the following reasons:

- The relationship between P values and confidence intervals is more clear with two-tailed P values.
- Two-tailed P values are larger (more conservative). Since many experiments do not completely comply with all the assumptions on which the statistical calculations are based, many P values are smaller than they ought to be. Using the larger two-tailed P value partially corrects for this.
- Some tests compare three or more groups, which makes the concept of tails inappropriate (more precisely, the P value has more than two tails). A two-tailed P value is more consistent with P values reported by these tests.
- Choosing one-tailed P values can put you in awkward situations. If you decided to calculate a one-tailed P value, what would you do if you observed a large difference in the opposite direction to the experimental hypothesis? To be honest, you should state that the P value is large and you found "no significant difference." But most people would find this hard. Instead, they'd be tempted to switch to a two-tailed P value, or stick with a one-tailed P value, but change the direction of the hypothesis. You avoid this temptation by choosing two-tailed P values in the first place.

When interpreting published P values, note whether they are calculated for one or two tails. If the author didn't say, the result is somewhat ambiguous. The terms *one-sided* and *two-sided* P values mean exactly the same thing as *one-tailed* and *two-tailed* P values.

EXAMPLE 10.1. COMPARING TWO PROPORTIONS FROM AN EXPERIMENTAL STUDY

Later in the book (Part VII), you'll learn how to calculate many common statistical tests. But you can interpret P values without knowing too much about the individual tests. This example and the next four show you real examples of P values and how to interpret them.

This example uses the same data as example 8.1. This study compared disease progression in asymptomatic people infected with HIV. The relative risk was 0.57. This means that disease progressed in 57% as many treated patients as placebo-treated patients.

The null hypothesis is that AZT does not alter the probability of disease progression. We want to calculate the P value that answers this question: If the null hypothesis were true, what is the chance that random sampling of subjects would result in incidence rates different (or more so) from what we observed? The P value depends on sample size and on how far the relative risk is away from 1.0.

You can calculate the P value using two different tests: Fisher's exact test or the chi-square test. Both tests will be described in Chapter 27, but you don't need to know much about the tests to interpret the results. The chi-square test is used more often, but only because it is easier to calculate by hand. Fisher's test calculates a more accurate P value and is preferred when computers do the calculating. The results from InStat are given in Table 10.1.

The P value is tiny, less than 0.0001. If you had chosen the chi-square test rather than Fisher's test, the P value would still have been less than 0.0001.

Interpreting the P value is straightforward: If the null hypothesis is true, there is less than a 0.01% chance of randomly picking subjects with such a large (or larger) difference in incidence rates. Note that InStat (like most statistical programs) also makes a statement about statistical significance. You'll learn what that means in the next chapter.

To interpret the P value you need to make the following assumptions:

• The subjects represent the population of all people who are infected with the HIV but have no symptoms who are or will be treated with AZT or placebo.
• Each subject was selected independently of the rest. Picking one subject should not influence the chance of picking anyone else.
• The only difference between the two groups is the treatment.

Are you ready to recommend AZT to these patients? Before deciding, remember that AZT is a toxic drug. You shouldn't base an overall conclusion from the study on one P value. Rather you should look at the benefit (which may be measured in several ways) and the risks or side effects of the treatment. You should also take into account other things you know about the AZT treatment from other studies.

EXAMPLE 10.2. COMPARING TWO PROPORTIONS FROM A CASE-CONTROL STUDY

This is a case-control study investigating the association between fleas and cat scratch fever. The results were shown in Table 9.1.

The odds ratio is 17.3. We want to calculate a P value. The null hypothesis is that there is no association between fleas and cat scratch disease, that the cats of cases are just as likely to have fleas as the cats of controls. The two-sided P value answers this question: If the null hypothesis were true, what is the chance of randomly picking subjects such that the odds ratio is 17.3 or greater or 0.058 (the reciprocal of 17.3) or lower?

To calculate a P value, you may use Fisher's exact test or the chi-square test. The same methods are used to analyze prospective and case-control studies. If a computer is doing the work, Fisher's test is your best choice. The results from InStat are shown

---

Table 10.1. Results from InStat Analysis of Example 10.1

Fisher's Exact Test

The two-sided P value is <0.0001, considered extremely significant.
There is a significant association between rows and columns.
Relative risk = 0.5718.
95% confidence interval: 0.4440 to 0.7363 (using the approximation of Katz).

Table 10.2. Results from InStat Analysis of Example 10.2

Fisher's Exact Test
The two-sided P value is <0.0001, considered extremely significant.
There is a significant association between rows and columns.
Odds ratio = 17.333
95% confidence interval: 5.506 to 54.563 (using the approximation of Woolf.)

in Table 10.2. Again the P value is less than 0.0001. If there were no association between fleas and cat scratch fever in the overall population, there is less than a 0.01% chance of randomly picking subjects with so much association.

To interpret the P value from a case-control study, you must accept these assumptions:

- The cases and controls are randomly selected from their respective populations, or at least are representative of those populations.
- Each subject was selected independently of the rest. Picking one subject should not influence the chance of picking anyone else.
- The controls do not differ systematically from the cases in any way, except for the absence of disease. It is very difficult to be sure you have satisfied this assumption.

EXAMPLE 10.3. COMPARING TWO MEANS WITH THE t TEST

In this study (same data as Example 7.1), the investigators compared stool output in babies treated conventionally with that in babies treated conventionally with the addition of bismuth salicylate. Control babies produced 260 ± 254 ml/kg of stool (N = 84, mean ± SD), while treated babies produced 182 ± 197 ml/kg of stool (N = 85). The difference between means is 78 ml/kg.

The null hypothesis is that the mean stool output in the overall population of babies with diarrhea treated with bismuth salicylate is equal to the mean stool output of babies treated conventionally. We need to calculate a P value that answers this question: If the null hypothesis is true, what is the probability of randomly choosing samples whose means are as different (or more different) as observed.

The unpaired t test calculates the P value. The test takes into account the size of the difference, the size of the samples, and the SD of the samples. The results from InStat are given in Table 10.3.

For this example, the two-tailed P value is 0.0269. If the drug is really ineffective, there is a 2.69% chance that two randomly selected samples of N = 84 and N = 85 would have means that are 78 ml/kg or further apart. This is a two-tailed P value, so it includes a 1.34% chance of randomly picking subjects such that the treatment reduces stool output by 78 ml/kg or more and a 1.34% of randomly picking subjects such that the treatment increases stool output by 78 ml/kg or more.

The t test is based on these familiar assumptions:

- The subjects are representative of all babies with diarrhea who have been or will be treated with oral rehydration or oral rehydration and bismuth salicylate.

Table 10.3. Results from InStat Analysis of Example 10.3

Unpaired t Test

Are the means of control and treated equal?
Mean difference = −78.000 (mean of column B minus mean of column A)
The 95% confidence interval of the difference: −146.99 to −9.013
t = 2.232 with 167 degrees of freedom.
The two-tailed P value is 0.0269, considered significant.

- The two groups are not matched or paired. (You should try to pair data whenever possible, but you need to use appropriate analyses. See Chapter 25.)
- Each subject was selected independently of the rest. Choosing one subject from the population should not influence the chance of choosing anyone else.
- The distribution of values in the overall population follows a roughly Gaussian distribution, and the two populations have identical SDs.

The last assumption presents a problem with these data. As discussed in Chapter 7, the data are very unlikely to come from a Gaussian distribution. The SD is about equal to the mean, yet stool output cannot be negative. There is no way this can happen with a Gaussian distribution. Since the sample size is so large, we can rely on the central limit theorem and ignore the assumption.

EXAMPLE 10.4. COMPARING MEANS OF PAIRED SAMPLES

This study (same as Example 7.2) measured the volume change of renal cysts when incubated in culture medium. The investigators studied nine cysts and measured the weight change (as a proxy for volume) of each. The average weight change was 0.50 grams with a standard error of 0.23.

The null hypothesis is that renal cysts are as likely to shrink as to grow in culture. More precisely, the null hypothesis is that the mean volume of renal cysts does not change in culture. We want to calculate a P value that answers this question: If the null hypothesis is true, what is the chance of randomly picking cysts with an average weight change of 0.50 grams or more? Use the paired t test to calculate the answer. The results from InStat are given in Table 10.4. The P value is 0.0614. If renal cysts on average do not change weight in culture, there is a 6.14% chance that nine randomly selected cysts would change weight as much or more than these did.

Table 10.4. Results from InStat Analysis of Example 10.4

Paired t Test for Renal Cysts Example

Does the mean change from BEFORE to AFTER equal 0.000?
Mean difference = 0.5000 (mean of paired differences)
The 95% confidence interval of the difference: −0.03038 to 1.030
t = 2.174 with 8 degrees of freedom.
The two-tailed P value is 0.0614, considered not quite significant.

EXAMPLE 10.5. COMPARING TWO SURVIVAL CURVES WITH THE LOG-RANK TEST

After a bone marrow transplant, the white cells from the donor can mount an immune response against the recipient. This condition, known as graft-versus-host disease (GVHD) can be quite serious. Immunosuppressive drugs are given in an attempt to prevent GVHD. Cyclosporine plus prednisone are commonly used. Chao et al.* compared those two drugs alone (two drugs) with those two drugs plus methotrexate (three drugs). Figure 10.3 shows their data.

They recorded time until GVHD and plotted the data as Kaplan-Meier survival curves. Note that the term *survival curve* is a bit misleading as the end point is onset of GVHD rather than death. The authors chose to plot the data going "uphill," showing the percent of patients who had GVHD by a certain time. Most investigators would have plotted these Kaplan-Meier curves going "downhill," showing the percentage of patients who had NOT had GVHD by a certain time. This is an artistic decision that does not affect the analyses.

The authors compared the two groups with the log-rank test (other investigators might use the Mantel-Haenszel test, which is nearly identical). You will learn a little bit about these tests in Chapter 33. Even though you have not yet learned about the test, you can understand the results.

As you would expect, the null hypothesis is that the two populations have identical survival curves and that the observed difference in our samples is due to chance. In other words, the null hypothesis is that treatment with three drugs is no better (and no worse) than treatment with two drugs in preventing GVHD. The P value from the log-rank test answers this question: If the null hypothesis is true, what is the probability of obtaining such different survival curves with randomly selected subjects?

The calculations are quite involved and should be left to computer programs. The authors report that the P value is 0.02. If addition of the third drug did not alter onset of GVHD, you'd see such a large difference in survival curves in only 2% of experiments of this size.

You've already learned about the assumptions that must be true to interpret a survival curve in Chapter 6. The log-rank test depends on those same assumptions (reviewed below):

- The subjects are representative of all bone-marrow transplant recipients.
- The subjects were chosen independently.
- Consistent criteria. The entry criteria and the definition of *survival* must be consistent during the course of the study.
- The survival of the censored subjects would be the same, on average, as the survival of the remaining subjects.

The data in Figure 10.3 show only one end point—the time to onset of GVHD. The investigators also compared other variables, such as renal function, liver function,

*NJ Chao, GM Schmidt, JC Niland et al. Cyclosporine, methotrexate and prednisone compared with cyclosporine and prednisone for prophylaxis of acute graft-versus-host disease. *N Engl J Med* 329:1225–1230, 1993.

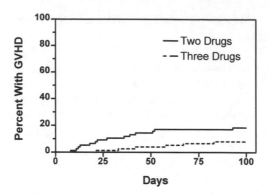

Figure 10.3. Data for Example 10.5. The investigators compared treating bone marrow transplants with two drugs or three drugs. The X axis shows time after the transplant, while the Y axis shows the percent of patients who have graft-versus-host disease (GVHD).

survival time, and recurrence of leukemia. Analysis of all these data suggest that addition of the third drug helps prevent GVHD without causing other problems. The authors recommendation to treat with three-drug therapy is based on analysis of all these variables, not just the survival curves.

SUMMARY

Most statistical tests calculate a P value. Even if you don't learn the details of all the tests it is essential that anyone reading the biomedical literature understand what a P value is (and what it isn't). A P value is simply a probability that answers the following question: If the null hypothesis were true (i.e., there is no difference between populations), what is the probability that random sampling (given the sample size actually used) would result in a difference as big or bigger than the one observed?

OBJECTIVES

1. You must be familiar with the following terms:
   • P value
   • Null hypothesis
   • One tailed test
   • Two tailed test
2. Whenever you see a P value reported in the biomedical literature, you should be able to identify the null hypothesis (even if you are not familiar with the statistical test being used) and pose the question that the P value answers.
3. You should know the difference between one- and two-tailed P values and know the arguments for and against using each.
4. Whenever you see a P value, you should be able to state the question it answers.

PROBLEMS

1. You wish to test the hypothesis that a coin toss is fair. You toss the coin six times and it lands on heads each time. What is the null hypothesis? What is the P value? Is a one- or two-tailed test more appropriate? What do you conclude?
2. You conduct an experiment and calculate that the two-tailed P value is 0.08. What is the one-tailed P value? What assumptions are you making?
3. (Same data as Problem 4 in Chapter 5 and Problem 1 in Chapter 7.) The serum levels of a hormone (the Y factor) was measured to be 93 ± 1.5 (mean $\pm$ SEM) in 100 nonpregnant women and 110 ± 2.3 (mean $\pm$ SEM) in 100 women in the first trimester of pregnancy. When the two groups are compared with a t test, the P value is less than 0.0001.

   A. Explain what the P value means in plain language.
   B. Review your answers to Problem 4 in Chapter 5 and Problem 1 in Chapter 7. Explain how the results complement one another.

4. (Same data as Problem 4 in Chapter 8.) Cohen et al. investigated the use of active cardiopulmonary resuscitation (CPR).* In standard CPR the resuscitator compresses the victim's chest to force the heart to pump blood to the brain (and elsewhere) and then lets go to let the chest expand. Active CPR is done with a suction device. This enables the resuscitator to pull up on the chest to expand it as well as pressing down to compress it. These investigators randomly assigned cardiac arrest patients to receive either standard or active CPR. Eighteen of 29 patients treated with active CPR were resuscitated. In contrast, 10 of 33 patients treated with standard CPR were resuscitated. Using Fisher's test, the two-sided P value is 0.0207. Explain what this means in plain language.

---

*TJ Cohen, BG Goldner, PC Maccaro, AP Ardito, S Trazzera, MB Cohen, SR Dibs. A comparison of active compression-decompression cardiopulmonary resuscitation with standard cardiopulmonary resuscitation for cardiac arrests occurring in the hospital. *N Engl J Med* 329:1918–1921, 1993.

Statistical Significance and Hypothesis Testing

In the previous chapter we treated the P value as a number and left the interpretation to the readers. Looked at this way, calculating a P value is analogous to calculating an average, ratio, or percentage. It is a useful way to summarize data to aid understanding and communication. This chapter shows how to make statements about statistical significance.

STATISTICAL HYPOTHESIS TESTING

When interpreting many kinds of data, you need to reach a decision. In a pilot experiment of a new drug, you need to decide whether the results are promising enough to merit a second experiment. In a phase III drug study, you need to decide whether the drug is effective and should be recommended for all patients. In a study comparing two surgical procedures, you need to decide which procedure to recommend.

Hypothesis testing is an approach that helps you make decisions after analyzing data. Follow these steps:

1. Assume that your samples are randomly selected from the population.
2. State the null hypothesis that the distribution of values in the two populations is the same.
3. Define a threshold value for declaring a P value significant. This threshold called the *significance level* of the test is denoted by α and is commonly set to 0.05.
4. Select an appropriate statistical test and calculate the P value.
5. If the P value is less than α, then conclude that the difference is *statistically significant* and decide to *reject the null hypothesis*. Otherwise conclude that the difference is not statistically significant and decide to not reject the null hypothesis.

Note that statisticians use the term *hypothesis testing* quite differently than scientists. Testing scientific hypotheses requires hard work involving many kinds of experiments. To test a new hypothesis, it is often necessary to design new experimental methodology and to design clever control experiments. Statistical hypothesis testing, in contrast, is easy. Just check whether one P value is above or below a threshold.

Converting a P value to the conclusion "significant" or "not significant" reminds me of the movie reviewers Siskel and Ebert. In their written reviews, they rate each movie on a scale (i.e., three and a half stars or B−). This is analogous to a P value. It is a concise way to summarize their opinions about the movie. When reviewing movies on television, they make a decision for you: see it (thumbs up) or don't see it (thumbs down).

The terminology of hypothesis testing is easiest to understand in the context of quality control. For example, let's assume that you run a brewery and that you have a warehouse full of the latest batch of beer. Before selling this batch, you need to test whether the batch meets various quality standards. Rather than test the entire warehouse full of beer (the population), you randomly choose some bottles (the sample) to compare to a "gold standard." The results can be expressed as a P value that answers the following question: If the new batch of beer is identical to the standard batch, what is the probability that a randomly selected sample of bottles would be as different from the standard as actually observed? If the P value is less than α (usually 0.05), then you reject the null hypothesis and reject the batch of beer (or at least do further tests to find out what is wrong). If the P value is greater than α, you do not reject the null hypothesis and do not reject the batch.

THE ADVANTAGES AND DISADVANTAGES OF USING THE PHRASE *STATISTICALLY SIGNIFICANT*

There are three advantages to using the phrase *statistically significant:*

- In some situations, it is necessary to reach a crisp decision from one experiment. Make one decision if the results are significant and the other decision if the results are not significant.
- With some statistical tests, it is difficult or impossible to obtain the exact P value, but it is possible to determine whether or not the P value exceeds α.
- People don't like ambiguity. The conclusion that "the results are statistically significant" is much more satisfying than the conclusion that "random sampling would create a difference this big or bigger in 3% of experiments if the null hypothesis were true."

The disadvantage to using the phrase *statistically significant* is that many people misinterpret it. Once some people have heard the word *significant,* they stop thinking about the data.

In biology and clinical medicine, it is not always necessary to reach a crisp decision from each P value. Instead, you can often base important decisions on several kinds of data, perhaps from several studies. Often, the best conclusion is wait and see. If it is not necessary to make a crisp decision from one experiment, there is no need to summarize the findings as being significant or not significant. Instead, just report the P value.

AN ANALOGY: INNOCENT UNTIL PROVEN GUILTY

Table 11.1 shows the analogy between the steps that a jury must follow to determine guilt (at least in the United States) and the steps that a scientist follows to determine statistical significance.

A jury reaches the verdict of guilty when the evidence is inconsistent with the assumption of innocence. Otherwise the jury reaches a verdict of not guilty. Note that a jury can never reach the verdict of innocent. The only choices are guilty or not guilty ("not proven" in Britain). A jury reaches a verdict of not guilty when the evidence is not inconsistent with the presumption of innocence. A jury does not have to be convinced that the defendant is innocent to reach a verdict of not guilty.

When performing a statistical test, you never conclude that you accept the null hypothesis. You reach the conclusion not significant whenever it is plausible that the data are consistent with the null hypothesis. A not significant result does not mean that the null hypothesis is true.

Journalists also evaluate evidence at a trial, and have different goals than jurors. As a journalist, you don't have to reach a verdict of guilty or not guilty. Instead your job is to summarize the proceedings. You want to present enough evidence so the readers can reach their own conclusions. You may include evidence unavailable to the jurors (evidence that wasn't presented at trial, or evidence that was ruled inadmissible). As a journalist, you don't have to reach any conclusion about the guilt of the defendant. If you do reach a conclusion, it is appropriate for it to be somewhat hazy if you think the evidence was inconclusive.

Table 11.1. Significant vs. Not Significant Decision Compared with Guilty vs. Not Guilty Decisions

Step	Trial by Jury	Statistical Significance
1.	Start with the presumption that the defendant is innocent.	Start with the presumption that the null hypothesis is true.
2.	Listen to factual evidence presented in the trial. Don't consider other data, such as newspaper stories you have read.	Base your conclusion only on data from this one experiment. Don't consider whether the hypothesis makes scientific or clinical sense. Don't consider any other data.
3.	Evaluate whether you believe the witnesses. Ignore testimony from witnesses who you think have lied.	Evaluate whether the experiment was performed correctly. Ignore flawed data.
4.	Think about whether the evidence is consistent with the assumption of innocence.	Calculate a P value.
5.	If the evidence is inconsistent with the assumption, then reject the assumption of innocence and declare the defendant to be guilty. Otherwise, reach a verdict of not guilty. As a juror, you can't conclude "maybe," can't ask for additional evidence, and can't say "wait and see."	If the P value is less than a preset threshold (typically 0.05), conclude that the data are inconsistent with the null hypothesis. Reject the null hypothesis, and declare the difference to be statistically significant. Otherwise, conclude that the difference is not significant. You can't conclude "maybe" or "wait and see."

I think that many scientists find themselves in the role of a journalist more often than they find themselves in the role of a juror. A scientist does not always need to evaluate data using the approach of statistical hypothesis testing. When presenting data, the most important thing is to present the methods and results clearly so the readers can reach their own conclusions. In many scientific situations, it is not necessary to reach a crisp conclusion of significant or not significant. Instead, it is sometimes appropriate to reach a more general conclusion that might be somewhat uncertain. And when reaching that conclusion, it is appropriate to consolidate data from several experiments and to take into account whether the hypothesis being tested fits theory and previous data.

TYPE I AND TYPE II ERRORS

When you conclude that results are significant or not significant, there are two ways you can be wrong:

- You can find that a result is statistically significant and reject the null hypothesis when in fact the null hypothesis is true. This is a Type I error. Convicting an innocent person is a Type I error. The probability of making a Type I error is α.
- You find that a result is not statistically significant and fail to reject the null hypothesis when the null hypothesis is in fact false. This is a Type II error. Failing to convict a guilty person is a Type II error.

We'll return to Type I and Type II errors in the next chapter.

CHOOSING AN APPROPRIATE VALUE FOR α

By tradition, α is usually set to equal 0.05. This cutoff is purely arbitrary, but is entrenched in the literature. It is amazing that scientists (who disagree about so much) all agree on such an arbitrary value! Ideally, the value of α should not be set by tradition but rather by the context of the experiment.

If you set α to a very low value, you will make few Type I errors. That means that if the null hypothesis is true, there is only a small chance that you will mistakenly call a result significant. However, there is also a larger chance that you will not find a significant difference even if the null hypothesis is false. In other words, reducing the value of α will decrease your chance of making a Type I error but increase the chance of a Type II error.

If you set α to a very large value, you will make many Type I errors. If the null hypothesis is true, there is a large chance that you will mistakenly find a significant difference. But there is a small chance of missing a real difference. In other words, increasing the value of α will increase your chance of making a Type I error but decrease the chance of a Type II error. The only way to reduce the chances of both a Type I error and a Type II error is to collect bigger samples. See Chapter 22.

You must balance the costs or consequences of Type I and Type II errors, and alter the value of α accordingly. Let's look at three examples:

- You are screening a new drug in an *in vitro* assay. If the experiment yields significant results, you will investigate the drug further. Otherwise you will stop investigating the drug. In this circumstance, the cost of a Type I error is a few additional experiments, and the cost of a Type II error is abandoning an effective drug. It makes sense to set α high to minimize Type II errors, even at the expense of additional Type I errors. I'd set α to 0.10 or 0.20 in this situation.
- You are conducting a phase III clinical trial of that drug to treat a disease (like hypertension) for which there are good existing therapies. If the results are significant, you will market the drug. If the results are not significant, work on the drug will cease. In this case, a Type I error results in treating future patients with a useless drug and depriving them of a good standard drug. A Type II error results in aborting development of a good drug for a condition that can be treated adequately with existing drugs. Thinking scientifically (not commercially), you want to set a low to minimize the risk of a Type I error, even at the expense of a higher chance of a Type II error. I'd set α to 0.01 in this situation.
- You are conducting a phase III clinical trial of that drug to treat a disease for which there is no good existing therapy. If the results are significant, you will market the drug. If the results are not significant, work on the drug will cease. In this case, a Type I error results in treating future patients with a useless drug instead of nothing. A Type II error results in cancelling development of a good drug for a condition that is currently not treatable. Here you want to set α high value because a Type I error isn't so bad but a Type II error would be awful. I'd set α to 0.10.

In this chapter, I've been deliberately vague when talking about the chances of a Type II error, and have not told you how to calculate that probability. You'll learn how to do that in Chapters 23 and 27.

Let's continue the analogy between statistical significance and the legal system. The balance between Type I and Type II errors depends on the type of trial. In the United States (and many other countries) a defendant in a criminal trial is considered innocent until proven guilty "beyond a reasonable doubt." This system is based on the belief that it is better to let many guilty people go free than to falsely convict one innocent person. The system is designed to avoid Type I errors in criminal trials, even at the expense of many Type II errors. You could say that α is set to a very low value. In civil trials, the court or jury finds for the plaintiff if the evidence shows that the plaintiff is "more likely than not" to be right. The thinking is that it is no worse to falsely find for the plaintiff than to falsely find for the defendant. The system attempts to equalize the chances of Type I and Type II errors in civil trials.

THE RELATIONSHIP BETWEEN α AND P VALUES

The P value and α are closely related. You calculate a P value from particular data. You set α in advance, based on the consequences of Type I and Type II errors. α is the threshold P value below which a difference is termed *statistically significant.*

THE RELATIONSHIP BETWEEN α AND CONFIDENCE INTERVALS

Although I have presented confidence intervals (CIs) and P values in separate sections of this book, the two are closely related. They both are based on the same assumptions and the same statistical principles. Ask yourself whether the 95% CI includes the value stated in the null hypothesis. If you are comparing two means, ask whether the 95% CI for the difference between the means includes 0. If you are analyzing a prospective study, ask whether the 95% CI for the relative risk includes 1.0. If the 95% CI includes the value stated in the null hypothesis, then the P value must be greater than 0.05 and the difference must be not significant. If the 95% CI does not include the null hypothesis, then the P value must be less than 0.05 and the result must be significant.

The rule in the previous paragraph works because 95% (CI) and 5% (threshold P value) add to 100%. Other pairs can be used as well. If the 99% CI includes the null hypothesis, then you can be sure that the P value must be greater than 0.01. If the 90% CI includes the null hypothesis, then you can be sure that the P value must be greater than 0.10.

STATISTICAL SIGNIFICANCE VERSUS SCIENTIFIC IMPORTANCE

A result is said to be statistically significant when the P value is less than a preset value of α. This means that the results are surprising and would not commonly occur if the null hypothesis were true. There are three possibilities:

- The null hypothesis is really true, and you observed the difference by coincidence. The P value tells you how rare the coincidence would be. If the null hypothesis really is true, α is the probability that you will happen to pick samples that result in a statistically significant result.
- The null hypothesis is really false (the populations are really different), and the difference is scientifically or clinically important.
- The null hypothesis is really false (the populations are really different), but the difference is so small as to be scientifically or clinically trivial.

The decision about scientific or clinical importance requires looking at the size of the difference (or the size of the relative risk or odds ratio). With large samples, even very small differences will be statistically significant. Even if these differences reflect true differences between the populations, they may not be interesting. You must interpret scientific or clinical importance by thinking about biology or medicine. For example, few would find a mean difference of 1 mmHg in blood pressure to be clinically interesting, no matter how low the P value. It is never enough to think about P values and significance. You must also think scientifically about the size of the difference.

SUMMARY

After calculating a P value, you "test a hypothesis" by comparing the P value with an arbitrary value α, usually set to 0.05. If the P value is less than α, the result is said to be "statistically significant" and the null hypothesis is said to be "rejected." If the P value is greater than α, the result is said to be "not statistically significant" and the null hypothesis is "not rejected." As used in the context of statistical hypothesis testing, the word *significant* has almost no relationship to the conventional meaning of the word to denote importance. Statistically significant results can be trivial. Differences that might turn out to be enormously important may result in "nonsignificant" P values. The scheme of hypothesis testing is useful in situations in which you must make a crisp decision from one experiment. In many biological or clinical studies, it is not always necessary to reach a crisp decision and it makes more sense to report the exact P value and avoid the term *significant.*

OBJECTIVES

1. You must be familiar with the following terms:
   - Hypothesis testing
   - Null hypothesis
   - α
   - Statistically significant
   - Type I error
   - Type II error
2. You must know the distinction between the statistical meaning of the terms *significant* and *hypothesis testing,* and the scientific meaning of the same terms.
3. You must know what *not significant* really means.

PROBLEMS

1. For Problems 3 and 4 of the last chapter, explain what a Type I and Type II error would mean.
2. Which of the following are inconsistent?
   A. Mean difference = 10. The 95% CI: -20 to 40. P = 0.45
   B. Mean difference = 10. The 95% CI: -5 to 15. P = 0.02
   C. Relative risk = 1.56. The 95% CI: 1.23 to 1.89. P = 0.013
   D. Relative risk = 1.56. The 95% CI: 0.81 to 2.12. P = 0.04
   E. Relative risk = 2.03. The 95% CI: 1.01 to 3.13. P < 0.001.

Interpreting Significant
and Not Significant P Values

THE TERM *SIGNIFICANT*

You've already learned that the term *statistically significant* has a simple meaning: The P value is less than a preset threshold value, α. That's it! In plain language, a result is statistically "significant" when the result would be surprising if there really were no differences between the overall populations.

It's easy to read far too much into the word *significant* because the statistical use of the word has a meaning entirely distinct from its usual meaning. Just because a difference is *statistically significant* does not mean that it is important or interesting. A statistically significant result may not be scientifically significant or clinically significant. And a difference that is not significant (in the first experiment) may turn out to be very important. This is important, so I'll repeat it: *Statistically significant results are not necessarily important or even interesting.*

EXTREMELY SIGNIFICANT RESULTS

Intuitively, you'd think that P = 0.004 is more significant than P = 0.04. Using the strict definitions of the terms, this is not correct. Once you have set a value for α, a result is either *significant* or *not significant*. It doesn't matter whether the P value is very close to α or far away. Many statisticians feel strongly about this, and think that the word *significant* should never be prefaced by an adjective. Most scientists are less rigid, and refer to *very significant* or *extremely significant* results when the P value is tiny.

When showing P values on graphs, investigators commonly use a "Michelin Guide" scale. *P < 0.05 (significant), **P < 0.01 (highly significant); ***P < 0.001 (extremely significant). When you read this kind of graph, make sure that you look at the key that defines the symbols, as different investigators use different threshold values.

BORDERLINE P VALUES

If you follow the strict paradigm of statistical hypothesis testing and set α to its conventional value of 0.05, then a P value of 0.049 denotes a significant difference and a P value of 0.051 denotes a not significant difference. This arbitrary distinction is unavoidable since the whole point of using the term *statistically significant* is to reach a crisp conclusion from every experiment without exception.

Rather than just looking at whether the result is significant or not, it is better to look at the actual P value. That way you'll know whether the P value is near α or far from it. When a P value is just slightly greater than α, some scientists refer to the result as *marginally significant* or *almost significant.*

When the two-tailed P value is between 0.05 and 0.10, it is tempting to switch to a one-tailed P value. The one-tailed P value is half the two-tailed P value and is less than 0.05, and the results become "significant" as if by magic. Obviously, this is not an appropriate reason to choose a one-tailed P value! The choice should be made before the data are collected.

One way to deal with borderline P values would be to choose between three decisions rather than two. Rather than decide whether a difference is significant or not significant, add a middle category of *inconclusive.* This approach is not commonly used.

THE TERM *NOT SIGNIFICANT*

If the P value is greater than a preset value of α, the difference is said to be *not significant.* This means that the data are not strong enough to persuade you to reject the null hypothesis. People often mistakenly interpret a high P value as proof that the null hypothesis is true. That is incorrect. A high P value does not prove the null hypothesis. This is an important point worth repeating: *A high P value does not prove the null hypothesis.* As you've already learned in Chapter 11, concluding that a difference is not statistically significant when the null hypothesis is, in fact, false is called a Type II error.

When you read that a result is not significant, don't stop thinking. There are two approaches you can use to evaluate the study. First, look at the confidence interval (CI). Second, ask about the power of the study to find a significant difference if it were there.

INTERPRETING *NOT SIGNIFICANT* RESULTS WITH CONFIDENCE INTERVALS

Example 12.1

Ewigman et al. investigated whether routine use of prenatal ultrasound would improve perinatal outcome.* They randomly divided a large group of pregnant women into two

---

*BG Ewigman, JP Crane, FD Frigoletto et al. Effect of prenatal ultrasound screening on perinatal outcome. *N Engl J Med* 329:821–827, 1993.

groups. One group received routine ultrasound sonogram exams twice during the pregnancy. The other group received sonograms only if there was a clinical reason to do so. The physicians caring for the women knew the results of the sonograms and cared for the patients accordingly. The investigators looked at several outcomes. Table 12.1 shows the total number of adverse events, defined as fetal or neonatal deaths of moderate to severe morbidity.

The null hypothesis is that the rate of adverse outcomes is identical in the two groups. In other words, the null hypothesis is that routine use of ultrasound neither prevents nor causes perinatal mortality or morbidity. The two-tailed P value is 0.86. The data provide no reason to reject the null hypothesis.

Before you can interpret the results, you need to know more. You need to know the 95% CI for the relative risk. For this study, the relative risk is 1.02, with the 95% CI ranging from 0.88 to 1.17. The null hypothesis can be restated as follows: In the entire population, the relative risk is 1.00. The data are consistent with the null hypothesis. This does not mean that the null hypothesis is true. Our CI tells us that the data are also consistent (within 95% confidence) with relative risks ranging from 0.88 to 1.17.

Different people might interpret these data in two ways:

> The confidence interval is very narrow, and is centered close to 1.0. These data convince me that routine use of ultrasound is neither helpful nor harmful. To reduce costs, I'll use ultrasound only when there is an identified problem.

> The confidence interval is narrow, but not all that narrow. There have been plenty of studies showing that ultrasound doesn't hurt the fetus, so I'll ignore the part of the confidence interval above 1.00. But ultrasound gives the obstetrician extra information to manage the pregnancy, and it makes sense that using this extra information will decrease the chance of a major problem. The confidence interval goes down to 0.88, a reduction of risk of 12%. If I were pregnant, I'd certainly want to use a risk-free technique that reduces the risk of a sick or dead baby by as much as 12%! The data certainly don't prove that routine ultrasound is beneficial, but the study leaves open the possibility that routine use of ultrasound might reduce the rate of truly awful events by as much as 12%. I think the study is inconclusive. Since ultrasound is not prohibitively expensive and appears to have no risk, I will keep using it routinely.

To interpret a not significant P value, you must look at the CI. If the entire span of the CI contains differences (or relative risks) that you consider to be trivial, then you can make a strong negative conclusion. If the CI is wide enough to include values you consider to be clinically or scientifically important, then the study is inconclusive. Different people will appropriately have different opinions about how large a difference

Table 12.1. Results of Example 12.1

	Adverse Outcome	Healthy Baby	Total
Routine sonograms	383	7,302	7,685
Sonograms only when indicated	373	7,223	7,596
Total	756	14,525	15,281

(or relative risk) is scientifically or clinically important and may interpret the same not significant study differently.

In interpreting the results of this example, you also need to think about benefits and risks that don't show up as a reduction of adverse outcomes. The ultrasound picture helps reassure parents that their baby is developing normally and gives them a picture to bond with and to show relatives. This can be valuable regardless of whether it reduces the chance of adverse outcomes. Although statistical analyses focus on one outcome at a time, you must consider all the outcomes when evaluating the results.

INTERPRETING *NOT SIGNIFICANT* P VALUES USING POWER ANALYSES*

A not significant P value does not mean that the null hypothesis is true. It simply means that your data are not strong enough to convince you that the null hypothesis is not true. As you have already learned, obtaining a not significant result when the null hypothesis is really false is called a Type II error. When you obtain (or read about) a not significant P value, you should ask yourself this question: "What is the chance that this is a Type II error?" That question can only be answered if you specify an alternative hypothesis—a difference Δ (or relative risk R) that you hypothesize exists in the overall population. Then you can ask, "If the true difference is Δ (or the true relative risk is R), what is the chance of obtaining a significant result in an experiment of this size? The answer is termed the *power* of the study.

Table 12.2 shows the power of example 12.1 for various hypothetical relative risks. This table uses the sample size of the example and sets $\alpha = 0.05$ and the risk in the control subjects to 5.0%. Calculating the power exactly is quite difficult, but there are several ways to calculate power approximately. You'll learn about the approximation used to create this table in Chapter 27. In Chapter 23 you'll learn how to calculate a similar table for studies that compare two means (rather than two proportions).

If the experimental treatment (routine ultrasound) reduced the risk by 25% (so the relative risk = 0.75), the study had a 97% power to detect a significant difference. If the treatment was really that good, then 97% of studies this size would wind up with a significant result, while 4% would come up with a not significant result. In contrast, the power of this study to detect a risk reduction of 5% (relative risk = 0.95) is only 11%. If the treatment truly reduced risk by 5%, only 11% of studies this size would come up with a significant result.

The numbers are particular to this study. The principle is universal. All studies have very little power to detect tiny differences and enormous power to detect large differences. If you increase the number of subjects, you will increase the power.

As you can see, calculations of power can aid interpretation of nonsignificant results. When reading biomedical research, however, you'll rarely encounter power calculations in papers that present not significant results. This is partly a matter of tradition, and it is partly because it is difficult to define the smallest difference or relative risk that you think it is important.

---

*This section is more advanced than the rest. You may skip it without loss of continuity.

Table 12.2. A Power
Analysis of Example 12.1

Relative Risk	Power
0.95	11%
0.90	30%
0.85	60%
0.80	84%
0.75	97%

SUMMARY

A result is statistically significant when the P value is less than the preset value of α. This means that the results would be surprising if the null hypothesis were really true. The statistical use of the term *significant* is quite different than the usual use of the word. Statistically significant results may or may not be scientifically or clinically interesting or important.

When results are statistically not significant it means that the results are not inconsistent with the null hypothesis. This does *not* mean that the null hypothesis is true. When interpreting results that are not significant, it helps to look at the extent of the CI and to calculate the power that the study would have found a significant result if the populations really were different (with a difference of a defined size).

Multiple Comparisons

Each example you've encountered so far was designed to answer one question. If you look through any medical journal, you'll discover that few papers present a single P value. It is far more common to see papers that present several—or several dozen—P values. It's not easy to interpret multiple P values.

COINCIDENCES

If your results are really due to a coincidence (the null hypothesis is true), the P value tells you how rare that coincidence would be. Interpreting multiple P values, therefore, is similar to interpreting multiple coincidences.

Example 13.1

In 1991, President Bush and his wife Barbara both developed hyperthyroidism due to Graves' disease. Could it be a coincidence, or did something cause Graves' disease in both? What is the chance that the president and his wife would both develop Graves' disease just by chance? It's hard to calculate that probability exactly, but it is less than 1 in a million. Because this would have been such a rare coincidence, they looked hard for a cause in the food, water, or air. No cause was ever found.

Was it really a one in a million coincidence? The problem with this "probability" is that the event had already happened before anyone thought to calculate the probability. To use a gambling analogy, the bets were placed after the ball stopped spinning.

A more appropriate question might be "what is the probability that a prominent person and his or her spouse would both develop the same disease this year?" By including all prominent people and all possible diseases, the probability is much higher. The coincidence no longer seems so strange. You could expand the question to include the next few years, and the answer would be higher still.

Example 13.2

Five children in a particular school got leukemia last year. Is that a coincidence? Or does the clustering of cases suggest the presence of an environmental toxin that caused the disease? That's a very difficult question to answer. It is tempting to estimate the answer to the question "what is the probability that five children in this particular school would all get leukemia this particular year?'' You could calculate (or at least estimate) the answer to that question if you knew the overall incidence rates of leukemia among children and the number of children enrolled in the school. The answer will be tiny. Everyone intuitively knows that and so is alarmed by the cluster of cases.

But you've asked the wrong question once you've already observed the cluster of cases. The school only came to your attention because of the cluster of cases, so you need to consider all the other schools and other diseases. The right question is ''what is the probability that five children in any school would develop the same severe disease in the same year?'' This is a harder question to answer, because you have to define the population of schools (this city or this state?), the time span you care about (one year or one decade?), and the severity of diseases to include (does asthma count?). Clearly the answer to this question is much higher than the answer to the previous one. When clusters occur, it is always worth investigating for known toxins and to be alert to other findings that might suggest a real problem. But most disease clusters are due to coincidence. It is surprising to find a cluster of one particular disease in one particular place at any one particular time. But chance alone will cause many clusters of various diseases in various places at various times.

Example 13.3

You go to a casino, and watch the roulette wheel. You notice that in the first 38 spins of a roulette wheel, the ball landed on red 25 times. By chance you'd only expect the ball to land on red 18 times (and 20 times on black, 0, or 00).* Using the binomial distribution, you can calculate that the chance of having the ball land on 25 or more red slots in the first 38 spins is about 2%. Are you going to place your bets on red?

A rare coincidence? Not really. You would have been just as surprised if the ball had landed 25 times on black, on an odd number, on a small number (1 to 18), on a large number (19 to 36), etc. The chance of seeing any one (or more) of these ''rare'' coincidences is far greater than 2%. It is only fair to calculate the probability of a rare coincidence if you define the coincidence before it happens. You have to place your bets before you know the outcome.

MULTIPLE INDEPENDENT QUESTIONS

Example 13.4

Hunter and colleagues investigated whether vitamin supplementation could reduce the risk of breast cancer.† The investigators sent dietary questionnaires to over 100,000 nurses in 1980. From the questionnaires, they determined the intake of vitamins A, C, and E and divided the women into quintiles for each vitamin (i.e., the first quintile

---

*An American roulette wheel has 18 red slots, 18 black slots, and 2 green slots labeled ''0'' and ''00.''

†DJ Hunter, JE Manson, GA Colditz, et al. A prospective study of the intake of vitamins C, E and A and the risk of breast cancer. *N Engl J Med* 329:234–240, 1993.

contains 20% of the women who consumed the smallest amount). They then followed these women for 8 years to determine the incidence rate of breast cancer. Using a test called the chi-square test for trend (which will be briefly discussed in Chapter 29) the investigators calculated a P value to test this null hypothesis: There is no linear trend between vitamin intake quintile and the incidence of breast cancer. There would be a linear trend if increasing vitamin intake was associated with increasing (or decreasing) incidence of breast cancer. There would not be a linear trend if the lowest and highest quintiles had a low incidence of breast cancer compared to the three middle quintiles. The authors determined a different P value for each vitamin. For vitamin C, $P = 0.60$; for vitamin E, $P = 0.07$; for vitamin A, $P = 0.001$.

Interpreting each P value is easy: If the null hypothesis is true, the P value is the chance that random selection of subjects would result in as large (or larger) a linear trend as was observed in this study. If the null hypothesis is true, there is a 5% chance of randomly selecting subjects such that the trend is significant. Here we are testing three independent null hypotheses (one for each vitamin tested). If all three null hypotheses are true, the chance that one or more of the P values will be significant is greater than 5%.

Table 13.1 shows what happens when you test several independent null hypotheses. In this example, we are testing three null hypotheses. If we used the traditional cutoff of $\alpha = 0.05$ for declaring each P value to be significant, the table shows that there would be a 14% chance of observing one or more significant P values, even if all three null hypotheses were true. To keep the overall chance at 5%, we need to lower our threshold for significance from 0.050 to 0.0170. With this criteria, the relationship between vitamin A intake and the incidence of breast cancer is statistically significant. The intakes of vitamins C and E are not significantly related to the incidence of breast cancer.

If you don't have access to the table, here is a quick way to approximate the value in the bottom row: Simply divide 0.05 by the number of comparisons. For this example, the threshold is 0.05/3 or 0.017—the same as the value in the table to three decimal points. If you tested seven hypotheses, the shortcut method calculates a threshold of 0.05/7 or .0071 (which is close to the exact value of 0.0073 on the table). If you make more than 10 comparisons, this shortcut method is not useful.

MULTIPLE GROUPS

Example 13.5

Hetland and coworkers were interested in hormonal changes in women runners.* Among their investigations, they measured the level of luteinizing hormone (LH) in nonrunners, recreational runners, and elite runners. Because hormone levels were not Gaussian, the investigators transformed their data to the logarithm of concentration

---

*ML Hetland, J Haarbo, C Christiansen, T Larsen. Running induces menstrual disturbances but bone mass is unaffected, except in amenorrheic women. *Am J Med* 95:53–60, 1993.

Table 13.1. Probability of Small P Value When Testing Many Null Hypotheses

Number of Independent Null Hypotheses (N)	2	3	4	5	6	7	8	9	10	20	50	100
Probability (P*) of obtaining one or more P values less than 0.05 by chance	10%	14%	19%	23%	26%	30%	34%	37%	40%	64%	92%	99%
α* (To keep overall α = 0.05, accept a P value as significant only if it is less than this value)	.0253	.0170	.0127	.0102	.0085	.0073	.0064	.0057	.0051	.0026	.0010	.0005

$P^* = 1 - 0.95^N$.

$\alpha^* = 1 - 0.95^{(1/N)}$.

This table assumes that you have set α to its usual value of 0.05. To calculate this table for other values, simply replace "0.95" in the two equations with "$(1 - \alpha)$".

Table 13.2. LH Levels in Three Groups of Women

Group	log(LH) ± SEM	N
Nonrunners	0.52 ± 0.027	88
Recreational runners	0.38 ± 0.034	89
Elite runners	0.40 ± 0.049	28

and performed all analyses on the transformed data. Although this sounds a bit dubious, it is a good thing to do, as it makes the population closer to a Gaussian distribution. The data are shown in Table 13.2.

The null hypothesis is that the mean concentration of LH is the same in all three populations. How do you calculate a P value testing this null hypothesis? Your first thought might be to calculate three t tests: one to compare nonrunners with recreational runners, another to compare nonrunners with elite runners, and yet another to compare recreational runners with elite runners. The problem with this approach is that it is difficult to interpret the P values. As you include more groups in the study, you increase the chance of observing one or more significant P values by chance. If the null hypothesis were true (all three populations have the same mean), there is a 5% chance that each particular t test would yield a significant P value. But with three comparisons, the chance that any one (or more) of them will be significant is far higher than 5%.

The authors used a test called *one-way analysis of variance* (ANOVA) to calculate a single P value answering this question: If the null hypothesis were true, what is the chance of randomly selecting subjects with means as far apart, or further, than observed in this study? The P value is determined from the scatter among means, the standard deviation (SD) within each group, and the size of the samples.

The answer is $P = 0.0039$. If the null hypothesis were true, there is only a 0.39% chance of randomly picking subjects and ending up with mean values so different. The authors concluded that the null hypothesis was probably not true.

Analysis of variance makes exactly the same assumptions as the t test. The subjects must be representative of a larger population. The data within each population must be distributed according to a Gaussian distribution with equal SDs. Each subject must be selected independently.

Next you want to know which group differs from which other group. But you shouldn't perform three ordinary t tests to find out. Instead, analysis of variance is followed by special tests designed for multiple comparisons. These tests are all named after the statistician(s) who developed them (Tukey, Newman–Keuls, Dunnett, Dunn, Duncan, and Bonferroni). The idea of all the tests is that if the global null hypothesis is true, there is only a 5% chance that any one or more of the comparisons will be statistically significant. The differences between the methods relate to the assumptions you are willing to make and to the number of comparisons you are interested in. You will learn a little bit about the differences between these tests in Chapter 30.

Using Tukey's test to compare each group with each other group with an overall $\alpha = 0.05$, we find that there is a statistically significant difference between the nonrunners and the recreational runners, but not between the nonrunners and the elite runners or between the recreational and elite runners.

MULTIPLE MEASUREMENTS TO ANSWER ONE QUESTION

Example 13.4 Continued

Recall that this study examined the possible relationship between vitamin intake and the incidence of breast cancer. You've already seen that they found a significant relationship between vitamin A intake and incidence of breast cancer. In addition to the three main analyses already mentioned, the authors analyzed their data in many other ways. If we focus only on vitamin A, the authors separately analyzed total vitamin A intake and performed vitamin A. They separately analyzed two different time periods (1980 to 1988 and 1984 to 1988). For each analysis, they analyzed the data using a simple test and again using a fancier test to adjust for known factors that influence the incidence of breast cancer (for example, number of children, age at first birth, age at menarche, family history of breast cancer). With two measures of vitamin A, two time periods, and two analytical methods they generated eight P values.

These eight P values don't test eight independent null hypotheses, so you shouldn't use the methods presented earlier in Table 13.1. The null hypotheses are interrelated—they are sort of measuring the same thing, but not quite. There are no good methods for dealing with this situation. Basically you need to look at the collection of P values and get an overall feel for what is going on. In this example, all four P values from the 1980 to 1988 analyses were tiny. The association between vitamin A intake and protection from breast cancer was substantial regardless of whether they looked at total or preformed vitamin A and whether or not they adjusted for other risk factors. In the 1984 to 1988 study, the associations all went in the same direction (more vitamin A = less breast cancer) but the P values were a bit larger. Putting all this together, the evidence is fairly persuasive.

A similar kind of problem comes up frequently in analyzing clinical trials. In many clinical trials, investigators measure clinical outcome using a variety of criteria. For example, in a study of a drug to treat sepsis, the main outcome is whether the patient died. Additionally, investigators may collect additional information on the patients who survived: how long they were in the intensive care unit, how long they required mechanical ventilation, and how many days they required treatment with vasopressors. All these outcomes are really measuring the same thing: how long the patient was severely ill. These data can lead to multiple P values but should not be corrected, as shown in previous sections, because the null hypotheses are not independent. To a large degree the various outcomes measure the same thing.

Although clinical studies often measure several outcomes, statistical methods don't deal with this situation very well. You should not make any formal corrections for multiple comparisons. Instead, you should informally integrate all the data before reaching any conclusions.

MULTIPLE SUBGROUPS

After analyzing the data in many studies, it is tempting to look at subgroups. Separately analyze the subjects by age group. Separately analyze the patients with severe disease and mild disease. The problem with doing separate analyses of subgroups is that the chance of making a Type I error (finding a statistically significant difference by chance) goes up.

Example 13.6

This problem was illustrated in a simulated study by Lee and coworkers.* They pretended to compare survival following two "treatments" for coronary artery disease. They studied a group of patients with coronary artery disease who they randomly divided into two groups. In a real study, they would give the two groups different treatments, and compare survival. In this simulated study, they treated the subjects identically† but analyzed the data as if the two random groups actually represented two distinct treatments. As expected, the survival of the two groups was indistinguishable.

They then divided the patients into six groups depending on whether they had disease in one, two, or three coronary arteries, and depending on whether the heart ventricle contracted normally or not. Since these are variables that are expected to affect survival of the patients, it made sense to evaluate the response to "treatment" separately in each of the six subgroups. Whereas they found no substantial difference in five of the subgroups, they found a striking result among the sickest patients. The patients with three-vessel disease who also had impaired ventricular contraction had much better survival under treatment B than treatment A. The difference between the two survival curves were statistically significant with a P value less than 0.025.

If this were a real study, it would be tempting to conclude that treatment B is superior for the sickest patients, and to recommend treatment B to those patients in the future. But this was not a real study, and the two "treatments" reflected only random assignment of patients. The two treatments were identical, so the observed difference was definitely due to chance. It is not surprising that the authors found one low P value out of six comparisons. Referring to Table 13.1, there is a 26% chance that one of six independent comparisons will have a P value less than 0.05, even if all null hypotheses are true. To reduce the overall chance of a Type I error to 0.05, you'd need to reduce α to 0.0085 when comparing six groups.

This is a difficult problem that comes up frequently. Beware of analyses of multiple subgroups as you are very likely to encounter small P values, even if all null hypotheses are true.

MULTIPLE COMPARISONS AND DATA DREDGING

In all the examples you've encountered in this chapter, you've been able to account for multiple comparisons because you know about all the comparisons the investigators made. You will be completely misled (and will reach the wrong conclusion) if the

---

*KL Lee, JF McNeer, CF Starmer, PJ Harris, RA Rosati. Clinical judgment and statistics. Lessons from a simulated randomized trial in coronary artery disease. *Circulation* 61:508–515, 1980.

†In fact, the study was done retrospectively. They looked back at hospital records to identify all patients who presented with coronary artery disease during a certain period. They divided these patients into two random groups, and then read the charts to see how long each patient lived.

investigator made many comparisons but only published the few that were significant. If the null hypothesis is true, a low P value means that a rare coincidence has occurred. But you can't evaluate the rarity of a coincidence unless you know how many different comparisons were made. As you've seen, if you test lots of null hypotheses the chance of observing one or more "significant" P values is far higher than 5%. If you test 100 independent null hypotheses that are all true, for example, you have a 99% chance of obtaining at least one significant P value. You will be completely misled if the investigators show you the significant P values but don't tell you about the others.

To avoid this situation, investigators should follow these rules:

• Analyses should be planned before the data are collected.
• All planned analyses should be completed and reported.

These rules are usually followed religiously for large formal clinical trials, especially when the data will be reviewed by the Food and Drug Administration. However, those rules are often ignored in more informal preliminary studies and in laboratory research. In many cases, the investigator never thought about how to perform the analyses until after perusing the data. Often looking at the data suggests new hypotheses to test.

It is difficult to know how to deal with analyses that don't follow those rules. If the investigators didn't decide what hypotheses to test until after they looked at the data, then they have implicitly performed many tests. When you read the paper, you need to figure out how many hypotheses the investigators really tested. Look at the number of variables, the number of groups, the number of time points, and the number of adjustments. Big studies can easily generate dozens or hundreds of P values. If the investigators implicitly tested many hypotheses, they are apt to find "significant" differences fairly often. To make sense of this kind of study, you need to look at the overall pattern of results and not interpret any individual P values too strongly.

You should always distinguish studies that test a hypothesis from studies that generate a hypothesis. Exploratory analyses of large databases can generate hundreds of P values, and scanning these can generate intriguing research hypotheses. After the hypothesis is generated, however, it is then necessary to test the hypotheses on a different set of data. Some investigators use half the data for exploration to define one or more hypotheses, and then test the hypotheses with the other half of the data. This is a terrific approach if plenty of data are available.

SUMMARY

Most scientific studies generate more than one P value. Some studies generate hundreds of P values. Interpreting multiple P values is difficult. If you make many comparisons, you expect some to have small P values just by chance. Therefore your interpretation of a small P value should be different when the P value is one of many. You'll encounter multiple P values in several situations: asking many independent questions, comparing multiple groups, measuring multiple end points, and reanalyzing data for multiple subgroups. You need to take into account the number of P values generated when interpreting the results. You'll be misled if the investigators calculated many P values, but only showed you the small ones.

PROBLEM

1. Assume that you are reviewing a manuscript prior to publication. The authors have tested many drugs to see if they increase enzyme activity. All P values were calculated using a paired t test. Each day they also perform a negative control, using just vehicle with no drug. After discussing significant changes caused by some drugs, they state that "in contrast, the vehicle control has consistently given a nonsignificant response (P > 0.05) in each of 250 experiments." Comment on these negative results. Why are they surprising?

IV

BAYESIAN LOGIC

When you interpret the results of an experiment, you need to consider more than just the P value. You also need to consider the experimental context, previous data, and theory. Bayesian logic allows you to integrate the current experimental data with what you knew before the experiment.

Since Bayesian logic can be difficult to understand in the context of interpreting P values, I first present the use of Bayesian logic in interpreting the results of clinical laboratory tests in Chapter 14. Then in Chapter 16 I explain how Bayesian logic is used in interpreting genetic data.

Interpreting Lab Tests: Introduction to Bayesian Thinking

Note to basic scientists: Don't skip this chapter because it appears to be too clinical. This chapter sets the stage for the discussion in the next two chapters.

What do laboratory tests have to do with P values? Understanding how to interpret "positive" and "negative" lab tests will help you understand how to interpret "significant" and "not significant" statistical tests.

THE ACCURACY OF A QUALITATIVE LAB TEST

We will consider first a test that yields a simple answer: positive or negative. Results can be tabulated on a two by two contingency table (Table 14.1). The rows represent the outcome of the test (positive or negative), and the columns indicate whether the disease is present or absent (based upon some other method that is completely accurate, perhaps the test of time). If the test is "positive," it may be true positive (TP), or it may be a false positive (FP) test in a person without the condition being tested for. If the test is "negative," it may be a true negative (TN) or it may be a false negative (FN) test in a person who does have the condition.

How accurate is the test? It is impossible to express the accuracy in one number. It takes at least two: sensitivity and specificity. An ideal test has very high sensitivity and very high specificity:

• The *sensitivity* is the fraction of all those with the disease who get a positive test result.
• The *specificity* is the fraction of those without the disease who get a negative test result.

$$\text{Sensitivity} = \frac{TP}{TP + FN} \qquad \text{Specificity} = \frac{TN}{TN + FP} \qquad (14.1)$$

It is easy to mix up sensitivity and specificity. Sensitivity measures how well the test identifies those with the disease, that is, how sensitive it is. If a test has a high sensitivity, it will pick up nearly everyone with the disease. Specificity measures how well the test excludes those who don't have the disease, that is, how specific it is. If a test has a very high specificity, it won't mistakenly give a positive result to many people without the disease.

Table 14.1. Accuracy of a Qualitative Lab Test

	Disease Present	Disease Absent
Test positive	TP	FP
Test negative	FN	TN

Sackett and colleagues have published a clever way to remember the difference between sensitivity and specificity.* Remember the meaning of sensitivity with this acronym SnNOut: If a test has high *sen*sitivity, a *n*egative test rules *out* the disorder (relatively few negative tests are false negative). Remember the meaning of specificity with this acronym: SpPIn. If a test has high *sp*ecificity, a *p*ositive test rules *in* the disorder (relatively few positive tests are false positive).

THE ACCURACY OF A QUANTITATIVE LAB TEST

Many lab tests report results on a continuous scale. For the purposes of this chapter, we will simplify things so that it reports either ''normal'' or ''abnormal.'' Figures 14.1 and 14.2 show the distribution of test results in patients with the condition being tested (dashed curve) and in those without the condition (solid curve). In Figure 14.1, the two distributions don't overlap, and it is easy to pick the cutoff value shown by the vertical dotted line. Every individual whose value is below that value (the left of the dotted line) does not have the condition, and every individual whose test value is above that threshold has it.

Figure 14.2 shows a more complicated situation where the two distributions overlap. Again, the solid curve represents patients without the condition and the dashed curve represents patients with the condition. Wherever you set the cutoff, some patients will be misclassified. The dark shaded area shows false positives, those patients classified as positive even though they don't have the disease. The lighter shaded area shows false negatives, those patients classified as negatives even though they really do have the disease.

Choosing a threshold requires you to make a tradeoff between sensitivity and specificity. If you increase the threshold (move the dotted line to the right), you will increase the specificity but decrease the sensitivity. You have fewer false positives but more false negatives. If you decrease the threshold (move the dotted line to the left), you will increase the sensitivity but decrease the specificity. You will have more false positives but fewer false negatives.

Choosing an appropriate threshold requires knowing the consequences (harm, cost) of false-positive and false-negative test results. Consider a screening test for a disease that is fatal if untreated but completely treatable. If the screening test is positive, it is followed by a more expensive test that is completely accurate and without risk.

*DL Sackett, RB Haynes, GH Guyatt, P Tugwell. *Clinical Epidemiology. A Basic Science for Clinical Medicine,* 2nd ed. Boston, Little Brown, 1991.

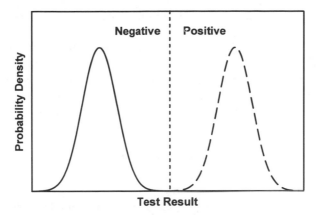

Figure 14.1. A perfect test. The solid line shows the distribution of values in people without the disease being tested for. The dashed curve shows the distribution of values in people with the disease. The vertical line shows the demarcation between normal and abnormal results. The two distributions do not overlap.

For this screening test you want to set the sensistivity very high, even at the expense of a low specificity. This ensures that you will have few false negatives but many false positives. That's OK. False positives aren't so bad, they just result in a need for a more expensive and more accurate (but safe) test. False-negative tests would be awful, as it means missing a case of a treatable fatal disease. Now let's consider another example, a screening test for an incurable noncommunicable disease. Here you want to avoid false positives (falsely telling a healthy person that she will die soon), while

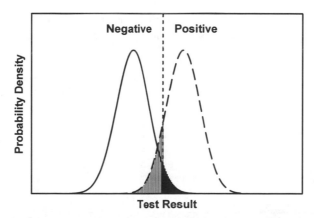

Figure 14.2. A typical test. As in Figure 14.1, the solid line shows the distribution of values in people without the disease, and the dashed curve shows the distribution of values in people with the disease. The two distributions overlap, so it is not obvious where to draw the line between negative and positive results. Our decision is shown as the dotted line. False-positive results are shown in the solid region: These are individuals without the disease who have a positive test result. False negatives are shown as the shaded area: These are individuals with the disease who have a negative test result.

false negatives aren't so bad (since you can't treat the disease anyway). In such a case, you would want the specificity to be high, even at the expense of a low sensitivity. It is impossible to make generalizations about the relative consequences of false-positive and false-negative tests, and so it is impossible to make generalizations about where to set the trade-off between sensitivity and specificity. Since the consequences of false-positive and false-negative tests are usually not directly comparable, value judgments are needed and different people will appropriately reach different conclusions regarding the best threshold value.

THE PREDICTIVE VALUE OF A TEST RESULT

Neither the specificity nor sensitivity answer the most important questions: If the test is positive, what is the chance that the patient really has the disease? If the test is negative, what is the chance that the patient really doesn't have the disease? The answers to those questions are quantified by the positive predictive value and negative predictive value:

$$\text{Positive predictive value} = \frac{\text{TP}}{\text{TP} + \text{FP}}.$$

$$\text{Negative predictive value} = \frac{\text{TN}}{\text{TN} + \text{FN}}. \tag{14.2}$$

The sensitivity and specificity are properties of the test. In contrast, the positive predictive value and negative predictive value are determined by the characteristics of the test and the prevalence of the disease in the population being studied. The lower the prevalence of the disease, the lower the ratio of true positives to false positives.

Look back at Figure 14.2. The two curves have equal areas, implying that there are equal numbers of tested patients with the condition and without the condition. In other words, Figure 14.2 assumes that the prevalence is 50%. In Figure 14.3, the

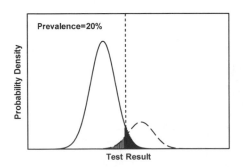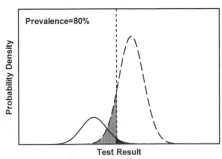

Figure 14.3. The effect of prevalence on the predictive value of the test. In Figures 14.1 and 14.2 the two curves had equal area, implying that half the tested population had the disease. The left half of this figure shows the results if the prevalence is 20%, while the right half shows the results if the prevalence is 80%. The fraction of all positive tests that are false positives is higher in the left panel. The fraction of all negative tests that are false negatives is higher in the right panel.

prevalence is changed to 20% (left panel) or 80% (right panel). As before, the solid curve represents people without the condition, and the dashed curve represents people with the condition. Any test result to the right of the dotted line is considered to be positive. This includes a portion of the area under the dashed curve (true positives) and a portion of the area under the solid curve (false positives).

The predictive values depend on the prevalence of the condition. A much larger fraction of positive tests are false positives in the left panel than in the right panel, so the predictive value of a positive test is lower in the left panel. Conversely, a much lower fraction of negative tests are false negative, so the predictive value of a negative test is much higher in the left panel.

CALCULATING THE PREDICTIVE VALUE OF A POSITIVE OR NEGATIVE TEST

Acute intermittent porphyria is an autosomal dominant disease that is difficult to diagnose clinically. It can be diagnosed by reduced levels of porphobilinogen deaminase. But the levels of the enzyme vary in both the normal population and in patients with porphyria, so the test does not lead to an exact diagnosis.

The sensitivity and specificity have been tabulated for various enzyme activities. Published data show that 82% of patients with porphyria have levels < 99 units (sensitivity = 82%) and that 3.7% of normal people have levels > 99 units (specificity = 100 − 3.7% = 96.3%). What is the likelihood that a patient with 98 units of enzyme activity has porphyria? The answer depends on who the patient is. We'll work through three examples.

Patient A

In this example, the test was done as a population screening test. Patient A has no particular risk for the disease. Porphyria is a rare disease with a prevalence of about 1 in 10,000. Since patient A does not have a family history of the disease, his risk of porphyria is 0.01%. Equivalently, we can say that the prior probability (prior to knowing the test result) is 0.01%. After knowing the test result, what is the probability that patient A has the disease? To calculate this probability, we need to fill in the blanks (A through I) in Table 14.2 for a large population similar to patient A.

Follow these steps to fill in the table. (Don't skip these steps. I show similar tables many times in this book, and you need to understand them.)

Table 14.2. Results of a Lab Test: Definitions of A through I

	Disease Present	Disease Absent	Total
Test positive	A	D	G
Test negative	B	E	H
Total	C	F	I

1. Assume a population of 1,000,000 (arbitrary). Enter the number 1,000,000 in position I. All we care about is ratios of values, so the total population size is arbitrary. I = 1,000,000.
2. Since the prevalence of the disease is 1/10,000, the total number in the disease present column is 0.0001 × 1,000,000 or 100. C = 100.
3. Subtract the 100 diseased people from the total of 1,000,000, leaving 999,900 disease absent people. F = 999,900.
4. Next calculate the number of people with disease present who also test positive. This equals the total number of people with the disease times the sensitivity. Recall that sensitivity is defined as the fraction of those with the disease whose test result is positive. So the number of people with disease and a positive test is 0.82 * 100 = 82. A = 82.
5. Next calculate the number of people without disease who test negative. This equals the total number of people without the disease (999,900) times the specificity (.963). Recall that the specificity is the fraction of those without the disease whose test result is negative. E = 962,904.
6. Calculate the number of people with the disease who test negative by subtraction: B = 100 − 82 = 18. So B = 18.
7. Calculate the number of people without the disease who test positive by subtraction: D = F − E = 36,996.
8. Calculate the two row totals. G = A + D = 37078. H = B + E = 962,922.

Table 14.3 is the completed table.

   If you screen 1 million people, you expect to find a test result less than 99 units in 37,078 people. Only 82 of these cases will have the disease. The predictive value of a positive test is only 82/37,078, which equals 0.22%. Only about 1 in about 500 of the positive tests indicate disease! The other 499 out of 500 positive tests are false positives. The test is not very helpful.

Patient B

This patient's brother has the disease. The disease is autosomal dominant, so there is a 50% chance that each sibling has the gene. The prior probability is 50%. Table 14.4 gives results you expect to see when you screen 1000 siblings of patients.

   You expect to find 429 positive tests. Of these individuals, 410 will actually have the disease and 19 will be false positives. The predictive value of a positive test is 410/429, which is about 96%. Only about 5% of the positive test are false positives.

Table 14.3. Porphyria Example A: Screening Test

Patient A	Disease Present	Disease Absent	Total
<99 units	82	36,996	37,078
>99 units	18	962,904	962,922
Total	100	999,900	1,000,000

Table 14.4. Porphyria Example B: Siblings of People with the Disease

	Disease Present	Disease Absent	Total
<99 units	410	19	429
>99 units	90	481	571
Total	500	500	1000

Patient C

This patient does not have a family history of porphyria, but he has all the right symptoms. You suspect that he has it. What pretest probability should you use? Base it on (informed) clinical judgment. In this case, you feel 30% sure that the patient has the disease. So the prior probability is 30%. Filling in the table is easy, as illustrated in Table 14.5.

If the test is positive, you can conclude that there is about a 246/272 = 90% chance that the patient has porphyria. Unlike the other examples, the "prior probability" in this example is a fairly fuzzy number. In the other two examples, the prior probability came from epidemiological data (prevalence) or from genetic theory. Here it just comes from clinical judgment. Since the prior probability is not an exact probability, the answer (90%) is also not exact. It is an estimate, but you can be fairly certain that patient C has porphyria. If the prior probability was really 20% or 40% (instead of 30%), you can calculate that the predictive value would be 85% or 94%.

The three patients had exactly the same test result, but the interpretation varies widely. To interpret the test result, it is not enough to know the sensitivity and specificity. You also need to know the prior probability (prior to seeing the test result) that the patient had the disease. For patient A, you obtained the prior probability from population prevalence. In patient B, you obtained the prior probability from genetic theory. In patient C, you estimated the prior probability using clinical intuition.

The predictive value of a positive test depends on who you are testing. Your interpretation of a positive test should depend partly on the characteristics of the test (as measured by sensitivity and specificity) but also on the prevalence of the disease in the group being tested. A test that is very useful for testing patients strongly suspected of having the disease (high prevalence) may turn out to be useless as a screening test in a more general population (low prevalence). Screening tests for rare diseases in the general population are only useful when both sensitivity and specificity are extremely high.

Table 14.5. Porphyria Example C: Clinical Suspicion

Patient C	Disease Present	Disease Absent	Total
<99 units	246	26	272
>99 units	54	674	728
Total	300	700	1000

Table 14.6. Porphyria Example D

	Disease Present	Disease Absent	Total
<79 units	219	2	221
>79 units	281	498	779
Total	500	500	1000

Patient D

Like patient B, this patient is the brother of an affected patient. But patient D's enzyme level is lower, 79 units. Since a low level is abnormal, the lower activity is associated with lower sensitivity and higher specificity. Fewer patients and many fewer normals have such a low enzyme level. For this level, the sensitivity is 43.8% and the specificity is 99.5%. If you test 1000 siblings of porphyria patients, the results you expect to find are given in Table 14.6.

You'd only expect to find 221 people whose enzyme level is so low, and 219 of those have the disease. The predictive value of a positive test is 219/222 or 98.6%. As you'd expect, a lower (more abnormal) test result has a higher predictive value.

BAYES' THEOREM

In these examples, calculating the predictive values using the tables took several steps. These can be combined into a single equation named for Thomas Bayes, an English clergyman who worked out the mathematics of conditional probability in the late 1700s. The equation can be written in terms of either probabilities or odds, but the equation is much simpler when expressed in terms of odds. Therefore, you need to review the difference between probability and odds before reading about Bayes theorem.

A REVIEW OF PROBABILITY AND ODDS

Likelihood can be expressed either as a probability or as odds.

- The *probability* that an event will occur is the fraction of times you expect to see that event in many trials.
- The *odds* are defined as the probability that the event will occur divided by the probability that the event will not occur.

A probability is a fraction and always ranges from 0 to 1. Odds range from 0 to infinity. Any probability can be expressed as odds. Any odds can be expressed as a probability. Convert between odds and probability with Equations 14.3 and 14.4:

$$\text{Odds} = \frac{\text{probability}}{1 - \text{probability}}. \tag{14.3}$$

$$\text{Probability} = \frac{\text{odds}}{1 + \text{odds}}. \tag{14.4}$$

If the probability is 0.50 or 50%, then the odds are 50:50 or 1. If you repeat the experiment often, you expect to observe the event (on average) in one out of two trials (probability = 1/2). That means you'll observe the event once for every time it fails to happen (odds = 1:1).

If the probability is 1/3, the odds equal $1/3/(1 - 1/3) = 1:2 = 0.5$. On average, you'll observe the event once in every three trials (probability = 1/3). That means you'll observe the event once for every two times it fails to happen (odds = 1:2).

BAYES' EQUATION

The Bayes' equation for clinical diagnosis is Equation 14.5:

$$\text{Post-test odds} = \text{pretest odds} \cdot \frac{\text{sensitivity}}{1 - \text{specificity}}. \qquad (14.5)$$

The post-test odds are the odds that a patient has the disease, taking into account both the test results and your prior knowledge about the patient. The pretest odds are the odds that the patient has the disease determined from information you know before running the test. The ratio sensitivity/(1 − specifity) is called the *likelihood ratio*. It is the probability of obtaining the positive test result in a patient with the disease (sensitivity) divided by the probability of obtaining a positive test result in a patient without the disease (1 − specificity). So Bayes' equation can be written in a simpler (and more general) form:

$$\text{Post-test odds} = \text{pretest odds} \cdot \text{likelihood ratio}. \qquad (14.6)$$

Using this equation we can rework the examples with intermittent porphyria. The test used in the example has a sensitivity of 82% and a specificity of 96.3%. Thus the likelihood ratio (sensitivity/1 − specificity) is $.82/(1.0 - .963) = 22.2$.* Analysis of patients A through C is shown in Table 14.7.

The first column shows the pretest probability, which came from epidemiological data (A), genetic theory (B), or clinical experience (C). The second column expresses the pretest probability as odds, calculated from the pretest probability using Equation 14.3. The third column was calculated from Equation 14.6. This is the odds that the patient has the disease, considering both the results of the test and the pretest odds. The last column converts this result back to a probability using Equation 14.4. The

Table 14.7. Intermittent Porphyria Examples Using Bayes' Equation

Patient	Pretest Probability	Pretest Odds	Post-Test Odds	Post-Test Probability
A	0.0001	0.0001	0.0022	0.0022
B	0.50	1.0000	22.2	0.957
C	0.30	0.4286	9.514	0.905

*If you express sensitivity and specificity as percents, rather than fractions, the likelihood ratio is defined as (sensitivity/100 − specificity).

results, of course, match those calculated earlier from the individual tables. Using Bayes' equation is more efficient.

SOME ADDITIONAL COMPLEXITIES

* Bayesian logic integrates the result of one lab test into the entire clinical picture. Many clinicians do a pretty good job of combining probabilities intuitively without performing any calculations and without knowing anything about Bayesian thinking. The formal Bayesian approach is more explicit and exact. Moreover, it clearly shows the necessity of knowing the prevalence of diseases and the sensitivity and specificity of tests.
* When thinking about the predictive value of tests, distinguish screening tests from confirmatory tests. It is OK if the quick and cheap screening tests turns up a lot of false positives, as long as the confirmatory test (often slower and more expensive) gives the correct result. When a positive screening test is followed by a confirmatory test, you really only care about the predictive value of the pair of tests.
* For some genetic diseases you need to distinguish the sensitivity to detect the genetic defect from the sensitivity to detect clinical disease. Some genetic diseases have poor penetrance, meaning that some people with the abnormal gene do not get the disease. A test that detects the gene with few false positives would produce a lot of false positives when assessed for its ability to detect the clinical disease.
* For many tests, sensitivity and specificity are tabulated for various values of the test. It is not necessary to pick a single threshold between positive and negative. Patient D described earlier demonstrates this point.
* The closer you can test for the real cause of the disease, the higher the sensitivity and specificity. If you tested for the abnormal gene sequence (rather than enzyme activity), the sensitivity and specificity would be extremely high (unless the penetrance is low). In this case, the test would probably be definitive, and there would be little need for Bayesian analyses.

SUMMARY

There are many ways of summarizing the accuracy of a diagnostic test. *Sensitivity* quantifies how well the test correctly detects the presence of the condition; *specificity* quantifies how well it correctly detects the absence of the condition.

The rates of false-negative and false-positive test results depend not only on the sensitivity and specificity of the test, but also on the prior probability that the subject has the disease. In some situations, you know the prior probability from population epidemiology. The prior probability is the prevalence of the disease. In other cases, you know the prior probability from genetic theory. In still other situations, you can estimate the prior probability from clinical experience.

Bayesian logic can be used to combine the result of the test with the prior probability to determine the probability that the patient has the condition. The Bayesian approach lets you combine the objective results of a test with your prior clinical suspicion to calculate the probability that a patient has a disease. Although formal

Bayesian analysis is seldom used in clinical settings, good clinicians intuitively combine these probabilities routinely.

OBJECTIVES

1. You must be familiar with the following terms:
   - Sensitivity
   - Specificity
   - False positives
   - False negatives
   - Predictive value
   - Bayes' equation
   - Bayesian logic
   - Likelihood ratio
2. You must understand why the rates of false positives and negatives depends on the prevalence of the condition being tested for.
3. Given the specificity, sensitivity, and prevalence, you should be able to calculate the rate of false positives and false negatives.
4. Using a book for reference, you should be able to calculate the probability that a patient has a disease if you are given the specificity and sensitivity of a test and the prior odds (or prior probability).

PROBLEMS

1. A test has a specificity of 92% and a sensitivity of 90%. Calculate the predictive values of positive and negative tests in a population in which 5% of the individuals have the disease.
2. A test has a specificity of 92% and a sensitivity of 99%. Calculate the predictive values of positive and negative tests in a population in which 0.1% of the individuals have the disease.
3. A woman wants to know if her only son is color blind. Her father is color blind, so she must be a carrier (because color blindness is a sex-linked trait). This means that, on average, half her sons will be color blind (she has no other sons). Her son is a smart toddler. But if you ask him the color of an object, his response seems random. He simply does not grasp the concept of color. Is he color blind? Or has he not yet figured out what people mean when they ask him about color? From your experience with other kids that age, you estimate that 75% of kids that age can answer correctly when asked about colors. Combine the genetic history and your estimate to determine the chance that this kid is color blind.
4. For patient C in the porphyria example, calculate the predictive value of the test if your clinical intuition told you that the prior probability was 75%.

Bayes and Statistical Significance

Setting the value of α, the threshold P value for determining significance, is similar to selecting the threshold value for a lab test that distinguishes "normal" from "abnormal." Recall from the previous chapter that selecting a threshold value for a lab test involves a trade-off between false positives and false negatives. Similarly, you learned in the previous chapter that selecting a value for α involves a tradeoff between Type I errors and Type II errors. The analogy is shown in Tables 15.1 and 15.2. You should memorize the differences between Type I and Type II errors. Those terms are sometimes mentioned in papers without being defined, and the distinction is not always clear from the context.

If a lab test measures the concentration of a chemical in the blood, you don't have to worry about false-positive and false-negative lab tests if you think about the actual concentration, rather than the conclusion positive or negative. Similarly, you don't have to worry about Type I and II errors if you think about the actual value of P (as a way to describe or summarize data) rather than the conclusion significant or not significant.

TYPE I ERRORS AND FALSE POSITIVES

You have made a Type I error when you reject the null hypothesis ($P < \alpha$) when the null hypothesis is really true. Note the subtle distinction between P values and α. Before collecting data, you choose the value of α, the threshold value below which you will deem a P value significant. Then you calculate the P value from your data.

The following statements summarize the analogy between specificity and P values:

Lab. If the patient really does not have the disease, what is the chance that the test will yield a positive result? The answer is 1 minus the specificity.

Statistics. If we assume that the two populations have identical means (or proportions), what is the chance that your study will find a statistically significant difference? The answer is α.

TYPE II ERRORS AND FALSE NEGATIVES

To define a Type II error, you must define your experimental hypothesis. To distinguish it from the null hypothesis, the experimental hypothesis is sometimes called the

Table 15.1. False Negatives and Positives in Diagnostic Tests

Diagnostic Test	Disease Is Really Present	Disease Is Really Absent
Test positive	No error (true positive)	False positive
Test negative	False negative	No error (true negative)

alternative hypothesis. It is not enough to say that the experimental hypothesis is that you expect to find a difference; you must define how large you expect the difference to be.

You have made a Type II error when you conclude that there is no significant difference between two means, when in fact the alternative hypothesis is true. The probability of making a Type II error is denoted by β and is sometimes called a *beta error.* The value of β depends on how large a difference you specify in the alternative hypothesis. If you are looking for a huge difference, the probability of making a Type II error is low. If you are looking for a tiny difference, then the probability of making a Type II error is high. Thus, one cannot think about β without defining the alternative hypothesis. This is done by deciding on a value for the minimum difference (or relative risk) that you think is clinically or scientifically important and worth detecting. This minimum difference is termed Δ (delta). Your choice of Δ depends on the scientific or clinical context. Statisticians or mathematicians can't help, the alternative hypothesis must be based on your scientific or clinical understanding. β is the probability of randomly selecting samples that result in a nonsignificant P value when the difference between population means equals Δ.

The *power* of a test is defined as 1 − β. The power is the probability of obtaining a significant difference when the difference between population means equals Δ. Like β, the power can only be defined once you have chosen a value for Δ. The larger the sample size, the greater the power. The lower you set α, the lower the power. Increasing Δ will increase the power, as it is easier to find a big difference than a small difference.

The following statements summarize the analogy between sensitivity and power:

Lab. If the patient really has a certain disease, what is the chance that the test will correctly give a positive result? The answer is the sensitivity. If the test can detect several diseases, the sensitivity of the test depends on which disease you are looking for.

Statistics. If there really is a difference (Δ) between population means (or proportions), what is the chance that analysis of randomly select subjects will result in a significant difference? The answer is the power, equal to one minus β. The answer depends on the size of the hypothesized difference, Δ.

Table 15.2. Type I and Type II Errors in Statistical Tests

Statistical Test	Populations Have Different Means (or Proportions)	Populations Have Identical Means (or Proportions)
Significant difference	No error	Type I error
No significant difference	Type II error	No error

PROBABILITY OF OBTAINING A FALSE-POSITIVE LAB RESULT: PROBABILITY THAT A SIGNIFICANT RESULT WILL OCCUR BY CHANCE

What is the probability of obtaining a false-positive lab result? This question is a bit ambiguous. It can be interpreted as two different questions:

• What fraction of all disease free individuals will have a positive test? This answer equals FP/(FP + TN), which is the same as one minus the specificity.
• What fraction of all positive test results are false positives? The answer is FP/(FP + TP). This is the conventional definition of the rate of false positives. As you learned in the previous chapter, this question can be answered only if you know the prevalence of the disease (or prior probability) in the population you are studying.

What is the probability of obtaining a statistically significant P value by chance? Again, this question is ambiguous. It can be interpreted as two different questions:

• If the null hypothesis is true, what fraction of experiments will yield a significant P value? Equivalently, if the null hypothesis is true, what is the probability of obtaining a statistically significant result ($P < \alpha$)? The answer is α, conventionally set to 5%.
• In what fraction of all experiments that result in significant P values is the null hypothesis true? Equivalently, if a result is statistically significant, what is the probability that the null hypothesis is true? The answer is not necessarily 5%. Conventional statistics cannot answer this question at all. Bayesian logic can answer the question, but only if you can define the prior probability that the null hypothesis is true. The next section discusses how to apply Bayesian logic to P values.

In each case (lab tests and statistical tests) the logic of the first question goes from population to sample, and the logic of the second goes from sample to population. When analyzing data, we are more interested in the second question.

THE PREDICTIVE VALUE OF SIGNIFICANT RESULTS: BAYES AND P VALUES

You perform a statistical test and obtain a significant result. Repeated from the last section, here is the question you wish to answer:

> In what fraction of all experiments that result in significant P values is the null hypothesis true? Equivalently, if the result is statistically significant, what is the probability that the null hypothesis is really true?

Here is an imaginary example. You are working at a drug company and are screening drugs as possible treatments for hypertension. You test the drugs in a group of animals. You have decided that you are interested in a mean decrease of blood pressure of 10 mmHg and are using large enough samples so that you have 80% power to find a significant difference ($\alpha = 0.05$) if the true difference between population means is 10 mmHg. (You will learn how to calculate the sample size in Chapter 22.)

You test a new drug and find a significant drop in mean blood pressure. You know that there are two possibilities. Either the drug really works to lower blood pressure, or the drug doesn't alter blood pressure at all and you just happened to get lower pressure readings on the treated animals. How likely are the two possibilities?

Since you set α to 0.05, you know that 5% of studies done with inactive drugs will demonstrate a significant drop in blood pressure. But that isn't the question you are asking. You want to know the answer to a different question: In what fraction of experiments in which you observe a significant drop in pressure is the drug really effective? The answer is not necessarily 5%. To calculate the answer you need to use Bayesian logic and need to think about the prior probability. The answer depends on what you knew about the drug before you started the experiment, expressed as the prior probability that the drug works. This point is illustrated in the following three examples.

Drug A

This drug is known to weakly block angiotensin receptors, but the affinity is low and the drug is unstable. From your experience with such drugs, you estimate that there is about a 10% chance that it will depress blood pressure. In other words, the prior probability that the drug works is 10%. What will happen if you test 1000 such drugs? The answer is shown in Table 15.3.

These are the steps you need to follow to create the table:

1. We are predicting the results of 1000 experiments with 1000 different drugs, so the grand total is 1000. This number is arbitrary, since all we care about are ratios.
2. Of those 1000 drugs we screen, we expect that 10% will really work. In other words, the prior probability equals 10%. So we place 10% of 1000 or 100 as the total of the first column, leaving 900 for the sum of the second column.
3. Of the 100 drugs that really work, we will obtain a significant result in 80% (because our experimental design has 80% power). So we place 80% of 100, or 80, into the top left cell of the table. This leaves 20 experiments with a drug that really works, but P > 0.05 so we conclude that the drug is not effective.
4. Of the 900 drugs that are really ineffective, we will by chance obtain a significant reduction in blood pressure in 5% (because we set α equal to 0.05). Thus the top cell in the second column is 5% $\times$ 900 or 45. That leaves 855 experiments in which the drug is ineffective, and we correctly observe no significant difference.
5. Determine the row totals by addition.

Out of 1000 tests of different drugs, we expect to obtain a significant difference (P < 0.05) in 125 of them. Of those, 80 drugs are really effective and 45 are not. When

Table 15.3. Statistical Significance When Testing Drug A

Drug A Prior Probability = 10%	Drug Really Works	Drug Is Really Ineffective	Total
Significant difference	80	45	125
No significant difference	20	855	875
Total	100	900	1000

you see a significant result for any particular drug, you can conclude that there is a 64% chance (80/125) that the drug is really effective and a 36% chance (45/125) that it is really ineffective.

Drug B

Here the pharmacology is much better characterized. Drug B blocks the right kinds of receptors with reasonable affinity and the drug is chemically stable. From your experience with such drugs, you estimate that the prior probability that the drug is effective equals 80%. What would happen if you tested 1000 such drugs? The answer is shown in Table 15.4.

If you test 1000 drugs like this one, you expect to see 650 significant results. Of those, 98.5% (640/650) will be truly effective. When you see a significant result for any particular drug, you can conclude that there is a 98.5% chance that it will really lower blood pressure and a 1.5% chance that it is really ineffective.

Drug C

This drug was randomly selected from the drug company's inventory of compounds. Nothing you know about this drug suggests that it affects blood pressure. Your best guess is that about 1% of such drugs will lower blood pressure. What would happen if you screen 1000 such drugs? The answer is shown in Table 15.5.

If you test 1000 drugs like this one, you expect to see 58 significant results. Of those, you expect that 14% (8/58) will be truly effective and that 86% (50/58) will be ineffective. When you see a significant result for any particular drug, you can conclude that there is a 14% chance that it will really lower blood pressure and an 85% chance that it is really ineffective.

These examples demonstrate that your interpretation of a significant result appropriately depends on what you knew about the drug before you started. You need to integrate the P value obtained from the experiment with the prior probability.

When you try to do the calculations with real data, you immediately encounter two problems:

• You don't know the prior probability. The best you can do is convert a subjective feeling of certainty into a "probability." If you are quite certain the experimental hypothesis is true, you might say that the prior probability is 0.99. If you are quite certain the experimental hypothesis is false, you might say that the prior probability is 0.01. If you think it could go either way, you can set the prior probability to 0.5.

Table 15.4. Statistical Significance When Testing Drug B

Drug B Prior Probability = 80%	Drug Really Works	Drug Is Really Ineffective	Total
Significant difference	640	10	650
No significant difference	160	190	350
Total	800	200	1000

Table 15.5. Statistical Significance When Testing Drug C

Drug C Prior Probability = 1%	Drug Really Works	Drug Is Really Ineffective	Total
Significant difference	8	50	58
No significant difference	2	940	942
Total	10	990	1000

- You don't know what value to give Δ, the smallest difference that you think is scientifically or clinically worth detecting. While it is usually difficult to choose an exact value, it is usually not too hard to estimate the value.

Despite these problems, it is often possible to make reasonable estimates for both the prior probability and Δ. It's OK that these values are estimated, so long as you treat the calculated probability as an estimate as well.

THE CONTROVERSY REGARDING BAYESIAN STATISTICS

It is possible to combine all the steps we took to create the tables into one simple equation called the *Bayes' equation*, as you saw in the last chapter. The entire approach discussed in the previous section is called *Bayesian thinking*. The Bayesian approach to interpreting P values is rarely used. If you knew the prior probability, applying Bayesian logic would be straightforward and not controversial. However, usually the prior probability is not a real probability but is rather just a subjective feeling. Some statisticians (Bayesians) think it is OK to convert these feelings to numbers ("99% sure" or "70% sure"), which they define as the prior probability. Other statisticians (frequentists) think that you should never equate subjective feelings with probabilities.

There are some situations where the prior probabilities are well defined. For example, see the discussion of genetic linkage in the next chapter. The prior probability that two genetic loci are linked is known, so Bayesian statistics are routinely used in analysis of genetic linkage. There is nothing controversial about using Bayesian logic when the prior probabilities are known precisely.

The Bayesian approach explains why you must interpret P values in the context of what you already know or believe, why you must think about biological plausibility when interpreting data. When theory changes, it is appropriate to change your perception of the prior probability and to change your interpretation of data. Accordingly, different people can appropriately and honestly reach different conclusions from the same data. All significant P values are not created equal.

APPLYING BAYESIAN THINKING INFORMALLY

When reading biomedical research, you'll rarely (if ever) see Bayesian calculations used to interpret P values. And few scientists use Bayesian calculations to help interpret P values. However, many scientists use Bayesian thinking in a more informal way without stating the prior probability explicitly and without performing any additional calculations. When reviewing three different studies, the thinking might go like this:

This study tested a hypothesis that is biologically sound and that is supported by previous data. The P value is 0.04, which is marginal. I have a choice of believing that the results are due to a coincidence that will happen 1 time in 25 under the null hypothesis, or of believing that the experimental hypothesis is true. Since the hypothesis makes so much sense, I'll believe it. The null hypothesis is probably false.

This study tested a hypothesis that makes no biological sense and has not been supported by any previous data. The P value is 0.04, which is lower than the usual threshold of 0.05, but not by very much. I have a choice of believing that the results are due to a coincidence that will happen 1 time in 25 under the null hypothesis, or of believing that the experimental hypothesis is true. Since the experimental hypothesis is so crazy, I find it easier to believe that the results are due to coincidence. The null hypothesis is probably true.

This study tested a hypothesis that makes no biological sense and has not been supported by any previous data. I'd be amazed if it turned out to be true. The P value is incredibly low (0.000001). I've looked through the details of the study and cannot identify any biases or flaws. These are reputable scientists, and I believe that they've reported their data honestly. I have a choice of believing that the results are due to a coincidence that will happen one time in a million under the null hypothesis or of believing that the experimental hypothesis is true. Even though the hypothesis seems crazy to me, the data force me to believe it. The null hypothesis is probably false.

You should interpret experimental data in the context of theory and previous data. That's why different people can legitimately reach different conclusions from the same data.

MULTIPLE COMPARISONS

Experienced clinicians do not get excited by occasional lab values that are marginally abnormal. If you perform many tests on a patient, it is not surprising that some are labeled "abnormal," and these may tell you little about the health of the patient. You need to consider the pattern of all the tests and not focus too much on any one particular test. If the test is quantitative, you also need to consider whether the test is just barely over the arbitrary line that divides normal from abnormal, or whether the result is really abnormal and far from the dividing line.

Similarly, experienced scientists do not get excited by occasional "significant" P values. If you calculate many P values, you expect some to be small and significant just by chance. When you interpret significant P values, you must take into account the total number of P values that were calculated. If you make multiple comparisons and calculate many P values, you expect to encounter some small P values just by chance. Chapter 13 discussed this problem in great detail.

SUMMARY

The analogy between diagnostic tests and statistical hypothesis tests is summarized in Table 15.6.

Table 15.6. Comparison Between Diagnostic Tests and Statistical Hypothesis Tests

	Lab Test	Statistical Hypothesis Test
Result	The result is a measurement, but it can be compared to a threshold and reported as "normal" or "abnormal."	The result is a P value, but it can be compared to a threshold and reported as "statistically significant" or "not statistically significant."
Scope	A lab test is performed for one individual and yields the diagnosis of positive or negative.	A P value is calculated from one experiment and yields the conclusion of significant or not significant.
Errors	A lab test can result in two kinds of errors: false positives and false negatives.	A statistical hypothesis test can result in two kinds of errors: Type I and Type II.
Threshold	You should choose the threshold between "normal" and "abnormal" based on the relative consequences of false-positive and false-negative diagnoses.	You should choose a value for α (the threshold between "not significant" and "significant" P values) based on the relative consequences of making a Type I or Type II error.
Accuracy	The accuracy of the lab test is expressed as two numbers: sensitivity and specificity.	The accuracy of the statistical test is expressed as two numbers: α and β (or power).
Interpretation	When interpreting the result of a lab test for a particular patient, you must integrate what is known about the accuracy of the laboratory test (sensitivity and specificity) with what is known about the patient (prevalence, or prior probability that the patient has disease). Bayesian logic combines these values precisely.	When interpreting the result of a statistical test of a particular hypothesis, you must integrate what is known about the accuracy of the statistical test (α and β) with what is known about the hypothesis (prior probability that the hypothesis is true). Bayesian logic combines these values precisely.
Multiple comparisons	If you perform many tests on one patient, you shouldn't be surprised to see occasional "abnormal" results. If you perform many tests, you need to look at overall patterns and not just individual results.	If you perform many statistical tests, you shouldn't be surprised to see occasional "significant" results. If you perform many tests, you need to look at overall patterns and not just at individual P values.

OBJECTIVES

1. You must be familiar with the following terms:
   - Type I error
   - Type II error
   - α error
   - β error
   - Power

2. You should be able to explain the analogy between false-positive and false-negative lab tests and Type II and Type I statistical errors.
3. You should understand why it is hard to answer this question: In what fraction of all experiments that result in a significant P value is the null hypothesis really true?
4. You should be able to explain why the answer to that question depends on the nature of the hypothesis being tested.
5. Given a prior probability and power, you should be able to calculate the predictive value of a statistically significant P value.

PROBLEMS

1. A student wants to determine whether treatment of cells with a particular hormone increases the number of a particular kind of receptors. She and her advisor agree that an increase of less than 100 receptors per cell is too small to care about. Based on the standard deviation of results you have observed in similar studies, she calculates the necessary sample size to have 90% power to detect an increase of 100 receptors per cell. She performs the experiment that number of times, pools the data, and obtains a P value of 0.04.

    The student thinks that the experiment makes a lot of sense and thought that the prior probability that her hypothesis was true was 60%. Her advisor is more skeptical and thought that the prior probability was only 5%.

    A. Combining the prior probability and the P value, what is the chance that these results are due to chance? Answer from both the student's perspective and that of the advisor.
    B. Explain why two people can interpret the same data differently.
    C. How would the advisor's perspective be different if the P value were 0.001 (and the power were still 90%)?

2. You go to Las Vegas on your 25th birthday, so bet on the number 25 in roulette. You win. You bet a second time, again on 25, and win again! A roulette wheel has 38 slots (1 to 36, 0, and 00), so there is a 1 in 38 chance that a particular spin will land on 25.

    A. Assuming that the roulette wheel is not biased, what is that chance that two consecutive spins will land on 25?
    B. If you were to spend a great deal of time watching roulette wheels, you would note two consecutive spins landing on 25 many times. What fraction of those times would be caused by chance? What fraction would be caused by an unfair roulette wheel?

Bayes' Theorem in Genetics

BAYES' THEOREM IN GENETIC COUNSELING

In genetic counseling you want to determine the probability that someone has a particular genetic trait.

Example 16.1

A woman wants to know her chances of being a carrier for Duchenne's muscular dystrophy, an X-linked recessive trait. Since her brother and maternal uncle both have the disease, it is clear that the gene runs in her family and is not a new mutation. From her family history, her mother must be a carrier and the woman had a 50% chance of inheriting the gene at birth.

Knowing that the woman has two sons without the disease decreases the chance that the woman is a carrier. Bayesian logic allows you to combine this evidence (two healthy sons) with the family history (50% chance of being a carrier). We'll first perform the calculations step by step with a table and then use Bayes' equation. Table 16.1 shows what you would expect to see if you were to examine many women with the same family history and two sons. The calculations are explained later.

To generate the table, follow these steps:

1. Set the grand total to 1000. This is arbitrary as we only care about ratios.
2. We know that half the women are carriers, so place 1/2 × 1000 or 500 into each column total.
3. If a woman is a carrier, there is a 1/4 chance (1/2 × 1/2) that both her sons would not have the disease. So place 1/4 × 500 = 125 in box A. That leaves 375 cases in box C.
4. If a woman is not a carrier, then none of her sons will have this disease (barring new mutations, which are very rare). So D = 0 and B = 500.
5. Compute the row totals.

Of the 1000 hypothetical women with two sons and this family history, 375 would have at least one son with the disease. We know that the woman in our example is not in this category. She is in the group of 625 women who have two sons without the disease. Of these 125 are carriers. So 125/625 = 20% of the women with two healthy sons are carriers. Thus we can say that the woman in our example has a 20% chance, or 1 in 5, of being a carrier.

Table 16.1. Calculations of Chance of Being a Carrier of Duchenne's Muscular Dystrophy in the Example

	Woman Is a Carrier	Woman Is Not a Carrier	Total
Both sons without disease	A = 125	B = 500	625
At least one son has the disease	C = 375	D = 0	375
Total	500	500	1000

From the laws of Mendelian genetics, we knew that her risk of being a carrier at birth was 1/2. Taking into account her two unaffected sons, using Bayesian logic lowers the risk to 1/5. Now let's use the Bayes' equation to streamline the calculations. Bayes' equation is as follows:

$$\text{Post-test odds} = \text{pretest odds} \cdot \text{likelihood ratio.} \qquad (16.1)$$

The likelihood ratio is the probability a carrier will have two unaffected sons divided by the probability that a noncarrier will have two unaffected sons. The probability that a carrier will have an unaffected son is 1/2. Therefore, the probability that both sons will be unaffected is $1/2 \times 1/2 = 1/4$ or 25%. The probability that a noncarrier will have two sons without this disease is 100% (barring new mutations, which are extremely rare). So the likelihood ratio is 25%/100% or 0.25.

From her family history, we know that this woman had a 50% chance of being a carrier at birth. This is the pretest probability. Therefore the pretest odds are 50:50 or 1.0. Multiply the pretest odds by the likelihood ratio to calculate the post-test odds, which equal 0.25 or 1:4. If you saw many people with the same family history as this woman, you'd see one carrier for every four noncarriers. Converting from odds to probability, the post-test probability is 20%. She has a 20% chance of being a carrier.

BAYES AND GENETIC LINKAGE

When two loci (genes or DNA sequences) are located near each other on the same chromosome, they are said to be linked. If the two loci are very close, crossing over or recombination between the two loci occurs rarely. Thus, alleles of linked loci tend to be inherited together. If the loci are further apart, recombination (a normal process) occurs more frequently. If the loci are very far apart, the two loci segregate independently just as if they were on different chromosomes.

Linkage is useful in genetic diagnosis and mapping. Since it is not possible to detect all abnormal genes directly, geneticists try to identify a marker gene (such as those for variable antigens or isozymes) or a variable DNA sequence that is linked to the disease gene. Once you know that the disease gene is linked to a marker, the presence of the marker (which you can identify) can then be used to predict the presence of the disease gene (which you cannot identify directly). This allows detection of genetic diseases prenatally or before they cause clinical problems. It also allows diagnosis of unaffected heterozygotes (carriers) who can pass the abnormal gene on to their children. This method works best for diseases caused by an abnormality of a single gene.

Before linkage can be useful in diagnosis, you need to identify a marker linked to the gene. This is usually done by screening lots of potential markers. How can you tell if a marker is linked to a disease gene? Geneticists study large families and observe how often the disease and marker are inherited together and how often there is recombination. If there are few recombination events between the marker and the disease, there are two possible explanations. One possibility is that the two are linked. The other possibility is that the two are not linked, but—just by coincidence—there were few recombination events.

Bayesian logic combines the experimental data with the prior probability of linkage to determine the probability that the gene is truly linked to the disease. To calculate Bayes' equation, we need to define the likelihood ratio in the context of linkage. When calculating the predictive values of lab tests in Chapter 14, we defined the likelihood ratio as sensitivity divided by one minus specificity—the probability that someone with the disease will have an abnormal test result divided by the probability that someone without the disease will have an abnormal test result. For studies of linkage, therefore, the likelihood ratio is the probability of obtaining the data if the genes really are linked* divided by the probability of observing those data if the genes are really not linked. The details of the calculations are beyond the scope of this book. When you read papers with linkage studies, you'll rarely see reference to the likelihood ratio. Instead you'll see the *lod score* (*log of odds*), which is simply the logarithm (base 10) of the likelihood ratio.

The higher the lod score, the stronger the evidence for linkage. A lod score of 3 means that the likelihood ratio equals 1000 (antilog of 3). This means that the data are 1000 times more likely to be observed if the marker is linked to the disease than if the marker is not linked.

To calculate the probability that the marker is linked to the gene requires accounting for the prior probability of linkage using Bayesian logic. Bayes' equation for linkage can be written as follows:

$$\text{Post-test odds of linkage} = \text{pretest odds of linkage} \cdot \text{likelihood ratio.}$$
$$\text{Post-test odds of linkage} = \text{pretest odds of linkage} \cdot 10^{lod}. \tag{16.2}$$

To calculate Bayes' equation, you must know the prior (or pretest) odds of linkage. Since there are 23 pairs of chromosomes, the chance that any particular randomly selected marker will be located on the same chromosome as the disease gene is 1/23 or 4.3%. But it is not enough to be on the same chromosome. To be linked to a disease, the marker must be close to the disease gene. So the prior probability that a random marker is linked to a particular gene must be less than 4.3%. In fact, genetic data tell us that 2% of randomly selected markers are linked to any particular disease gene.† Converting to odds, the pretest odds of linkage are about 0.02. The values presented here assume that the marker was randomly selected, as is often the case. If you pick a marker known to be on the same chromosome as the disease, then the pretest odds of linkage are higher.

*This can only be calculated once you specify a hypothetical genetic distance θ.

†For these calculations, we define linkage to mean that the probability of recombination is 30% or less.

Let's assume that a lod score equals 3. What is the probability that the marker and disease are truly linked? The post-test odds equal the pretest odds (0.02) times the likelihood ratio ($10^3 = 1000$), which is 20. Converting to a probability, the post-test probability equals 20/21 (Equation 14.4), which is about 95%. If you observe a lod score of 3.0, you will conclude that the marker and gene are linked. When you make that conclusion, there is a 95% chance that you will be correct, leaving a 5% chance that you will be wrong.

If a lod score equals or exceeds 3, geneticists usually conclude that the marker and disease are linked. If a lod score is less than or equal to -3, geneticists conclude that the marker and disease are not linked. See Problem 2 to calculate the probability that this conclusion is wrong. If the lod score is between -3 and 3, geneticists conclude that the evidence is not conclusive.

PROBLEMS

1. In Example 16.1, assume that the woman had three unaffected sons. What is the probability that she is a carrier?
2. If the lod score is -3, what is the probability that the marker is linked to the disease?
3. It would be possible to calculate a P value from linkage data. Explain in plain language what it would mean.
4. You perform a t test and obtain a P value of 0.032. You used enough subjects to ensure that the experiment had a 80% power to detect a specified difference between population means with $P < 0.05$. Does it make sense to calculate a likelihood ratio? If so, calculate the ratio and explain what it means.

V

CORRELATION AND REGRESSION

17

Correlation

INTRODUCING THE CORRELATION COEFFICIENT

Example 17.1

Borkman et al.* wanted to understand why insulin sensitivity varies so much among individuals. They hypothesized that the lipid composition of the cell membranes of skeletal muscle affected the sensitivity of the muscle for insulin. They tested the hypothesis in two studies. We'll look only at the second study of 13 healthy young men. They determined insulin sensitivity in each with a glucose clamp study. To do this they infused insulin at a standard rate (adjusting for size differences) and determined how much glucose they needed to infuse to keep the blood glucose level constant. Insulin causes the muscles to take up glucose and thus causes the level of glucose in the blood to fall. The amount of glucose infused to keep a constant blood level is thus a measure of insulin sensitivity. They only needed to infuse a lot of glucose when the muscle was very sensitive to insulin. They also took a small muscle biopsy from each subject and measured its fatty acid composition. We'll focus on the fraction of polyunsaturated fatty acids that have between 20 and 22 carbon atoms (%C20–22). The authors show the data as a graph, from which I read off the approximate values in Table 17.1. Table 17.1 is sorted in order of %C20–22, not in the order the subjects were studied. These data are graphed in Figure 17.1.

Note that both variables are scattered. The mean of the insulin-sensitivity index is 284 and the standard deviation (SD) is 114 mg/m²/min. The coefficient of variation is 114/284, which equals 40.1%. This is quite high. The authors knew that there would be a great deal of variability, and that is why they explored the causes of the variability. There is also reasonable variability in the content of fatty acids. The %C20–22 is 20.7% and the SD is 2.4%. So the coefficient of variation is 11.6%. If you don't look at the graph carefully, you could be misled. The X axis does not start at 0, so you get the impression that the variability is greater than it actually is.

Looking at the graph, there is an obvious relationship between the two variables. In general, individuals whose muscles have more C20–22 polyunsaturated fatty acids

*M Borkman, LH Sorlien, DA Pan, AB Jenkins, DJ Chisholm, LV Campbell. The relation between insulin sensitivity and the fatty-acid composition of skeletal-muscle phospholipids. *N Engl J Med* 328:238–244, 1993.

Table 17.1. Correlation Between %C20–22 and
Insulin Sensitivity

% C20–22 Polyunsaturated Fatty Acids	Insulin Sensitivity $(mg/m^2/min)$
17.9	250
18.3	220
18.3	145
18.4	115
18.4	230
20.2	200
20.3	330
21.8	400
21.9	370
22.1	260
23.1	270
24.2	530
24.4	375

also have greater sensitivity to insulin. The two variables vary together—statisticians
say that there is a lot of *covariation* or a lot of *correlation.*

The direction and magnitude of the linear correlation can be quantified with a
correlation coefficient, abbreviated r. Its value can range from −1 to 1. If the correlation coefficient is 0, then the two variables do not vary together at all. If the correlation coefficient is positive, the two variables tend to increase or decrease together.
If the correlation coefficient is negative, the two variables are inversely related, that
is, as one variable tends to decrease, the other one tends to increase. If the correlation

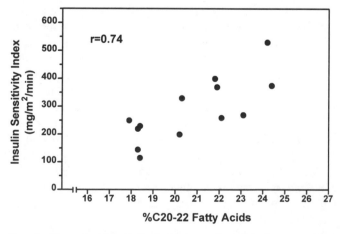

Figure 17.1. Data for Example 17.1. Each circle shows the results for one subject. The subjects
whose muscles have a higher percentage of C20–22 fatty acids tend to have a higher sensitivity
to insulin.

coefficient is 1 or -1, the two variables vary together completely, that is, a graph of the data points forms a straight line.

In the example the two variables increase together, so the correlation coefficient must be positive. But there is some scatter, so the correlation coefficient must be less than 1.0. In fact, the correlation coefficient equals 0.77.* As always, we'd like to make inferences about the correlation coefficient for the entire population. We know the correlation coefficient for this particular sample of 13 men. Using an equation given at the end of the chapter, we can calculate that the 95% confidence interval (CI) for the correlation coefficient ranges from 0.38 to 0.93. We can be 95% sure that the overall correlation coefficient lies within this range. Even the low end of the CI has a strong positive correlation. So we can be quite confident that there is a strong correlation in the entire population.

You can calculate a P value from these data. The null hypothesis is that there is no correlation in the overall population. The two-tailed P value answers this question: If the null hypothesis were true, what is the chance that 13 randomly picked subjects would have a r greater than 0.77 or less than -0.77?

For this example the P value is 0.0021. If there really were no relationship between insulin sensitivity and the %C20–22 fatty acids, there is only a 0.21% chance of randomly observing such a strong correlation in an experiment of this size.

INTERPRETING r

Why do the two variables correlate so well? There are four possible explanations:

- The lipid content of the membranes determines insulin sensitivity.
- The insulin sensitivity of the membranes somehow affects lipid content.
- Both insulin sensitivity and lipid content are under the control of some other factor (perhaps a hormone).
- The two variables don't correlate in the population at all, and the observed correlation in our sample was a coincidence.

You can never rule out the last possibility, but the P value tells you how rare the coincidence would be. In this example, you would observe a correlation that strong (or stronger) in 0.21% of experiments if there is no correlation in the overall population.

You cannot decide between the first three possibilities by analyzing only these data, you need to perform additional experiments where you manipulate one of the variables. The authors, of course, want to believe the first possibility. While most people immediately think of the first two possibilities, the third possibility is often ignored. But correlation does not necessarily imply causality. Two variables can be correlated because both are influenced by the same third variable. Height and weight are correlated quite well, but height does not cause weight or vice versa. Infant mortality in various countries is negatively correlated with the number of telephones per capita, but buying telephones will not make kids live longer. Instead increased wealth (and

---

*I calculated all values from the data read off the graph in the paper. Since this is not completely accurate, the calculations shown in the paper are slightly different.

thus increased purchases of telephones) relates to better plumbing, better nutrition, less crowded living conditions, more vaccinations, etc.

INTERPRETING r^2

The square the correlation coefficient is an easier value to interpret than r. For the example, $r^2 = 0.59$. While r^2 is sometimes called the *coefficient of determination,* most scientists simply refer to it as "r squared." Because r is always between -1 and 1, r^2 is always between 0 and 1 and is smaller than r. More precisely, $r^2 \leq |r|$.

If you can accept the assumptions listed in the next section, you can interpret r^2 as the fraction of the variance that is shared between the two variables. In the example, 59% of the variability in insulin tolerance is associated with variability in lipid content. Knowing the lipid content of the membranes lets you explain 59% of the variance in the insulin sensitivity. That leaves 41% of the variance that is explained by other factors or by measurement error. X and Y are symmetrical in correlation analysis, so you can also say that 59% of the variability in lipid content is associated with variability in insulin tolerance. By squaring both ends of the 95% CI for r, we find that the 95% CI for r^2 is 0.14 to 0.86. We can be 95% sure that in the overall population somewhere between 14% and 86% of the variance in insulin sensitivity is associated with variance in membrane lipid content.

ASSUMPTIONS

You can calculate the correlation coefficient from any set of data, and it is a useful descriptor of the data. However, you cannot make inferences from the correlation coefficient (and P value) unless the following assumptions are true:

- *Subjects are randomly selected* from, or at least representative of, a larger population.
- *Paired samples.* Each subject (or each experimental unit) must have both X and Y values.
- *Independent observations.* Sampling one member of the population should not influence your chances of sampling anyone else. The relationship between all the subjects should be the same. In this example, the assumption of independence would be violated is some of the subjects are related (i.e., siblings). It would also be violated if the investigator purposely chose some people with diabetes and some without, or if the investigator measured each subject on two occasions and treated the values as two separate data points.
- *X and Y values must be measured independently.* The correlation calculations are not meaningful if the values of X and Y are somehow intertwined. For example, it would not be meaningful to calculate the correlation between a midterm exam score and the overall course score, as the midterm exam is one of the components of the course score.
- *X values were measured and not controlled.* If you systematically controlled the X variable (i.e., concentration, dose, or time), you should calculate linear regression

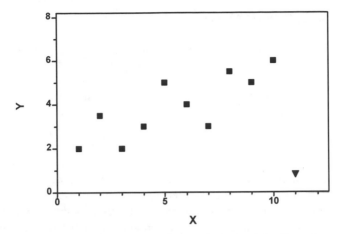

Figure 17.2. Effects of an outlier. In correlation, a single point far from the rest (an outlier) can have an enormous impact. If you analyze all 11 points (including the triangle), you'll find that $r - 0.020$ and $P = 0.38$. This analysis suggests that there is no correlation between X and Y. If you only analyze the 10 squares, $r = 0.82$ and $P = 0.0046$. This analysis provides strong evidence that X and Y are correlated. The presence or absence of a single point (the triangle) completely changes the conclusion.

rather than correlation (see Chapter 19). You will get the same value for r^2 and the P value. The confidence interval of r cannot be interpreted if the experimenter controls the value of X.

- *Gaussian distribution.* The X and Y values must each be sampled from populations that follow a Gaussian distribution, at least approximately.
- *All covariation must be linear.* The correlation coefficient would not be meaningful, for example, if Y increases as X increases up to a certain point but the Y decreases as X increases further.

OUTLIERS

Calculation of the correlation coefficient can depend heavily on one outlying point; change or delete that point and the analysis may be quite different. An example showing the influence of a single outlying point is shown in Figure 17.2. If you analyze the 10 squares only, $r = 0.81$ and $P = 0.005$. If you analyze all 11 data points (including the triangle), $r = 0.29$ and $P = 0.382$. Including or excluding just one point completely changes the results.

Outliers can influence all statistical calculations, but especially in correlation. You should look at graphs of the data before reaching any conclusion from correlation coefficients. Don't dismiss outliers as "bad" points that mess up the analysis. It is possible that the outliers are the most interesting observations in the study!

SPEARMAN RANK CORRELATION

You can't interpret the correlation coefficient unless you are willing to assume that the distribution of both X and Y in the overall population is Gaussian. What if you can't support that assumption?

As we'll see in later chapters, this problem comes up often in many contexts. Many statistical tests are based on the assumption that the data are sampled from Gaussian populations. Alternative tests are needed when that assumption cannot be supported. One alternative approach would be to make some other assumption about the distribution of values in the population, but this approach is not used often. If you are not willing to assume a Gaussian distribution, you're rarely willing to assume other distributions. And even if you were, statistical tests are not commonly available to deal with other distributions. A better approach is to use a method that does not make any assumption about how values in the population are distributed. Statistical methods that do not make assumptions about the distribution of the population (or at least not restrictive assumptions) are called *nonparametric* tests.

Most nonparametric ideas are based on a simple idea. List the values in order from low to high, and assign each value a rank. Base all further analyses on the ranks. By analyzing ranks rather than values, you don't need to care about the distribution of the population. One nonparametric method for quantifying correlation is called the *Spearman rank correlation.* Spearman rank correlation is based on the same assumptions as ordinary (Pearson) correlations listed earlier, with the exception that rank correlation does not assume Gaussian distributions. For the insulin sensitivity example, $r_s = 0.74$ with a 95% CI ranging from 0.31 to 0.92.

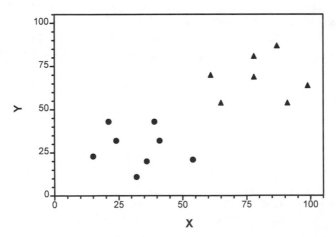

Figure 17.3. Don't combine two populations in correlation. If you analyze all the data, you'll find that $r = 0.72$ and $P = 0.0009$. This appears to be strong evidence that X and Y are correlated. But you've really sampled from two different populations, shown as circles and triangles. If you analyzed each sample separately, you'd find very low correlation coefficients and P values greater than 0.50. There is no evidence that X and Y are correlated. Combining two populations created the illusion of correlation.

DON'T COMBINE TWO POPULATIONS IN CORRELATION

To interpret the results of linear correlation or regression, you must assume that all data points are sampled from one population. If they combine two populations, you can easily be misled. Figure 17.3 shows an example. When all the data are examined, the correlation coefficient is 0.72. The X and Y variables appear to show an extremely strong correlation and the P value is 0.0009. However, the data are really composed of two different populations, shown as circles and triangles. In each population, the X and Y variables correlate very poorly, with r = −0.16 and r = 0.05, respectively. Each P value is greater than 0.50. The correct conclusion from these data is that the two populations (circles and triangles) differ substantially in both variables (denoted by X and Y axes). However, the data do not show any correlation between X and Y among individuals. If you didn't know that the study combined two populations, you would reach a very misleading conclusion.

CALCULATING THE CORRELATION COEFFICIENT*

The equation for calculating r is programmed into many 20 dollar calculators, so you'll rarely need to calculate it yourself. The equation can be expressed in several ways. Equation 17.1 is fairly intuitive:

$$r = \frac{\sum_{i=1}^{N} \left[\frac{(X_i - \overline{X})}{s_x} \cdot \frac{(Y_i - \overline{Y})}{s_y} \right]}{(N - 1)}. \tag{17.1}$$

Here $\overline{X}$ is the mean X value, s_x is the standard deviation of all X values, and N is the number of data points. The mean X and mean Y values define a point at "the center of gravity" of the data. The position of each point is compared to that center. The horizontal distance $(X_i - \overline{X})$ is positive for points to the right of the center and negative for points to the left. The vertical distance $(Y_i - \overline{Y})$ is positive for points above the center and negative for points below.

The distances are standardized by dividing by the SD of X or Y. The quotient is the number of SDs that each point is away from the mean. Dividing a distance by the SD cancels out the units so it doesn't matter what units X and Y are expressed in. The two standardized distances are multiplied in the numerator of the equation. The product is positive for points that are northeast (product of two positive numbers) or southwest (product of two negative numbers) of the center, and negative for points that are to the northwest or southeast (product of a negative and a positive number). If X and Y are not correlated, then the positive products will approximately balance the negative ones, and the correlation coefficient will be close to 0. If X and Y are correlated, the positive and negative products won't balance, and the correlation coefficient will be far from 0. The magnitude of the numerator depends on the number of data points. Finally, account for sample size by dividing by N − 1. N is the number of XY pairs.

---

*This section contains the equations you need to calculate statistics yourself. You may skip it without loss of continuity.

Note that X and Y are totally symmetrical in the definition of the correlation coefficient. It doesn't matter which variable you label X and which you label Y.

THE 95% CI OF A CORRELATION COEFFICIENT*

Correlation coefficients (like most summary statistics) are more informative when expressed as a CI. You have calculated r from data in one sample and can be 95% sure that the population value of the correlation coefficient lies within the 95% CI.

As r can never be larger than 1.0 or smaller than -1.0, the CI is usually asymmetrical and thus a bit tricky to calculate. If this is not done by your computer program, use Equation 17.2. This equation gives an accurate confidence interval only if N > 10.

$$\text{Define: } z = 0.5 \times \ln\left(\frac{1 + r}{1 - r}\right) \quad z_L = z - \frac{1.96}{\sqrt{N - 3}} \quad z_U = z + \frac{1.96}{\sqrt{N - 3}}.$$

$$(17.2)$$

$$\text{Confidence interval of r: } \frac{e^{2z_L} - 1}{e^{2z_L} + 1} \text{ to } \frac{e^{2z_U} - 1}{e^{2z_U} + 1}.$$

CALCULATING THE SPEARMAN CORRELATION COEFFICIENT*

First, separately rank the X and Y values. The smallest value gets a rank of 1. Then calculate the correlation coefficient between the X ranks and the Y ranks using Equation 17.1. The resulting coefficient is called r_S. You must use a special method to determine the 95% CI of r_S. Equation 17.2 is a reasonable approximation if you have more than 10 subjects.

CALCULATING A P VALUE FROM CORRELATION COEFFICIENTS*

The relationship between r and the P value depends on N, the number of XY pairs. For example to reach significance (P < 0.05) r must be larger than 0.88 with 5 data points, greater than 0.63 with 10 data points, greater than 0.44 with 20 data points, but only greater than 0.20 with 100 data points.

If you have 10 or more XY data pairs (N $\geq$ 10), you can use Equation 17.3, which calculates z from r.

$$z = \frac{0.5 \cdot \ln\left(\frac{1 + r}{1 - r}\right)}{\sqrt{1 / (N - 3)}}.$$

$$(17.3)$$

*This section contains the equations you need to calculate statistics yourself. You may skip it without loss of continuity.

The value z is from a standard Gaussian distribution. The P value is determined by answering this question: In a Gaussian distribution with mean = 0 and SD = 1, what fraction of the population has a value greater than z or less than −z? You can then find the answer (the two-tailed P value) from the last column of Table A5.2 in the Appendix. (Note that ln is the abbreviation for natural logarithm.)

You can determine the P value more accurately (even when N < 10) using Equation 17.4, which converts r into t. Although you haven't learned about the t distribution yet, you can use Table A5.4 in the Appendix to find the P value. The number of degrees of freedom equals N − 2, and this determines which column in the table to use. Find the row corresponding to the closest value of t and read the P value.

$$t = r \cdot \sqrt{\frac{N-2}{1-r^2}} \qquad df = N - 2. \qquad (17.4)$$

SUMMARY

The correlation between two variables can be quantified by the correlation coefficient, r. It is easier to interpret the square of the correlation coefficient, r^2. It is the proportion of the variance in one variable that is "explained" by variance in the other.

You can calculate a P value to test the significance of a correlation. The P value answers this question: If there is no correlation overall, what is the chance that randomly chosen subjects will correlate as well (or better) than observed?

If you are not willing to assume that both variables distribute according to a Gaussian distribution (at least approximately), then you can use a nonparametric form of correlation. The most commonly used method of nonparametric correlation is called Spearman correlation.

OBJECTIVES

1. You should be familiar with the following terms:
   - Correlation coefficient
   - Coefficient of determination
   - Spearman rank correlation
   - Nonparametric
   - Pearson correlation
   - Outliers
2. Without looking at a book you must know the meaning of correlation coefficient, r, and r^2.
3. Using books, calculators, and computers, you should be able to calculate the correlation coefficient between two variables. You should also be able to obtain the appropriate P value.
4. You should be able to recognize data for which the Spearman rank correlation is appropriate.
5. You should be able to estimate the correlation coefficient from an XY graph.

PROBLEMS

1. In Example 17.1, how should the investigators have analyzed the data if they had measured the insulin sensitivity and %C20–22 twice in each subject?
2. The P value in Example 17.1 was two tailed. What is the one-tailed P value? What does it mean?
3. Do X and Y have to be measured in the same units to calculate a correlation coefficient? Can they be measured in the same units?
4. What is the P value if r = 0.5 with N = 10? What is the P value if r = 0.5 with N = 100?
5. Can you calculate the correlation coefficient if all X values are the same? If all Y values are the same?
6. Golino et al. investigated the effects of serotonin released during coronary angioplasty.* After angioplasty (inflating a balloon positioned inside to coronary artery to force open a blocked artery) they measured the degree of vasoconstriction in the next few minutes as the percent change in cross-sectional area (monitored by angiograms). They also measured the amount of serotonin in blood sampled from the coronary sinus. The data for eight patients are shown (I read these values off the graph in the publication, so they may not be exactly correct). To make the serotonin levels follow a distribution closer to a Gaussian distribution, the authors calculated the logarithm of serotonin. Calculate the correlation between the logarithm of serotonin levels and the percent change in cross-sectional area. Interpret what it means.

Serotonin (ng/ml)	% Change in Cross-Sectional Area
2.0	4.0
5.0	7.0
6.0	28.0
10.0	26.0
15.0	30.0
60.0	34.0
65.0	35.0
165.0	42.0

*P Golino, F Piscione, CR Benedict, et al. Local effect of serotonin released during coronary angioplasty. N Engl J Med 330:523–528, 1994.

An Introduction to Regression

All the statistical tests you have encountered so far are designed to compare two or more groups with each other. Are the means the same? Are the proportions the same? Are the survival curves the same? Are two variables correlated? A whole other class of statistical tests have been developed to answer a different question: How well do the data fit a theoretical model?

WHAT IS A MODEL?

A model is a mathematical abstraction that is an analogy of events in the real world. Many models are written as an equation that defines a value you want to predict (Y) from one or more variables (X) that you know.

Figure 18.1 shows three theoretical models. The first model is that insulin levels increase linearly with body weight. The second is that pulse rate increases sigmoidally with increasing concentrations of norepinephrine. The third is that the number of open ion channels decreases exponentially with time. As you'll see, you can also write models that predict the odds ratio or relative risk from one or more variables.

WHY BOTHER WITH MODELS?

Regression techniques can be used for several purposes:

- *Looking for a trend.* This is the simplest form of regression. It is used to ask questions such as these: Do fasting insulin levels tend to increase with age? Has the incidence of breast cancer tended to increase over the last 20 years?
- *Adjusting for a confounding variable.* Regression techniques can answer questions such as this: Did the new treatment alter the incidence of ulcers after adjusting for age and aspirin consumption?
- *Curve fitting.* Regression techniques can answer questions such as these? What is the EC_{50} of a new drug?* What is the rate constant for closing of an ion channel?
- *Prediction.* Regression techniques can answer questions such as these: How can you predict the risk of a myocardial infarction from knowing someone's age, blood pressure, and cholesterol level? How can you predict success in medical school from college grades, exam scores, and interview scores?
- *Standard curve.* Regression is used to analyze many assays. The assay is run with known concentrations of the substance being measured. Regression is used to fit a line or curve to the graph of concentration versus assay response (which might be optical density, radioactivity, fluorescence, etc.). That line or curve can then be used to determine the concentration from the response obtained with unknown samples.

*The EC_{50} is the concentration needed to achieve a half-maximal effect.

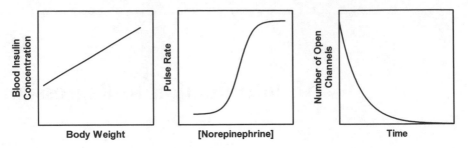

Figure 18.1. Three regression models. The left graph shows a model that blood insulin increases linearly with body weight. The middle graph shows a model that pulse rate varies sigmoidally with norepinephrine concentration. The right graph shows a model that the number of open channels decreases exponentially with time.

DIFFERENT KINDS OF REGRESSION

Regression includes a large family of techniques.

Simple Linear Regression

This is the most common form of regression. The outcome variable is a measurement. There is a single X variable. A graph of X versus Y is a straight line. The next chapter discusses linear regression.

Multiple Linear Regression

Here Y is still a measured variable (not a proportion and not a survival time), but there are two or more X variables. Multiple regression is used to determine the influence of one X variable while adjusting for the effects of the other. Multiple regression (discussed in Chapter 31) is also used to find an equation to predict future outcomes.

Logistic Regression

Here Y is a binary variable (or proportion) such as infected/not infected, or cancer/no cancer. There may be only one X variable, but logistic regression is more commonly used with several X values. Logistic regression is discussed in Chapter 32.

Proportional Hazards Regression

Here the outcome is survival time. There may be only one X variable, but proportional hazards regression (discussed in Chapter 33) is more commonly used with several.

Nonlinear Regression

Again Y is a measured variable, and there is a single X variable. But a graph of X versus Y is curved. Nonlinear regression is discussed in Chapter 34.

Simple Linear Regression

The most commonly used regression method is simple linear regression. You've probably already seen this method used to find the "best line" through a graph of data points. In this chapter, we'll first work through an example and then go back and explore the principles and assumptions in more detail.

AN EXAMPLE OF LINEAR REGRESSION

We'll continue Example 17.1. Recall that the investigators were curious to understand why insulin sensitivity varies so much between individuals. They measured insulin sensitivity in 13 men, and also measured the lipid content of muscle obtained at biopsy. You've already seen that the two variables (insulin sensitivity, and the fraction of the fatty acids that unsaturated with 20–22 carbon atoms, %C20–22) correlate substantially. The correlation coefficient r is 0.77. It is easier to interpret r^2 which equals 0.59. This means that 59% of the variance in insulin sensitivity can be accounted for by the linear relationship between insulin sensitivity and %C20–22.

The investigators used linear regression to fit a line through the graph so they could find out how much the insulin sensitivity increases for every percentage point increase in %C20–22. The program InStat produced Figure 19.1 and Table 19.1.

Figure 19.1 shows the best-fit regression line. *Best-fit* means any other line would be further away from the data points. More precisely, with any other line the sum of the squares of the vertical distances of the points from the regression line would have been larger. Later in the chapter you'll learn why the method minimizes the *square* of the distances.

Figure 19.1 also shows the 95% confidence interval for the regression line as two dotted curves. The interpretation is familiar to you. The best-fit line determined from this particular sample of subjects is unlikely to really be the best-fit line for the entire population. If the assumptions of linear regression are true (we'll discuss these later), you can be 95% sure that the overall best-fit regression line lies somewhere within the space enclosed by the two dotted curves. Figure 19.2 shows five possible regression lines (solid) that lie within the 95% confidence interval (dotted). Even though the borders of the 95% confidence region are curved, we are not allowing for the possibility of a curved (nonlinear) relationship between X and Y.

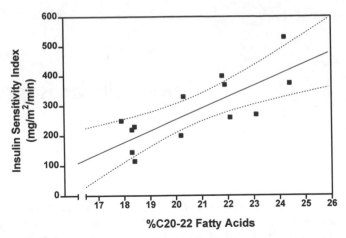

Figure 19.1. The results of linear regression. The best-fit line is shown as a solid line. The 95% CIs for that line are shown as dotted curves.

Now let's turn to Table 19.1 which shows the numerical output of InStat.

- The slope is 37.2. This means that when %C20–22 increases by 1.0, the average insulin sensitivity increases by 37.2 mg/m²/min. The program also reports the 95% CI for the slope, and its interpretation is familiar. The slope calculated from this particular sample is unlikely to equal the true slope in the overall population. But we can be 95% sure that the true slope is within the range of the 95% CI, between 16.7 and 57.7 mg/m²/min.
- The Y-intercept is −486.5. This is the value of Y when X is zero. Taking this number at face value, this says that when the membranes have no C20–22 fatty acids, the insulin sensitivity would be −486.5. This is not biologically possible, as the sensitivity is the amount of glucose needed to maintain a constant blood level and so cannot be negative. We'll discuss this problem later in the chapter.

Table 19.1. Output for InStat for Example 17.1

| | Linear Regression | | | |
| | Number of points = 13 | | | |
Parameter	Expected Value	Standard Error	Lower 95% CI	Upper 95% CI
Slope	37.208	9.296	16.747	57.668
Y intercept	−486.54	193.72	−912.91	−60.173
X intercept	13.076			

r squared = 0.5929
Standard deviation of residuals from line (Sy.x) = 75.895

Test: Is the slope significantly different from zero?
F = 16.021
The P value is 0.0021, considered very significant.

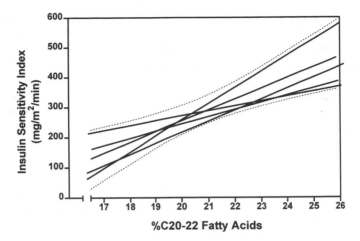

Figure 19.2. Meaning of the confidence intervals of a regression line. The 95% CIs for the regression line from our example are shown as dotted curves. If we can accept the assumptions of linear regression, we can be 95% sure that the true best-fit line lies within these confidence limits. For example, five potential best-fit lines are shown (solid lines). All are within the CI.

- The X-intercept is the value of X when Y equals zero. Taken at face value it says that when the %C20–22 equals 13.076, the muscles will have no sensitivity to insulin. Since we have no data points where %C20–22 is near 13, there is no way to know whether this is really true.
- Next, the output shows that r^2 equals 0.5929. This is the same value determined by linear correlation in Chapter 17. It means that 59% of all the variance in insulin sensitivity can be accounted for by the linear regression model—that 59% of the variance in insulin sensitivity can be explained by variance in %C20–22. The remaining 41% of the variance may be due to other factors, to measurement errors, or to a nonlinear relationship between insulin sensitivity and %C20–22.
- The standard deviation of the residuals equals 75.9 mg/m²/min. This is the standard deviation of the vertical distances of the data points from the regression line.
- Finally, the program reports the P value testing the null hypothesis that there really is no linear relationship between insulin sensitivity and %C20–22. If the null hypothesis were true, the best-fit line in the overall population would be horizontal with a slope of zero. The P value answers this question: If that null hypothesis were true, what is the chance that linear regression of data from a random sample of subjects would have a slope as far from zero (or further) than we actually observed? In this example, the P value is tiny, so we conclude that the null hypothesis is very unlikely to be true and that the observed relationship is unlikely to be due to a coincidence of sampling.

COMPARISON OF LINEAR REGRESSION AND CORRELATION

We have now analyzed the data of Example 17.1 both as an example of correlation and of regression. The values of r^2 and the P value were the same with both analyses.

It made sense to interpret the CI of the correlation coefficient (Chapter 17) only because the experimenters measured X, and because we are willing to assume that both insulin sensitivity and %C20–22 approximate Gaussian distributions. You cannot interpret (and shouldn't calculate) the CI of the correlation coefficient if the experimenters manipulated X.

It made sense to interpret the linear regression line only because we were able to decide which variable was X and which was Y. The investigators hypothesized that the lipid content of the membranes influenced insulin sensitivity, and so defined %C20–22 to be X and insulin sensitivity to be Y. The results of linear regression (but not correlation) would be different if the definitions of X and Y were swapped.

In many cases, it makes sense to calculate only linear regression or only correlation but not both. In Example 17.1 it made sense to perform both sets of calculations.

THE LINEAR REGRESSION MODEL

The whole point of linear regression is to fit an ideal mathematical model to your data. To understand linear regression, therefore, you must understand the model. Don't be scared by the word *model*. It's not that complicated.

Recall that a line is defined by Equation 19.1:

$$Y = \text{intercept} + \text{slope} \cdot X = \alpha + \beta \cdot X. \tag{19.1}$$

The Y-intercept, α, is the place where the line crosses the Y axis. To see this, note that when X equals 0 the equation calculates that Y equals α. The slope, β, is the change in Y for every unit change in X.*

Equation 19.1 is not quite sufficient to model linear regression. The problem is that it places every data point exactly on the line. We need to add a random component to the linear regression model to account for variability (Equation 19.2):

$$Y = \text{intercept} + \text{slope} \cdot X + \text{scatter} = \alpha + \beta \cdot X + \varepsilon. \tag{19.2}$$

Epsilon (ε) represents random variability. The only way to make this model useful is to make some assumptions about the random variable. We'll assume that the random factor follows a Gaussian distribution with a mean of 0. This means any particular value is just as likely to be above the line as below it, but is more likely to be close to the line than far from it. The random variable is often referred to as *error*. As used in this statistical context, the term *error* refers to any random variability, whether caused by experimental imprecision or biological variation.

α and β represent single values, whereas ε has a different value for each subject. Figure 19.3 shows simulated data that follow Equations 19.1 and 19.2.

---

*You've already seen that the Greek letters α and β are also used to represent the probabilities of Type I and Type II errors. There is no relationship between the intercept and slope of a line, and the probabilities of Type I and Type II errors. Don't confuse the two uses of the variables α and β.

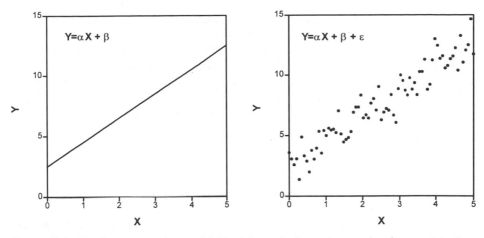

Figure 19.3. The linear regression model. The left graph shows the equation for a straight line. The right graph adds random error. The error follows a Gaussian distribution. Its standard deviation is the same for all parts of the line (it does not vary with X).

THE ASSUMPTIONS OF LINEAR REGRESSION

The linear regression model is built around these assumptions:

- X and Y are asymmetrical. The model predicts Y values from X. X is often a variable you control, such as time or concentration. Or X can be a variable that logically or chronologically precedes Y.
- The relationship between X and Y can be graphed as a straight line. In many experiments, the relationship between X and Y is curved, but linear regression only deals with linear relationships. You can calculate linear regression even when the relationship is curved, but the results are unlikely to be helpful.
- The equation defines a line that extends infinitely in both directions. No matter how high or how low a value of X you propose, the equation can predict a Y value. This assumption is rarely reasonable with biological data. But we can salvage the model by restricting ourselves to using the predictions of the model only within a defined range of X values. Thus we only need to assume that the relationship between X and Y is linear within that range. In the example, we know that the model cannot be accurate over a broad range of X values. At some values of X, the model predicts that Y would be negative, a biological impossibility. In fact, the Y-intercept is negative. But the linear regression model is useful within the range of X values actually observed in the experiment.
- The variability of values around the line follows a Gaussian distribution. Even though no biological variable follows a Gaussian distribution exactly, it is sufficient that the variation be approximately Gaussian.
- The standard deviation (SD) of the variability is the same everywhere. In other words, ε has the same SD everywhere, regardless of the value of X. The assumption that the SD is the same everywhere is termed *homoscedasticity*.

- The model only provides for variability in the Y variable. The model assumes that you know X exactly. This is rarely the case, but it is sufficient to assume that any imprecision in measuring X is very small compared to the variability in Y. Certainly, uncertainty in measuring Y should not alter your assessment of X.
- Each subject (or each XY data pair) was randomly sampled from the population. At a minimum, we assume that our subjects are representative of the entire population.
- Each subject (or each XY data pair) was selected independently. Picking one subject from the population should not influence the chance of picking anyone else.

LINEAR REGRESSION AS A LEAST SQUARES METHOD*

How does the linear regression procedure find the "best" values of α and β to make the regression model (Equation 19.2) fit the data? Linear regression defines the "best line" as the line that minimizes the sum of the squares of the vertical distances of the points from the line. This means that the sum would be higher for any other line.

Why minimize the square of the distances? The simplest answer is that this avoids the problem of negative distances. But why not minimize the absolute value of the distances? A simple answer is that distances are squared because it is better to have two points sort of close to the line (say five units each) than to have one very close (one unit) and one further (nine units). Another answer is that the criteria of minimizing the square of the distances assures that a unique line is defined from any set of data points. A method that minimized the absolute value of the distances would not always generate a unique answer.

Here is another way to understand why linear regression minimizes the square of the distances. The vertical distance of each point from the regression line is called the *residual.* Linear regression finds the line that minimizes the SD of the residuals. When calculating the SD, you sum the squared deviations. So to minimize the SD, you minimize the sum of the squared deviations.

Note that linear regression does not really find the line that "comes closest" to the points, since it looks only at vertical distances (parallel to the Y axis). This also means that linear regression calculations are not symmetrical with respect to X and Y. Switching the labels "X" and "Y" will produce a different regression line (unless the data are perfect, with all points lying directly on the line). This makes sense, as the whole point is to find the line that best predicts Y from X. The line that best predicts X from Y is usually different.

An extreme example makes this more clear. Consider data for which X and Y are not correlated at all. You know X and have to predict Y. Your best bet is to predict that Y equals the mean of all Y values for all values of X. The linear regression line for predicting Y from X, therefore, is a horizontal line through the mean Y value. In contrast, the best line to predict X from Y would be a vertical line through the mean of all X values, 90° different.

---

*This section is more advanced than the rest. You may skip it without loss of continuity.

THE MEANING OF r^2

You've already learned about r^2 in the context of linear correlation. It is a fraction between 0 and 1, and has no units. When r^2 equals 0, there is no linear relationship between X and Y. In this case, the best-fit line is horizontal (slope = 0), so knowing X does not help you predict Y. When $r^2 = 1$, all points lie exactly on a straight line with no scatter.

You can think of r^2 as the fraction of the total variance in Y that can be "explained" by the linear regression model. It is the fraction of the total variance in Y that is accounted for by linear regression with X. The value of r^2 (unlike the regression line itself) would be the same if X and Y were swapped. So r^2 is also the fraction of the variance in X that is explained by variation in Y. Alternatively, r^2 is the fraction of the variation that is shared between X and Y.

Statisticians sometimes call r^2 the *coefficient of determination,* but scientists call it *r squared.*

You may find it easier to interpret r^2 after you understand how it is calculated. It is calculated from two values, s_y and s_e. The first, s_y, is the SD of all Y values without regard to X. The second, s_e, is the SD of the residuals. It is the SD of the vertical distances of the points from the regression line. If the data are not correlated at all, the best-fit line is horizontal and s_e equals s_y. In all other cases s_e is smaller than s_y. Since the variance is the square of the SD, s_y^2 is the total variance of Y and s_e^2 is the variance unexplained by the regression model. Equation 19.3 defines r^2.

$$r^2 = \frac{s_y^2 - s_e^2}{s_y^2} = \frac{\text{total variance} - \text{unexplained variance}}{\text{total variance}} = \frac{\text{explained variance}}{\text{total variance}}.$$

(19.3)

Let's look at the two extreme cases. With completely random data, the SD of the data points around the regression line is the same as the SD of the Y values, so s_e equals s_y, and $r^2 = 0$. With perfectly linear data, the SD of the points around the regression line is 0. In this case, s_e equals 0 and r^2 equals 1.0.

MAXIMUM LIKELIHOOD

A puzzling aspect of linear regression is that we minimize the sum of the squares of the distances of the points from the line, rather than minimize the sum of the absolute distances. Why minimize the square of the differences? A rigorous answer is that the regression line determined by the least-squares method is identical to the line determined by maximum likelihood calculations. The next paragraph gives you a feel for how maximum likelihood calculations work.

For any particular model, it is possible to calculate the probability (likelihood) of obtaining any particular set of data. The regression model must specify particular values for the slope and intercept of a hypothetical line, as well as for the scatter of the data around that line. The model defines the entire population. For any particular model, many different samples of data could be obtained, but some data sets are more likely than others. Thus it is possible to calculate the conditional probability (likelihood) that answers this question: If the model were true, what is the chance of obtaining the particular set of data that we collected? Different models will give different answers. If you try models with all possible combinations of slope and intercept, it is possible to find the regression line that has the maximum likelihood of producing the observed set of data. In other words, this method finds the values for the variables that are most likely to be true.

Restricting ourselves to models in which the scatter of data around the line is Gaussian, the maximum-likelihood method yields identical values for slope and intercept as does the least-squares method. Least-square calculations are easier and thus have become standard. Least-square calculations are inappropriate for logistic and proportional-hazard regression, and maximum-likelihood calculations are necessary.

GRAPHING RESIDUALS

Linear regression assumes that the scatter of Y values around the regression line follow a Gaussian distribution with a SD that is the same for all values of X. The best way to informally test these assumptions is to inspect a residual plot. Figure 19.4 shows a residual plot from the example. In such a plot the X values are unchanged, but the Y values are replaced by the distance between the points and the line (keeping track of sign). A residual is positive when the point lies above the line, and is negative when the point lies below the regression line. Residuals equal the actual Y value minus the predicted Y value.

When inspecting a graph of residuals, ask yourself these questions:

• Does the scatter appear to be Gaussian, with many points close to the horizontal line at Y = 0, fewer points far away, and no outliers much further away? If your answer is No, linear regression may not be appropriate.

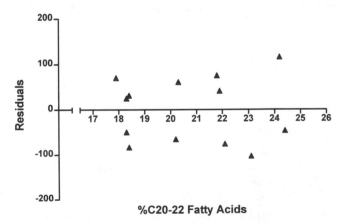

Figure 19.4. A residual plot. The X axis is the same as it is in Figure 19.2 The Y axis shows the distance of each point from the best-fit line. If the line went directly through a point, that point would have Y = 0 on the residual plot. Positive residuals show that the point is above the line; negative residuals show that the point is below the line.

- Does the average distance of the points from the Y = 0 tend to increase (or decrease) as you move along the graph? If you see such a trend, linear regression may not be appropriate.
- Are there large clusters of adjacent points all above the line or all below the line? Again, if you see such clusters, linear regression would not be appropriate.

USING THE REGRESSION LINE AS A STANDARD CURVE TO FIND NEW VALUES OF Y FROM X

Knowing the best-fit regression line, it is easy to calculate Y from X. Just plug X into the regression equation and calculate Y. The regression line extends infinitely in both directions, and nothing but common sense stops you from attempting to make predictions beyond the range of X values that encompass the original data points. But the relationship between X and Y may change beyond this range. Mark Twain pointed out the absurdity of extrapolation:

> In the space of one hundred and seventy-six years, the Lower Mississippi has shortened itself two hundred and forty-two miles. This is an average of a trifle over one mile and a third per year. Therefore, any calm person, who is not blind or idiotic, can see that in the Old Oölitic Silurian Period, just a million years ago next November, the Lower Mississippi was upward of one million three hundred thousand miles long, and stuck out over the Gulf of Mexico like a fishing rod. And by the same token, any person can see that seven hundred and forty-two years from now, the lower Mississippi will be only a mile and three-quarter long. . .
>
> Life on the Mississippi

Quantifying the accuracy of the predicted Y value is harder. We want to calculate the 95% prediction interval, that is, the range of values that contains 95% of new points. This is different from the CI shown earlier. The prediction interval must include uncertainty in the position of the regression line (quantified by the 95% CI of the regression line) and the scatter of points around the line. Thus the prediction interval is much wider than the CI.

The distinction between CI and the prediction interval is analogous to the difference between standard error and SD, and has already been discussed in Chapter 5. The prediction intervals are always wider than the CIs. As the number of data points increases, the CIs grow narrower, while the prediction intervals stay about the same. Like the CI, the prediction intervals are curved.

THE REGRESSION FALLACY

When interpreting the results of linear regression, make sure that the X and Y axes represent separate measurements. Otherwise, you can be mislead by a problem called *the regression fallacy*. Here is an example.

Figure 19.5 shows computer generated data simulating an experiment where blood pressure was measured before and after an experimental intervention. The left panel

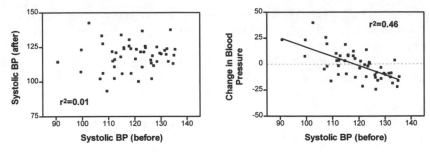

Figure 19.5. The regression fallacy. The left panel shows simulated data showing blood pressure before and after an intervention. All values were sampled from a Gaussian distribution with a mean of 120 and a SD of 10. There is no correlation between the two sets of values. The right panel shows the same data after some calculations. The Y axis now shows the change in blood pressure (after − before). There appears to be a strong correlation and the best-fit line (shown) has a slope far from horizontal. Subjects whose pressure was low originally tended to increase their pressure (left side of the graph). Subjects whose pressure was high originally tended to decrease their pressure (right side of graph). If these were real data, you might be intrigued by this finding. But there is no finding, and these are random data. Because the values plotted on the Y axis include the values shown on the X axis, linear regression is not appropriate for these data.

shows a graph of the data. Each point represents an individual whose blood pressure was measured before (X axis) and after (Y axis) an intervention. The data are entirely random, and there is no trend whatsoever. Each value was sampled from a Gaussian distribution with mean = 120 and SD = 10. As expected, the best-fit regression line is horizontal. While blood pressure levels varied between measurements, there was no systematic effect of the treatment. The right panel shows the same data. But now the Y axis shows the change in blood pressure (after − before). Notice the striking linear relationship. Individuals who initially had low pressures tended to increase; individuals with high pressures tended to decrease. This is entirely an artifact of data analysis and tells you nothing about the effect of the treatment, only about the stability of the blood pressure levels between treatments.

These are not real data. But the figures clearly make the point that it is very misleading to plot a change in a variable versus the initial value of the variable. Attributing a significant correlation on such a graph to an experimental intervention is termed the *regression fallacy*. Such a plot should not be analyzed by linear regression because these data (so presented) violate one of the assumptions of linear regression, that the X and Y values were determined independently. Here, instead, the X value is used in the calculation of the Y value.

CALCULATING LINEAR REGRESSION*

Linear regression calculations are best left to computer programs and calculators. The equations defining the slope and intercept can be expressed in many forms. The

*This section contains the equations you need to calculate statistics yourself. You may skip it without loss of continuity.

equations shown here were chosen for clarity. Other equations are more efficient for hand calculations.

Slope and Intercept

Equation 19.5 calculates the slope. Note that the equation is not symmetrical with respect to X and Y. We call the best fit slope b, to distinguish it from β, the ideal slope in the population.

$$\text{Slope} = b = \frac{\Sigma(X_i - \overline{X})(Y_i - \overline{Y})}{\Sigma(X_i - \overline{X})^2}. \tag{19.5}$$

The standard error (SE) of the slope is defined by Equation 19.6:

$$\text{SE of slope} = s_b = \frac{s_e}{s_X \cdot (N - 1)}. \tag{19.6}$$

This equation makes sense. If the scatter of the points around the line (s_e) gets larger, the uncertainty in the exact value of the slope also gets larger. If you have collected data over a larger range of X values, the SD of X (s_x) will increase, resulting in a more precise slope. Finally, if you collect more data (increase N) you will know the slope more exactly. The units of the SE of the slope are the same as the units of the slope: the Y units divided by the X units.

The best-fit line always goes through the point defined by the average of all X values and the average of all Y values. Thus the Y intercept can be calculated from Equation 19.7:

$$\text{Y intercept} = a = (\text{mean Y}) - \text{slope (mean X)} = \overline{Y} - b \cdot \overline{X}. \tag{19.7}$$

Goodness of Fit

One way to quantify goodness of fit is to calculate the SD of the residuals (or ''errors''), as shown in Equation 19.8.

$$s_e = \sqrt{\frac{\Sigma(Y_i - \text{predicted Y})^2}{N - 2}}. \tag{19.8}$$

The numerator of the fraction sums the square of the vertical distances of the points from the line. In other words, it sums the square of the vertical distance of the actual value of Y from the value predicted from the regression line. That is the quantity that linear regression minimizes. The denominator is $N - 2$. This makes sense because you had to calculate the slope and intercept of the regression line before you could calculate the predicted Y values. If you knew the slope and intercept and $N - 2$ of the XY pairs, you could calculate what the other two XY pairs must be. Therefore there are $N - 2$ degrees of freedom, and the average of the squared deviations is determined by dividing by $N - 2$. You already learned how to calculate r^2 from s_y and s_e (Equation 19.3).

Confidence Intervals and Predictions Intervals of the Line

The 95% CI of the slope is calculated from Equation 19.9.

$$95\% \text{ CI of slope} = \text{slope} - t^* \cdot s_b \text{ to slope} + t^* \cdot s_b. \tag{19.9}$$

To find the critical value of t^*, use Table A5.3 in the Appendix. The number of degrees of freedom equals the number of XY pairs minus two. To determine the CI of the Y position of the linear regression line at any position X, use Equation 19.10. You can be 95% certain that the true population regression line will lie within these limits for all values of X. In this equation (and the next) Y_{reg} is the Y value of the regression line at the particular value of X you have chosen:

$$Y_{reg} = X \cdot \text{slope} + Y \text{ intercept}.$$

95% CI of regression line

$$= \left[Y_{reg} - s_e \cdot t^* \sqrt{\frac{1}{N} + \frac{(X - \overline{X})^2}{s_x^2}} \right] \text{ to } \left[Y_{reg} + s_e \cdot t^* \sqrt{\frac{1}{N} + \frac{(X - \overline{X})^2}{s_x^2}} \right]. \tag{19.10}$$

To determine the prediction interval at any position X, use Equation 19.11.

95% prediction interval

$$= \left[Y_{reg} - s_e \cdot t^* \sqrt{1 + \frac{1}{N} + \frac{(X - \overline{X})^2}{s_x^2}} \right] \text{ to } \left[Y_{reg} + s_e \cdot t^* \sqrt{1 + \frac{1}{N} + \frac{(X - \overline{X})^2}{s_x^2}} \right]. \tag{19.11}$$

You can be 95% certain that all new data points you collect will lie within this prediction interval. The prediction intervals are much wider than the CIs. This is because the prediction interval must include uncertainty in the position of the best-fit line (as calculated by the CI) as well as the scatter of data around the regression line.

P Value

It is possible to calculate a P value answering this question: If the slope of the best fit line in the overall population equals 0, what is the chance of picking subjects in which the slope is as far or further from 0 than we actually observed? To calculate the P value, first determine the slope and the SE of the slope. Then calculate t using Equation 19.12 and find the P value using Table A5.4 in the Appendix.

$$t = \frac{\text{slope}}{\text{SE of slope}} = \frac{b}{s_b} \qquad df = N - 2. \tag{19.12}$$

The SE of the slope takes into account the scatter of the data around the line and the number of data points. If you have many data points, the SE of the slope will be small. Therefore, the value of t (and thus the P value) depends on how steep the slope is (the value of b), the amount of data scatter, and the number of data points.

The P value can be interpreted in terms of r^2 rather than slope. The P value also answers this question. If the best-fit regression line in the overall population is horizontal, what is the chance that we would randomly select subjects and end up with a value of r^2 as large or larger than we actually observed?

SUMMARY

Linear regression finds the "best" straight line that goes near the data points. More precisely, linear regression finds the line that minimizes the sum of the square of the vertical distances of the points from the line.

OBJECTIVES

1. You should be familiar with the following terms:
   - Regression   - r^2
   - Model   - Least squares
   - Intercept   - Residual
   - Slope
2. You should know the assumptions of linear regression and know that linear regression calculations will be misleading with nonlinear data. You should know that regression calculations are not symmetrical with respect to X and Y.
3. Using book, calculator, and computer, you should be able to perform linear regression calculations.
4. You should be able to interpret the output of linear regression programs.
5. You should understand why CIs of regression lines are depicted by curves and why prediction bands are wider than confidence intervals.

PROBLEMS

1. Will the regression line be the same if you exchange X and Y? How about the correlation coefficient?
2. Why are the 95% CIs of a regression line curved?
3. Do the X and Y axes have to have the same units to perform linear regression?
4. How many P values can be generated from a simple linear regression?
5. The results of a protein assay are presented in the following table. Chemicals are added to tubes that contain various amounts of protein. The reaction forms a blue color. Tubes with higher concentrations of protein become a darker blue. The darkness of the blue color is measured as optical density.

Concentration (micrograms)	Optical Density
0	0
4	0.017
8	0.087
12	0.116
16	0.172
Unknown 1	0.097
Unknown 2	0.123

    A. Calculate the best-fit linear regression line through the standard curve (five known concentrations).

    B. Read the unknown concentrations from the standard curve.

6. What is r^2 if all points have the same X value? What about if all points have the same Y value?

7. Sketch some examples of residual plots for data that do not meet the assumptions of linear regression.

    A. Scatter increases as X increases.

    B. Data form a curve, not a straight line.

8. Can r^2 ever be 0? Negative?

9. Do you need more than one Y value for each X value to calculate linear regression? Does it help?

VI

DESIGNING CLINICAL STUDIES

Note to basic scientists: You may skip Chapters 20 and 21 without loss in continuity. Don't skip Chapter 22 about calculating sample size—it applies to basic as well as clinical research.

20

The Design of Clinical Trials

Note to basic scientists: You may skip this chapter without loss of continuity.

There are lots of ways to do medical research. Some research can be done by examining existing records such as death certificates or medical charts. Other kinds of research are done with animals. Many clinical studies are done as case-control studies (as discussed in Chapter 9). This chapter focuses on clinical trials. These are medical experiments where some patients are given one treatment, other patients are given another treatment, and the results are compared.

While medicine as a healing profession has existed for millennia, the idea that medical treatments should be tested experimentally is fairly new. There were a handful of medical experiments published in the 19th and early 20th century, but it is really only since 1940 that clinical research became well established. Now it is commonly accepted that new treatments must be tested before they are widely used, and that the test must be carefully controlled to avoid biases. While anecdotal or observational evidence can be used to generate a hypothesis, we do experiments to test the hypothesis. This is a very straightforward idea but one that is historically new to medicine. The idea has yet to penetrate very far in other fields (such as education).

Before a new drug treatment can be used clinically, it must be tested in a defined series of steps. The initial work, of course, is preclinical. Many properties of the drug can be defined using animals and cell cultures. After that, clinical research on new drugs is defined by four phases:

- Phase 1 is the initial testing of a drug on humans. The drug is given to a few dozen people to learn enough about its actions, metabolism, and side effects to design a valid phase 2 study.
- Phase 2 studies are controlled experiments to determine if the drug is effective for a specific purpose and to determine the common side effects. Phase 2 studies typically enroll no more than several hundred patients.
- Phase 3 studies are larger studies involving hundreds to thousands of patients. The goal is to learn more about the effectiveness and safety of the drug and to compare it to other treatments. If the phase 3 studies are successful, the drug will be approved for general use. Information gathered in the phase 3 studies are included in the package insert so physicians will know how to prescribe the drug appropriately.

- Phase 4 studies monitor the effectiveness and safety of a drug after it has been released for general use. Phase 4 studies also are designed to learn about new uses of the drug.

The same steps ought to be taken to test new therapy, whether a drug, diet, exercise, or procedure. Because medical or surgical procedures are not subject to government regulations like drugs, however, new procedures are rarely tested so rigorously. The rest of this chapter deals mostly with phase 3 trials.

DESIGNING THE STUDY

Who Are the Subjects?

Before starting a study, the investigators need to decide which patients they will study. If you use rigid enough entry criteria, you can be quite sure that an identified extraneous variable cannot influence your results. For example, if you only accept patients between 50 and 55 years old into the study, you can be quite sure that age difference cannot confound the results. But you don't want to be too rigid, or you'll never be able to recruit enough subjects. And even if you could get enough subjects, the results of the study would apply only to that narrowly defined population and may not be of general interest.

The Need for a Comparison Group

The initial uses of a new drug or therapy are uncontrolled. An investigator tries it and observes what happens. Uncontrolled trials are useful in the early stages of a new therapy. The history of medical research has demonstrated over and over the need for side-by-side comparisons of the new drug with existing drug or placebo. People tend to be enthusiastic about new therapies, and uncontrolled trials often reach optimistic conclusions that are not supported by later research. It is too easy to be misled when therapies are evaluated without a comparison group. To properly test a new therapy, you need to compare it with an existing standard therapy or with placebo (no active therapy).

It is tempting to want to compare the outcome of patients taking the new drug with results previously recorded for patients taking another drug. The advantage of this approach is obvious: All the new patients get the new drug. The problem with historical comparison groups, however, is that you can never be sure that the two groups of patients didn't differ in some other way. Interpreting the results of such a comparison is difficult.

Instead of comparing two large groups of subjects, you can select matched pairs of subjects. The pair should be of similar age and also be matched for other variables that might confound the results, such as disease severity, location, ethnic group, or weight. These data should be analyzed using special methods for matched data.

Random Assignment of Treatment

The decision as to which patient gets which treatment should be made randomly. This simple idea is in part responsible for the incredible achievements of clinical research in the last few decades. If you decide in any other way, the two groups may differ in important ways.

For example, one alternative to randomization would be to give patients whose last name begins with A to M one treatment and patients whose name begins with N to Z the other treatment. Another alternative might be to give patients in the morning clinic one treatment and patients in the afternoon clinic the other treatment. These are not good alternatives to randomization, as the two groups might differ in an important way. Since the last name is determined in part by one's ethnic origin, the first example might end up with more people of Italian ancestry in the first group and more of Oriental ancestry in the other. In the second example, patients who choose morning or afternoon appointments may differ in important ways. Or perhaps physicians will instruct patients to make an appointment for either morning or afternoon to get a particular treatment. To avoid these kind of problems, treatments should be assigned randomly.

Don't confuse the two very different uses of the concept of *randomization* in statistics. (1) Statistical analysis is based on the assumption that the data were collected from a sample that is randomly selected from (or at least representative of) a larger population. (2) Good experimental design randomly assigns a treatment to each subject. The phrase *randomized* clinical trials refers to the latter definition.

Sometimes investigators first divide the subjects into subgroups and then randomize within each subgroup. This is called *stratified randomization*. For example, you might divide the subjects into four age groups. Within each age group, you use a randomization scheme that ensures an equal number of subjects are allocated to each treatment. This technique ensures differences in age cannot affect the results. While stratification is very useful, it is impractical to stratify patients for every conceivable variable that might affect the result.

Rather than conduct a randomized experiment to compare two treatments, it is much easier to just observe the outcome in patients who happened to receive one treatment versus the other. But the results of this kind of comparison can never be definitive. Without knowing that the two groups of patients were similar before the treatment, you can never be sure that differences in outcome are due to differences in treatment.

The Blind Treating the Blind

When possible, clinical studies should be *double blind*. The first blind is that the physicians treating the patients should not know which patient gets which treatment. This avoids the possibility that they will treat the two groups of patients differently, or that they will interpret clinical data in a biased manner depending on which treatment the patient received. In some protocols (surgery vs. drug) it is impossible for the treating physicians to be blind as to treatment. Instead, it is possible to blind the people who collect or interpret data, for example, radiologists or pathologists.

The second blind is that the patients should not know which treatment they receive. Response to a treatment depends, in part, on psychological factors, and blinding the patient prevents psychological factors from confounding the results.

In some cases, studies are *triple blind,* which means that the people doing the data analysis don't know which treatment group is which until the very end of the analysis. This prevents the statisticians from introducing their biases when deciding exactly how to do the analysis.

Compulsive Controls

The two treatment groups must be treated identically, except for the treatment being compared. In clinical studies this is taken care of by randomly allocating treatment and performing the study double blind. Sometimes, special care must be taken to treat the two groups similarly. If laboratory tests must be done to monitor therapy in one group, then the same tests (or at least the same blood drawing) should be done in the other. If the results of the test frequently require an adjustment of the dose in one group, then the ''dose'' of the other group (often placebo) should be altered as well.

Crossover Design

In crossover experiments, each subject serves as his or her own control. This study design is useful for drug therapy that can be evaluated fairly quickly and for conditions that don't progress rapidly. Crossover designs cannot usually be used to study surgical procedures (you can't take away the operation) or for cancer chemotherapy (the disease progresses too rapidly) but are very useful for treatment of chronic conditions such as arthritis or asthma. So as not to introduce a new confounding variable (time or treatment order), the order of the two treatments should be determined randomly for each patient. Some patients get treatment A first; others get treatment B first.

There are two problems with crossover designs. One problem is that the effect of the first treatment may persist into the time period of the second treatment. Studies usually include a washout period between the two treatments to prevent carryover. Another problem is that subjects may drop out of the study before the second treatment.

Intention to Treat

After a subject has been randomly assigned to receive a particular treatment, he or she may not get it. How do you analyze data from these subjects? The obvious answer is to consider the treatment the patient actually got rather than the treatment he or she was supposed to get, but this leads to problems.

For example, assume that we are comparing medical (drugs) with surgical (coronary bypass) treatment of patients with heart disease. One of the patients assigned to surgery gets pneumonia, and so the operation is cancelled. The patient did not get the treatment (surgery) he was assigned to get, but rather got the other treatment. Now say he dies of pneumonia. How do you analyze that death? The obvious answer is that the patient received medical treatment (drugs) not surgery, so the death should be counted as a death among the medically treated patients. But that would bias the results. The reason the operation was cancelled is that the patient was sick. If you take those patients out of the surgery group, then you are removing the sickest patients from one group and adding them to the other. The two groups are no longer comparable.

Another patient agreed to join the study, knowing that she may get surgery and may get drugs. She is assigned to receive drugs. After thinking it over, the patient decides she wants surgery after all. How do you analyze the outcome from this patient?

She didn't get the treatment she was assigned to get. Ethically, you have to allow the patient to get the treatment she wants. Scientifically, you don't want to consider this patient's outcome as part of the surgery group. If every subject chooses his or her own treatment, then the two groups will differ in many ways, and the comparison of the outcomes will be meaningless.

Dealing with data from these patients is difficult. There is no perfect way to deal with the problems. Most investigators adopt the conservative policy named *intention to treat.* This means that data are analyzed assuming that every patient received the assigned treatment. Even if the patient never received the treatment, the data from that patient are pooled with the patients who actually got the treatment. This is termed a *conservative policy,* because it makes it harder to find significant effects.

Another approach would be to simply not analyze data from any patient who didn't actually get the assigned treatment. This can also introduce biases. In the first example above, it would take the sickest patients out of the surgery group but leave similar patients in the medical group. Now if the surgery group does better overall, you won't know whether that was because the surgical treatment was superior or whether it was because the surgical patients were healthier to start with.

In many papers, investigators report the analyses calculated two ways: First, they use the intention-to-treat policy. Then they reanalyze their data again after eliminating all those patients from the analysis. If the two methods yield different conclusions, you should be suspicious of either conclusion. If the two analyses reach the same conclusion, you can be confident that the data from patients who didn't actually receive the assigned treatment did not affect the results much. In many studies, only a small fraction of patients fail to get the assigned treatment, so it doesn't matter too much how those patients are handled.

THE NEED FOR A STUDY PROTOCOL

Before a clinical study can begin, all aspects of the study are written down in a document that can be anywhere from several dozen to several hundred pages. This document is called the *study protocol.* There are several reasons to be so formal:

• Before the study can begin, it must be approved by one or more committees, as discussed later. The committees need to see a complete protocol in order to decide whether to approve.
• Many studies are performed in several medical centers at one time. This allows the study to include many patients in a relatively short period of time. A detailed study protocol is needed to ensure that the study is run the same way at all medical centers.
• Most clinical studies collect reams of data. Most studies measure the patient's response to treatment by monitoring several variables (end points). Most studies collect enough information to subdivide the subjects in many different ways. Most studies last long enough to allow you to analyze the data for several time intervals. Armed with a powerful computer program, several end points, several subgroups, several time intervals, and several confounding variables to adjust for, you can churn out hundreds of P values. Just by chance, you are quite likely to find that some of these are "significant." These mean nothing, as you expect 1 in 20 P values to be significant

just by chance. In order to interpret a P value properly, you need to decide in advance how many you will calculate and how you will adjust for the many comparisons. Good experimental protocols, therefore, specify how the data will be analyzed, what specific end points will be examined, and what P value will be considered significant. See Chapter 13 to learn more about interpreting multiple P values.

- It is not always obvious how you should analyze all the data in a study. What do you do when a patient gets killed in a car crash halfway through? Does that count as a death or as a withdrawal? What do you do when some of the tests are done several weeks late—include the data even though it is late, or exclude that patient's data from the analysis? What do you do when a lab result is too high to be believable—exclude it or include it? Study protocols specify the answers to these questions in advance. Otherwise, the people analyzing the data can introduce their own biases.

WHAT IS IN THE STUDY PROTOCOL

The study protocol spells out every detail of the study. Some points that are covered follow:

- What is the rationale of the study?
- Who will be studied? How will the patients be selected? What patients will be rejected? Typically study protocols define a narrow group of patients and exclude patients with other simultaneous illnesses. By restricting the study in this way, variability is reduced and the power of the study to detect differences is increased. But if the patient group is too narrow, it can be hard to know how to apply the results of the study to other patients. It is important to decide in advance who will be included and who won't. Otherwise it is too tempting afterwards to selectively exclude certain kinds of patients in order to make the results come out as desired.
- How will treatments be assigned? When possible, the assignment of the patient to a treatment should be random.
- What is the comparison group? If there is no established treatment, then the experimental treatment is compared to placebo. If there is a good established treatment, then the experimental treatment ought to be compared to the standard therapy.
- Is the study blind? Double blind?
- How many subjects? Justify. Chapter 22 discusses how you can determine the appropriate number of subjects.
- Exactly what treatment will the various groups get? What dose? When will dosages be adjusted? When will drugs be stopped? How are drugs stored?
- What variables will be monitored? What data will be collected and when?
- How will side effects be managed? What side effects have been observed before?
- Under what circumstances will subjects be dropped from the trial? How will those data be analyzed?
- Will there be interim analyses? If the one treatment is clearly superior or clearly harmful, when will the trial be terminated?
- How is the consent form worded?
- What data will be collected and how will the data be analyzed?

HUMAN SUBJECT COMMITTEES AND INFORMED CONSENT

Clinical investigators in all countries agree that research on humans must be based on these simple principles:

- The care of individual patients takes precedence over the clinical trial. Each subject must be treated appropriately. You cannot deprive a patient of a treatment known to work for the sake of a trial, unless you have strong reason to believe that the experimental treatment *might* be better. It is only ethical to choose a treatment randomly when the people planning the study and administering the drug can honestly say that they are not sure which of the treatments (or placebo) is best.
- The anticipated benefits of the trial must balance the anticipated risks. The principle is clear, but it is sometimes hard to apply. One problem is that many benefits and risks are unknown. Another problem is that future patients stand to gain most of the benefits, while current patients must take all the risk.
- Each patient in the trial must voluntarily decide to take part. Before asking the subject to participate, the investigator should explain the study, answer questions, and describe the alternative treatments. The subject's decision to participate should be an informed decision. Patients should not be coerced into participating, but defining coercion is difficult. Studies are no longer conducted on prisoners because prisoners may feel subtly coerced into participating. Some might argue that studies should not be conducted on medical students for the same reason.
- No subject is enrolled in the experiment until she or he formally agrees by signing a consent document. This document explains the purpose of the trial (including an explanation of randomization and placebo when appropriate) and lists anticipated (or potential) benefits and risks to the patient. Finally, the consent document must note that participation is voluntary and that patient may withdraw at any time without denying care to which they would otherwise be entitled. Consent documents should be written in plain language, without medical jargon. Typically consent forms are two or three pages long. Human subject committees allow investigators to bypass informed consent in unusual circumstances (infants, comatose patients, emergency treatments) or to obtain consent from guardians or family members.
- The subject must be informed if additional information becomes available during the course of the trial that might influence the decision to remain in the trial.
- Research protocols must be approved by a committee of clinicians, scientists, and others not involved in the research. This review process guards against overenthusiastic researchers performing unsafe studies. In most countries, this review process is mandated by law. The review committee is often called the Human Subjects Committee or Institutional Review Board. If the research involves a new drug, it must first be approved (in the United States) by the Food and Drug Administration. If it involves radioactive isotopes or recombinant DNA, it must also be approved by specialized committees.

ETHICS OF PLACEBOS

Sometimes it seems like the patients who are randomized to receive placebo are getting a bad deal. They are sick, and instead of getting an experimental treatment they get an inactive drug. It is only ethical for investigators to include a placebo in a trial, when there is doubt whether any standard therapy would substantially benefit the patient. When standard therapies are available and clearly effective, the control patients should get the standard therapy rather than placebo.

Just because a treatment is standard, logical, and accepted doesn't mean that it helps patients. There are plenty of examples in medical history of therapies that were once considered to be conventional that we now know harm patients. George Washington was probabily killed by blood letting, a standard therapy of his day. Another example is oxygen for premature babies. Some thought oxygen might be toxic and cause a form of blindness (retrolental fibroplasia). To test this idea, they performed several controlled experiments. At the time, some thought it unethical to reduce the oxygen (the established therapy) delivered to premature babies with lung disease. It turns out that the overuse of oxygen was causing blindness. The conventional therapy was harming the patient.

HOW IS THE POPULATION DEFINED?

When reading the results of a clinical study, you first need to think about samples and populations. Ideally, the subjects in the study were randomly selected from a defined population of patients. This almost never is the case. Instead, the sample of patients in the study is supposed to be representative of a larger population. What population? The answer, unfortunately, is often poorly defined.

Papers often give detailed descriptions of the studies inclusion and exclusion criteria. For example, the study might include only patients whose white cell count is below a certain threshold but exclude those who also have liver disease. Usually the list of inclusion and exclusion criteria fills a long paragraph. But despite the apparent detail, these criteria only tell part of the story. They are used to decide whether a particular patient can be included in the trial. They often don't tell you which group of patients could possibly have been considered for the study in the first place. If the study was done in an inner-city hospital, the results are likely to be different than if the study was done in a suburban hospital. If the study was done in Sweden, the results may not be helpful to a doctor participating in Mexico. If the study was performed in a tertiary referral hospital, the patients (by definition) are likely to be unusual (or they wouldn't have been referred), and the results may be quite different than they would be in a community hospital.

The whole point of statistical analysis is to extrapolate results from a sample to a larger population. If you can't define the population, you can't interpret the statistics.

REVIEWING DATA FROM CLINICAL TRIALS

When reviewing the statistical analyses, you need to distinguish between several kinds of analyses:

• *Data checking analysis.* Before doing any real analyses, investigators should first check that the data are reasonable and try to check for errors in data entry. Here are some questions that investigators should ask: Are there any impossible values (negative blood pressures)? Does the number of subjects in the computer database match the number of patients actually studied? Are lab values in the expected range? Are dates consistent (discharge dates should follow admission dates, etc.)? These sorts of analyses are rarely shown in scientific papers.

- *Demographic analysis.* Most clinical studies show tables comparing baseline characteristics of the treatment groups. For example, these tables will show the average age, the fraction of women, the average blood pressure. When patients are randomly divided into two groups, it is possible that the two groups differ substantially from each other. The point of the demographic analyses is to convince you that this didn't happen. Additionally, these analyses show you the composition of the different ''strata'' in the trial (perhaps age groups).

- *Main analysis.* Good studies are clearly designed to ask one main question and to specify in advance how the data will be analyzed to answer that question. It is important to define the main analysis when the study is designed and *not to alter the main analysis* after looking at the data. If you look at the data enough ways, you are likely to stumble on some ''significant'' findings just by chance. If the investigators changed their main question after looking at the data, you can't know how to interpret the P value.

- *Interim analyses.* In many studies, the data are analyzed during the course of the study. This is done to protect the safety of patients. If one treatment is clearly much better than the other, it is not ethical to continue the trial. The problem is multiple comparisons. If you reanalyze the data too often, you are too likely to find a ''significant'' difference just be chance. The probability of a Type I error exceeds the value of α (usually 5%). In most trials, the threshold α for significance in interim analyses is much lower (often 1%) than it is for the main analysis.

- *Secondary analyses.* In addition to the main analysis, most clinical studies contain many secondary analyses, either to look at additional outcome variables or to look at subsets of patients. You need to be a bit skeptical when looking at secondary analyses. The problem, again, is multiple comparisons. If you divide the data enough ways and look at enough end points, some ''significant'' differences are sure to emerge by chance. After all, you expect one in twenty P values to be less than 0.05 even if all null hypotheses are true. See Chapter 13 for further discussion.

  The statistical issue is pretty clear: The more comparisons you do, the more apt you are to find spurious findings. But one also must use clinical and scientific judgment. Multiple secondary analyses can often paint a clearer clinical picture than just one main analysis. The point is to look at all the secondary analyses as a group and not to focus on a particular analysis that seems intriguing without placing it in the context of the others.

- *Meta-analyses.* Investigators sometimes pool together results from several different studies.

21

Clinical Trials where N = 1

Most clinical trials involve many patients, and the number of patients is abbreviated N. A study with N = 40 enrolled 40 patients. Some clinical questions can be answered by doing a formal study using only a single patient, N = 1.

The goal of an N = 1 trial is modest. You are not trying to learn the secrets of nature and are not trying to gather data that will help future patients. You just want to know which of several alternatives is best for this patient. Is the current medication any better than placebo? Is dose A better than dose B? Does a generic pill work as well as a brand name pill?

It only makes sense to conduct an N = 1 trial when there is considerable doubt as to which therapy is best. Perhaps neither patient nor clinician thinks that the current therapy is working. Do a N = 1 trial comparing drug and placebo to find out. Perhaps the patient insists on taking a treatment that the physician thinks is worthless. Do a N = 1 trial to find out if the treatment is better than the alternative.

A N = 1 trial is only feasible in the following circumstances:

- The disease is chronic, so the treatment will be continued for a long time.
- The patient's course is fairly stable, so the severity of the disease won't change during the study.
- The treatments work quickly, and the effects are rapidly reversible.
- The patient understands the nature of the experiment and wants to participate.
- The effectiveness of the therapy can be quantified either by a laboratory measurement or by a clinical scale (i.e., quality of life). Ideally, you will record several variables that measure the effectiveness of the treatment and also record any side effects.
- A pharmacist is available to prepare the unlabelled medications and keep track of which is which.

Drugs should be administered in a double-blind manner according to a random schedule.

Analyze the data with any appropriate method; there is nothing special about the analysis of a one-patient trial. When analyzing the data, N refers to the number of measurements, not the number of subjects (which is 1). The study will be more powerful if the outcome is a measurement rather than a binary variable.

When interpreting the results, remember that the "population" in this study is all possible responses in this particular patient. You are generalizing from the responses observed in the sample of data you collected to the responses you would expect to see over the next few years in this particular patient. Until you repeat the N = 1 trial on several patients, you should not generalize the findings to other patients.

It is difficult to get enough data to have much power of achieving significance with the traditional cutoff of 0.05. Since the consequences of a Type I error are usually not severe with an N = 1 trial, α is sometimes raised to 0.10.

EXAMPLE

You think that taking vitamins makes you feel more alert. But you aren't sure whether it is a placebo effect or a real effect of the vitamins. You arrange with a pharmacist to prepare 10 vials with a week's supply of a multivitamin pill and 10 vials with a week's supply of placebo. The pharmacist randomly scrambles the vials and tells you which vial to use which week. Each week you rate your "alertness" on a scale of 0 (tired and distracted) to 10 (wide awake and alert).

Before you collect the data, you decide how you want to analyze the data and how you will interpret the results. Since the data are expressed as a rating scale, rather than a true measurement, it does not make sense to think of the data as being sampled from a Gaussian population. So you decide to use a test that does not make that assumption. There is no pairing between placebo weeks and vitamin weeks, so you can't use a test that assumes pairing or matching. So you decide to analyze the data with the Mann-Whitney test, a nonparametric test that compares two groups. See Chapter 24 for details.

In deciding how to set α, you need to consider the consequences of a Type I or Type II error. Making a Type I error (concluding that the vitamins are effective when they really aren't) isn't so bad because there is really no risk, the expense is minimal,

Table 21.1. Results for Vitamins and Alertness Example

Week	Preparation	Score
1	Placebo	4
2	Placebo	5
3	Vitamin	7
4	Placebo	8
5	Vitamin	9
6	Vitamin	7
7	Placebo	5
8	Vitamin	8
9	Placebo	9
10	Vitamin	8
11	Placebo	7
12	Placebo	5
13	Placebo	7
14	Placebo	6
15	Vitamin	7
16	Vitamin	6
17	Placebo	5
18	Vitamin	6
19	Vitamin	7
20	Vitamin	9

and only one person (yourself) is affected. Making a Type II error is not all that bad, but you are intrigued by the hypothesis and don't want to miss a real finding. You decide to set α to 0.10 and to make your decision accordingly.

You take the pills every day for 20 weeks and tabulate the results in Table 21.1. You plug the numbers into a computer program that can perform the Mann-Whitney test. The answer is that the two-tailed P value equals 0.063. If overall alertness is not affected by taking vitamins versus placebo, there is a 6.3% chance of obtaining a difference as large as you did in an experiment of this size.

Since you previously decided to set α equal to 0.10, you conclude that the data provide convincing evidence that taking vitamins increases your alertness. Since this study only involved one subject (yourself), you should not generalize the findings. And since the P value is fairly high (above the usual cutoff of 0.05) and is based on a small sample size, you can't be too sure the conclusion is correct.

Choosing an Appropriate Sample Size

Before starting a study, investigators must decide how many subjects to include. You want enough subjects to get valid results but not so many as to make the study unfeasible. Sample size can be determined by "gut feel," and that works fine so long as you have an experienced gut. It is better to calculate size using statistical principles, as explained in this chapter.

This book has repeatedly emphasized two different (but complementary) approaches to statistics: calculating confidence intervals and calculating P values. Each of these approaches can be used to determine sample size.

CONFIDENCE INTERVALS

Previous chapters have shown you how to calculate confidence intervals (CIs) for many kinds of data. In all cases, the width of the CI depends on sample size. Everything else being equal, a larger sample size makes the CI narrower.

If you can state the desired width of the CI, you can calculate how many subjects you need.

One Mean

Let's start with the simplest case. You plan to measure a variable in N subjects and calculate the mean and 95% CI of the mean. The larger you set N, the narrower the CI. How large must you make N to reduce the width of the 95% CI to a certain precision? The answer is shown in Equations 22.1. Note that we define precision* to equal half the width of the 95% CI. In other words, the precision is the distance the CI extends on either side of the sample mean.

$$\text{Precision} = t^* \cdot \frac{\text{SD}}{\sqrt{N}}.$$

$$t^* \approx 2. \tag{22.1}$$

$$N \approx 4 \left(\frac{\text{SD}}{\text{precision}} \right)^2.$$

*The word *precision* is sometimes defined in other ways.

The only trick is that N depends upon the critical value of t (t*), which itself depends on N. The way out of this loop is to realize that t* (for 95% confidence) is close to 2 unless the sample size is very small, so we set t* = 2. Some books present fancier equations that attempt to correct for the fact that the true value of t* might be higher. I don't think the discrepancy is worth worrying about, because the calculations are based on an estimated value of standard deviation (SD) and are only supposed to calculate an estimated sample size.

Example 22.1. We estimate that the SD of blood pressure is 10 mmHg. We wish to measure the blood pressure of a large enough sample so that we can define the mean value with a precision of 5 mmHg. How large must our sample be? The answer is 16. If we collected many samples of 16, on average the precision would equal 5 mmHg (assuming that the SD was estimated correctly). In any particular sample, the precision is equally likely to be less than or greater than 5.

If you want to use the equation, you may have difficulty estimating the value of the SD. Usually this can be done by inspecting previous data. Alternatively, you can run a pilot experiment first. Because the actual SD may not equal the value you estimated, the calculated sample size may not be exactly right. That's OK, as long as you realize that the equations are not precise but *estimate* an *approximate* sample size.

Difference Between Two Means

If you want to look at the difference between two means, you need more subjects as uncertainty in each mean contributes toward uncertainty in their difference. Use Equation 22.2 to determine the necessary sample size for *each* group from the estimated SD (assuming the two populations have equal SDs) and the desired precision (half-width of the 95% CI of the difference):

$$N \approx 8 \left(\frac{SD}{precision} \right)^2. \tag{22.2}$$

Continuing the example, now we want to measure blood pressure in samples of two populations and select large enough samples so that the 95% CI for the difference between the two means has a precision equal to 5 mmHg. Again we assume that the SD of blood pressures is about 10 mmHg and further assume that the two populations have equal SDs. Plugging the numbers into the equation, we need approximately 32 subjects in each group.

One Proportion

How many subjects are needed to determine a proportion with a specified precision? Equation 22.3 is a simple rearrangement of Equation 2.1.

$$\text{Precision} = 1.96 \cdot \sqrt{\frac{p(1-p)}{N}}.$$

$$N \approx \frac{4 \cdot p(1-p)}{precision^2}. \tag{22.3}$$

If you can estimate the proportion (p) and specify the desired precision, the sample size is readily calculated. If you can't estimate the value of p, set it equal to 0.5. The quantity $p(1 - p)$ is maximum when $p = 0.5$, and so setting $p = 0.5$ is a worst-case assumption that may overestimate the needed sample size.

When reading or watching to news, you've undoubtedly heard the phrase "this poll has a margin of error of 3%." How many subjects were used in such a poll? Pollsters use the term *margin of error* the same way we use the word *precision*. Set precision to .03 and $p = 0.5$, and N is 1111. Indeed, many polls use a sample size of about that size. If p is set to any value other than 0.5, fewer subjects would be needed.

Two Proportions

The number of subjects needed in each group to determine the difference between two proportions with a specified precision is as follows:

$$N \approx \frac{8 \cdot p_{av}(1 - p_{av})}{(\text{precision})^2}. \tag{22.4}$$

In Equation 22.4, the precision is the desired half-width of the 95% CI, p_{av} is the anticipated average of the two proportions, and N is the number of subjects needed in *each* group.

Example 22.2. You know that the incidence of hypertension in one population is about 0.10 and suspect that the incidence in a second population is 0.16. How large a sample size do you need to use in order for the half-width of the 95% CI of the difference between proportions to be 0.02? Set p_{av} to the average of .10 and .16, or 0.13. Set precision to 0.02. You need 2262 subjects in each group.

If you have difficulty estimating p_{av}, you can calculate a worst-case sample size by setting p_{av} equal to 0.5. Any other value would require a smaller sample size.

General Comments on Calculating Sample Size for Estimation

The preceding sections gave equations for estimating necessary sample size in several situations. If your assumptions are correct, the calculated sample size is large enough so that, on average, the half-width of the CI will equal the precision you specify. The entire CI spans a distance equal to twice the precision. For any particular experiment, you have a 50% chance that the CI will be wider and a 50% chance that it will be narrower. You will need to double or triple that sample size in order to have a 95% chance that the half-width of the CI will be narrower than the precision you specify.

The calculated sample sizes are the number of subjects you will need at the end of the experiment. It is usually wise to start with a larger number to allow for subjects who drop out and for experimental problems.

HYPOTHESIS TESTING

The previous sections showed you how to calculate the sample size needed to ensure that a CI has a specified width. It is far more common to calculate the sample size needed to achieve statistical significance, as explained now.

As you might imagine, the required sample size depends on these variables:

- α, the threshold for significance. Typically α is set to 0.05. If you set a more rigid criteria for significance by setting α to a smaller value (say 0.01), you will need a larger sample size.
- β, the probability of missing an important difference or making a Type II error. If you lower β to get more power, you will need more subjects. Sample size is often calculated for 90% power or 80% power (equivalent to setting β to 0.10 or 0.20).
- Δ, the minimum difference between population means that you wish to detect as significant. It takes more subjects to find a small difference than a large one.
- SD, the estimated standard deviation of each group (obviously not relevant when you are comparing two proportions).

Comparing Two Means

The sample size needed to compare two means is calculated using Equation 22.5, where N is the number of subjects in *each* group. As you'd expect, the necessary sample size increases if the data are more scattered (larger SD), if you set harsher criteria for α and β, or if you want to detect smaller differences (lower Δ).

$$N \approx \frac{2 \cdot SD^2 \cdot (z_\alpha + z_\beta)^2}{\Delta^2}. \tag{22.5}$$

The calculations are based on the assumption that the two populations have equal SDs. The estimated SD can come from previous data or from a pilot study. By the time you are ready to organize a formal study, you ought to know enough about the variable you are measuring to estimate its SD. Note that you don't have to separately estimate SD and Δ, it is sufficient to estimate their ratio. In some contexts, you'll find it easier to estimate the ratio than the individual values.

If you conduct a study using the calculated sample size, and the difference in population means really does equal Δ, what will you find? If you were to perform many such studies, you would detect a significant difference in some studies but not in others. The proportion of studies in which you will find a difference that is statistically significant with $P < \alpha$ equals $1 - \beta$. In other words, there is a $1 - \beta$ chance of obtaining a significant result if you use the calculated sample size and your assumptions are correct.

The term $(z_\alpha + z_\beta)^2$ is sometimes referred to as the *power index*. Values of this term are listed in Table 22.1. This table shows the square of the sum of the two z values. As the value is already squared, don't make the mistake of squaring again.

z_α is the critical value for the z distribution for the desired value of α. You may choose either the one- or two-sided distribution depending on how you plan to report the P values. If you can justify a one-sided P value, you will need fewer subjects. Z_β is the critical value of the z distribution for the desired value of β.*

---

*A one-tailed distribution is always used for β, even if you plan to report a two-tailed P value. If there really is a difference between populations, you make a Type II error when your samples means are so close together that the result is not significant. The other tail is when your sample means are much further apart than the population means. This tail is not important when calculating sample size.

Table 22.1. Values of the Power Index

α		Power Index $= (z_\alpha + z_\beta)^2$				
1-sided	2-sided	$\beta = .01$ Power = 99%	$\beta = 0.05$ Power = 95%	$\beta = 0.10$ Power = 90%	$\beta = .20$ Power = 80%	$\beta = .50$ Power = 50%
0.05	0.10	15.8	10.9	8.6	6.2	2.7
0.025	0.05	18.3	13.0	10.5	7.9	3.8
0.005	0.01	23.9	17.8	14.9	11.7	6.6

Before trying to calculate the power index for different values of z_α or z_β, first try to reproduce Table 22.1. Depending on which table you use, you may need to look up $z_{1-\beta}$, $z_{1-\alpha}$ or $z_{1-\alpha/2}$.

Example 22.3. We know that the SD of blood pressure in our population is about 10 mmHg. How many subjects do we need to have 80% power to detect a difference between means of 5 mmHg with P < 0.05 (two-sided)? The answer is about 63 subjects in each group.

Computer programs sometimes calculate sample size for several values of α and β. Table 22.2 is the output of InStat for the previous example.

Comparing Two Proportions

Necessary sample size for comparing two proportions can be estimated from Equation 22.6:

$$N \approx \frac{2 \cdot p_{av}(1 - p_{av})(z_\alpha + z_\beta)^2}{\Delta^2}. \tag{22.6}$$

Here p_{av} is the anticipated average proportion, Δ is the difference between the two proportions that you are looking for, and N is the number of subjects needed in *each* group. You can use this equation both for cross-sectional, experimental prospective and case-control studies. In cross-sectional studies, the proportions are the prevalences in the two groups. In prospective and experimental studies, the proportions are the

Table 22.2. Calculation of Sample Size for Example 22.3

Input

    Minimum difference you wish to detect as significant: 5
    Estimated standard deviation of each population: 10

Results

Power	β	α (two-sided) 0.10	0.05	0.02	0.01
80%	0.20	50	63	81	94
90%	0.10	69	85	105	120
95%	0.05	87	105	127	143

The values are the number of subjects needed in each group.

incidence rates in the two groups. In case-control studies, the proportions are the fraction of cases and controls exposed to the risk factor.

Equation 22.6 is simple to calculate. Other books give more complicated equations that attempt to calculate sample size more accurately. The difference between the simpler and more complicated equations is usually not too important. No matter how fancy the equation, the calculated sample size is no more than an estimate, since the value of p_{av} is an estimate and the values of α, β, and Δ are fairly arbitrary.

Example 22.4. You know that the incidence of hypertension in one population is about 0.10 and suspect that the incidence in a second population is higher. How many subjects do you need to have 90% power to detect an increase of 0.06 (to an incidence of 0.16) with P < 0.05? Set $\alpha = 0.05$ (two-sided), $\beta = .10$, $p_{av} = .13$, and $\Delta = 0.6$, and use the equation provided above. The answer is 660 in each group. The computer program InStat uses a more sophisticated equation and reports that 692 subjects are needed in each group (Table 22.3).

Comparing Two Survival Curves

Example 22.5. You know that 20% of patients treated for a certain kind of tumor will die within 3 years. An alternative treatment may be better. How many subjects do you need to have a 90% power of detecting an improvement to 10% with P < 0.05 (two sided)? Use the techniques of the previous section for comparing two proportions. Set $p_1 = .10$, $p_2 = .20$, $p_{av} = 0.15$, $\alpha = 0.05$ (two sided), $\beta = 0.90$, and the equation calculates a sample size of 268 in each group. The more accurate equation programmed into InStat calculates a sample size of 286 per group.

The discussion in the previous paragraph assumed that the only outcome you care about is whether each individual survived 3 years. Thus you can analyze the data by comparing two proportions. There are two problems with this analysis. First, most studies accrue patients over time, so you won't follow all patients for the same length of time. This problem tends to increase the number of subjects you will need. Second, you will know more than just whether or not each subject survived to 3 years, you will know exactly how long each subject survived. If you display the data as full survival curves and compare curves with an appropriate method (log-rank test), you

Table 22.3. Calculation of Sample Size for Example 22.4

Input
  Minimum difference you wish to detect as significant: 0.06
  Estimated value of the smaller proportion: 0.10

Results

Power	β	\(\alpha\) (two-sided) 0.10	0.05	0.02	0.01
80%	0.20	421	525	663	766
90%	0.10	570	692	850	967
95%	0.05	711	847	1022	1150

The values are the number of subjects needed in each group.

can get by with fewer subjects. It is not easy to determine how many fewer subjects are needed, as it depends on study design (how long are patients accrued?) and on assumptions about their survival rates.

Unequal Sample Sizes

Equations 22.5 and 22.6 assume that both groups will have the same number of subjects. In some situations, it makes sense to have more subjects in one group than another. It is possible to compensate for fewer subjects in one group by adding additional subjects in the other group. But it is not an even trade-off, and the total number of subjects must increase. For example, you may decrease the number of subjects in one group by 25% if you double the number in the other group, and you may decrease the number in one group by 40% if you quadruple the number in the other group. There is not much point in increasing the sample size in one group to more than about four times the calculated value. No matter how many subjects you include in the larger group, you can't reduce the number of subjects in the other group to less than half the number that would be needed with equal size samples.

INTERPRETING A STATEMENT REGARDING SAMPLE SIZE AND POWER

Papers often include a statement something like this one: "We chose to study 313 subjects in each group in order to have 80% power of detecting a 33% reduction in the recurrence rate from a baseline rate of 30% with a significance level of 0.05 (two tailed)." This sentence defines α, β, and Δ. In this example $\alpha = 0.05$ (the significance level), $\beta = 0.20$ (100% minus the power), $p_1 = .30$, and $\Delta = 0.10$ (33% of p_1). Let's review again the meaning of α and β.

- $\alpha = 0.05$. If the null hypothesis is true, there is a 5% chance of obtaining a significant result and a 95% chance of obtaining a not significant result in a study of this size.
- $\beta = 0.20$. The investigators assume that the recurrence rate is normally 30% and hypothesize that the new treatment reduces the recurrence rate by one third. If this hypothesis is true, there is a 80% chance of obtaining a significant result and a 20% chance of obtaining a not significant result in a study with 313 subjects in each group.

Don't be confused by the analogy between β errors and false-negative lab results. The 80% power does *not* refer to individual subjects. Some people misinterpret 80% power to mean this: "80% of the subjects would be improved by the treatment and 20% would not be." This is *incorrect*. Power refers to the fraction of studies that would report significant P values, not the fraction of patients who benefit from the treatment.

Although a statement regarding sample size calculations sounds very precise ("we calculated that we need 313 subjects in each group"), in fact they are just "ballpark" estimates. Consider these problems:

- The equations are based on a few simplifying assumptions. Accordingly, the resulting values of N should be considered to be an estimate. More sophisticated equations usually calculate slightly larger sample sizes than the equations shown here.

- The calculations tell you how many subjects you need at the end of the study. You will usually need to begin the study with additional subjects to allow for dropouts and experimental problems.
- The values of α and β are arbitrary. Ideally the values should be based on the relative consequences of Type I and Type II errors, but more often α and β are simply set to conventional values.
- The value of Δ is arbitrary. Ideally, Δ is the smallest difference that would be clinically (or scientifically) important. In practice, this value is hard to define.
- Sample size calculations assume that you will only measure and analyze one outcome. Common sense tells you that you should appraise all relevant clinical variables, and most clinical investigators do so. While it seems like this should increase the power of a study, current statistical methods are not well designed to cope with multiple outcome variables.
- Although it always sounds like the investigator calculated sample size from α, β, and Δ, often the process went the other way. The investigator chose the sample size and then calculated values of α, β, and Δ that would justify the sample size. Often the process is iterative. The investigator specified ideal values for α, β, and Δ, and was horrified at the enormous number of subjects required. The investigator then altered those values and recalculated N. This process was repeated until N sounded "reasonable."

SEQUENTIAL STUDIES

In all the examples presented in this chapter, we assume that you choose a sample size before the study begins, study all the subjects, and then analyze the data. An alternate approach seems far more appealing—add new patients to the study over time and periodically reanalyze the data. If the result is significant, stop. Otherwise, recruit some new subjects. This approach should not be used (unless special analyses are employed; see the next paragraph). The problem is that the probability of a Type I error with such a design is far greater than 5%. By stopping the study if the results happen to be significant but continuing otherwise, you have biased the results towards significance. The P value from such a study design cannot be interpreted. In fact, if you kept going long enough, every experiment would eventually reach significance just by chance, although some might take a *very* long time to do so.

   Statisticians have devised special study designs that are designed to be reanalyzed with every pair of patients. Such methods are only useful when the outcome for each patient is determined in a short time. With a sequential analysis, the data are reanalyzed (using special techniques) after each pair of patients (one patient gets treatment A, the other treatment B). The study continues until the analysis reaches a conclusion that the one treatment is significantly better than the other, or that the two are indistinguishable. The investigator needs to specify α and β before the study begins. Data can be plotted on special graph paper that automates the analysis. Because the analysis is repeated with each pair of subjects, you don't need to calculate sample size before you begin.

   Other special methods are designed to allow periodic reanalysis after every batch of subjects. For example, you might test a new drug on 18 patients. If it works on none, stop. If it works on any, then continue testing on more patients. These methods take into account the multiple analyses.

SUMMARY

Before beginning a study, investigators need to know how many subjects to include. You need enough subjects to get valid results, but want to avoid wasting time and money by using too many subjects. There are two approaches you can use to calculate the needed sample size. One approach asks how many subjects are needed so that the confidence interval (or the confidence interval of a difference) has a desired width. The other approach asks how many subjects are needed so that you have a specified power to obtain a significant difference (with a specified value of α) given a specified experimental hypothesis. In order to use either of these methods, you need to estimate the standard deviation of the values (if the outcome is a measurement) or the estimated value of the proportion (if the outcome is binomial).

OBJECTIVES

- Without using books or computers, you should know what information is needed to calculate sample size in various circumstances.
- Using books, calculator, or computer, you should be able to determine the sample size needed to determine a mean or proportion to within a specified tolerance. You should also be able to calculate the necessary sample size to find a specified difference with specified power.

PROBLEMS

1. You are preparing a grant for a study that will test whether a new drug treatment lowers blood pressure substantially. For previous experience, you know that 15 rats in each group is enough. Prepare a convincing power analysis to justify that sample size.
2. The overall incidence of a disease is 1 in 10,000. You think that a risk factor increases the risk. How many subjects do you need if you want to have 95% power to detect a relative risk as small as 1.1 in a prospective study?
3. How large a sample would you need in an election poll to determine the percent voting for each candidate to within 1%? What assumptions are you making?
4. Can a study ever have 100% power?
5. About 10% of patients die during a certain operation. A new technique may reduce the death rate. You are planning a study to compare the two procedures. You will randomly assign the new and standard technique, and compare the operative death rate.
   A. How many subjects do you need to have 95% power to detect a 10% reduction in death rate (α = 0.05)?

B. How many subjects do you need to have 60% power to detect a 50% reduction in death rate ($\alpha = 0.10$)?

C. Is it ethical to compare a new technique (that you think is better) with a standard technique?

6. Lymphocytes contain beta-adrenergic receptors. Epinephrine binds to these receptors and modulates immune responses. It is possible to count the average number of receptors on human lymphocytes using a small blood sample. You wish to test the hypothesis that people with asthma have fewer receptors. By reading various papers, you learn that there are about 1000 receptors per cell and that the coefficient of variation in a normal population is about 25%.

A. How many asthmatic subjects do you need to determine the receptor number to plus or minus 100 receptors per cell with 90% confidence?

B. You want to compare a group of normal subjects with a group of asthmatics. How many subjects do you need in each group to have 80% power to detect a mean decrease of 10% of the receptor using $\alpha = 0.05$?

C. How many subjects do you need in each group to have 95% power to detect a mean difference in receptor number of 5% with $\alpha = 0.01$?

7. You read the following in a journal:

Before starting the study, we calculated that with a power of 80% and a significance level of 5%, 130 patients would be required in each group to demonstrate a 15-percentage-point reduction in mortality (from the expected rate of 33 percent to 18 percent).*

A. Explain in plain language what this means.

B. Show the calculations, if possible.

*MA Hayes, AC Timmins, EHS Yau, M Palazzo, CJ Hinds, D Watson. Elevation of systemic oxygen delivery in the treatment of critically ill patients. *N Engl J Med* 33:1717–1722, 1994.

VII

COMMON STATISTICAL
TESTS

Even if you use computer programs to calculate statistical tests, it is helpful to know how the tests work. This part explains the most commonly used tests in reasonable detail. The next part explains more advanced tests in less detail.

23

Comparing Two Groups: Unpaired t Test

Example 23.1

To keep the calculations simple, we will use these fake data comparing blood pressures in a sample of five medical students from the first- (MS1) and second- (MS2) year class. You've already seen these data in Chapter 10 (Table 23.1).

MS1: 120, 80, 90, 110, 95

MS2: 105, 130, 145, 125, 115

Now we want to calculate a P value that answers this question: If the two populations have identical means, what is the probability that the difference between the means of two random samples of five subjects will be as large (or larger) than the difference actually observed?

There are several methods to calculate this P value, and this example is continued in the next chapter. For this chapter, we'll make the assumption that the two populations follow a Gaussian distribution and will calculate the P value with a t test.

The P value obviously depends on the size of the difference. If the difference is large, the P value must be small. But large compared to what? The t test compares the size of the difference between means with the standard error (SE) of that difference, which you have already learned to calculate in Chapter 7. The SE of the difference combines the standard deviations (SDs) of the groups and the number of data points.

Equation 23.1 defines the t ratio, simply called:

$$t = \frac{\text{difference between means}}{\text{SE of difference}} \qquad df = N_{total} - 2. \qquad (23.1)$$

If the two groups are the same size, calculate the SE of the difference using Equation 23.2. Otherwise use Equation 23.3. (You've already seen these equations in Chapter 7.)

$$\text{SE of difference (equal N)} = \sqrt{SEM_a^2 + SEM_b^2}. \qquad (23.2)$$

$$\text{Pooled SD} = \sqrt{\frac{(N_a - 1) \cdot SD_a^2 + (N_b - 1) \cdot SD_b^2}{N_a + N_b - 2}}.$$

$$(23.3)$$

$$\text{SE of difference} = \text{pooled SD}\sqrt{\frac{1}{N_a} + \frac{1}{N_b}}.$$

For this example, t = 25/9.84 = 2.538.

Table 23.1. Example 23.1 Blood Pressure Data

	Mean	SD	SEM	95% Confidence Interval
A: (MS1)	99	15.97	7.14	79.17 to 118.82
B: (MS2)	124	15.17	6.78	105.17 to 142.83
Difference (B-A)	25		9.84	2.29 to 47.71

SOME NOTES ON t

• The sign of t only tells you which mean is larger. In this example, t is positive because the difference was defined as the mean of group B minus the mean of group A. It would have been negative had the difference been defined as the mean of group A minus the mean of group B.
• Because t is a ratio, it doesn't have any units.
• You don't need raw data to calculate t. You can do all the calculations from the mean, standard error of the mean (or SD), and sample size of each group.

OBTAINING THE P VALUE FROM t, USING A TABLE

If you don't use a computer program that calculates an exact P value, you need to determine the P value using a table.

To convert t to a P value, you have to know three things:

• What is the value of t? If it is negative, remove the minus sign (i.e., convert it to its absolute value). As t gets higher, P gets smaller.
• Do you want to calculate a one- or two-tailed P value? For reasons already mentioned, this book always uses two-tailed P values.
• How many degrees of freedom (df) are there? For a two-sample t test, df is two less than the total number of subjects, in this case df = 8. If you haven't studied statistics intensively, the rules for calculating df are not always intuitive, and you need to just learn them or look them up. For any particular value of t, P gets smaller as df gets bigger.

A portion of a t table is shown in Table 23.2. Look across the line for df = 8. Under $\alpha = 0.05$ is the value 2.306. Our t is higher than this, so our P value is less than 0.05. Under $\alpha = 0.01$ is the value 3.355. Since this is higher than our value of t, our P value is not this low. Thus the P value for example 23.1 is less than 0.05 but greater than 0.01.

The table shown here, and the longer Table A5.5 in the Appendix are similar to tables in many other books. Table A5.4 in the Appendix is less standard. In this table, the columns represent df, the rows are various values of t, and you read the P value directly. With this table, we can bracket the P value between 0.037 (t = 2.5) and 0.032 (t = 2.6). As we have already seen, a computer program can calculate the P value with more precision; in this case the two-tailed P value is 0.035.

Although consulting statistical tables should be no more complicated than reading an airline timetable or a newspaper's financial page, the inconsistent use of terminology makes the use of statistical tables more challenging. You should find several statistics

Table 23.2. Determining the P Value from t

df	α (two-tailed)						
	0.250	0.200	0.150	0.100	0.050	0.010	0.001
6	1.273	1.440	1.650	1.943	2.447	3.707	5.959
7	1.254	1.415	1.617	1.895	2.365	3.499	5.408
8	1.240	1.397	1.592	1.860	2.306	3.355	5.041
9	1.230	1.383	1.574	1.833	2.262	3.250	4.781
10	1.221	1.372	1.559	1.812	2.228	3.169	4.587
15	1.197	1.341	1.517	1.753	2.131	2.947	4.073
20	1.185	1.325	1.497	1.725	2.086	2.845	3.850
100	1.157	1.290	1.451	1.660	1.984	2.626	3.390

books and look up the P value for this example. Some potentially confusing points are as follows: Some books label the columns as P values instead of α values. In this context, α and P mean almost the same thing. The P value is calculated for particular data; the α value is a threshold P value below which you term the difference *significant.* Other books label the columns with one minus the P value. In this case you need to look in the column labeled .95 to get the critical value for P − 0.05. In some books, it is difficult to tell whether the P values are one or two tailed.

THE t DISTRIBUTION AND THE MEANING OF THE P VALUE*

If you repeated the experiment many times, you would get many different values for t. Statisticians have figured out what this distribution looks like when the null hypothesis is true. This probability distribution depends on the number of df.

Our example had five subjects in each group, and so had 8 df (5 + 5 − 2). Figure 23.1 shows the distribution of t with 8 df assuming that the null hypothesis is true. This graph shows the probability distribution of all possible values of t that could be obtained if the null hypothesis were true. The area under the curve represents the results of all possible experiments. In our particular experiment, t = 2.54. All values of t greater than 2.54 or less than −2.54 are shaded. The two-tailed P value is the fraction of all possible experiments (under the null hypothesis) that would lead to t > 2.54 or t < −2.54. In other words, the P value is the fraction of the total area under the curve that is shaded. By inspection you can see that each tail is a small fraction of the total. Exact calculations reveal that each tail represents 1.7% of all possible experiments (again, assuming that the null hypothesis were true). The two-tailed P value is therefore the sum of those two tails, which is 3.4% or 0.034.

The probability distribution of t shown in Figure 23.1 is for 8 df. If you have more data points, and thus more df, the probability distribution is more compact. Figure 23.2 shows the probability distribution for t under the null hypothesis for 2, 6, and 1000 df. With many df, the distribution is almost identical to a Gaussian distribution.

*This section is more advanced than the rest. You may skip it without loss of continuity.

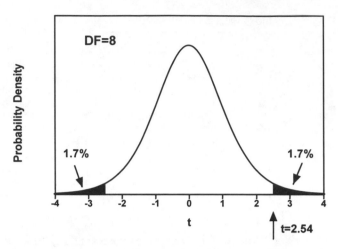

Figure 23.1. The t distribution for eight degrees of freedom (DF). The figure shows the probability distribution for t assuming 8 degrees of freedom (total of ten subjects in both groups) assuming the null hypothesis is really true. In the example, t = 2.54. All values of t greater than 2.54 and less than −2.54 are solid. The area of these solid areas as a fraction of the total area under the curve is 3.4%, which is the P value.

ASSUMPTIONS OF A t TEST

The t test is based on these assumptions:

• The samples are randomly selected from, or at least representative of, the larger populations.

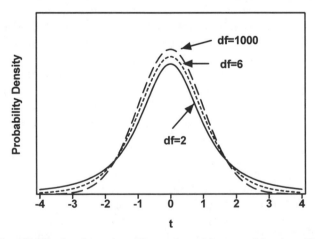

Figure 23.2. The t distribution depends on the number of degrees of freedom. This figure shows the distribution of t under the null hypothesis for df = 2, df = 6, and df = 1000. When there are many degrees of freedom, the t distribution is virtually identical to the Gaussian (z) distribution. With fewer degrees of freedom, the t distribution is wider.

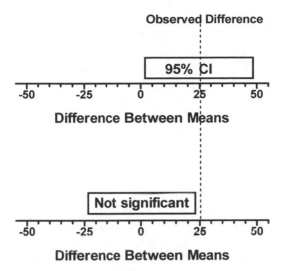

Figure 23.3. The relationship between CIs and statistical significance. The width of the 95% CI equals the width of the range of differences that would be not significant ($\alpha = 0.05$). In this example, the 95% CI does not include 0 and the range of not significant values does not include the observed difference.

- The two samples were obtained independently. If the subjects were matched, or if the two samples represent before and after measurements, then the paired t test should be used instead.
- The observations within each sample were obtained independently. Choosing any one member of the population does not alter the chance of sampling any one else.
- The data are sampled from populations that approximate a Gaussian distribution. With large samples, this assumption is not too important.
- The SD of the two populations must be identical. If you are unwilling to assume that the SDs are equal, look in a more advanced book for modifications of the t test that do not require this assumption.

THE RELATIONSHIP BETWEEN CONFIDENCE INTERVALS AND HYPOTHESIS TESTING*

Confidence intervals and hypothesis testing are closely related. Figure 23.3 shows the relationship between the CI and hypothesis testing for the blood pressure example. The top part of the figure shows the 95% CI for the difference between means. The calculations were explained in Chapter 7. The CI is centered on the observed difference between sample means and extends in each direction by a half-width defined by Equation 23.4:

$$\text{Half-width} = t^* \cdot \text{SE of difference.} \tag{23.4}$$

*This section is more advanced than the rest. You may skip it without loss of continuity.

For this example, the critical value of t is 2.306 and the SE of the difference is 9.84. So the 95% CI extends a distance of 22.69 on either side of the observed mean difference of 25. Thus the 95% CI box extends from 2.31 to 47.69. We can be 95% sure that the true difference between population means lies within this range, which is shown as a box in the top part of the figure.

The bottom half of the figure shows the results of hypothesis testing. Given the sample size and scatter, any difference between means inside the box would be not significant, and any difference outside the box would be statistically significant. Determining where to draw the box is easy. We can rearrange Equation 23.1 to Equation 23.5:

$$\text{Difference between means} = t^* \cdot \text{SE of difference.} \tag{23.5}$$

From Table 23.2, we know that any value of t between -2.306 and 2.306 will yield a P value (two-tailed) greater than 0.05 (assuming 8 df). From our data we know the SE of the difference is 9.84. So a difference between sample means of 2.306 $\times$ 9.84 = 22.69 (in either direction) would be right on the border of significance. The not significant box extends from a mean difference of -22.69 to 22.69. Given the sample size and scatter, any difference between means of less than -22.69 or greater than 22.69 would be statistically significant.

As you can see graphically (Figure 23.3) and algebraically (Equations 23.3 and 23.4), the two boxes in Figure 23.3 are the same size. The width of the 95% CI equals the range of differences that are not significant.

In this example, the 95% CI does not include 0, and the not significant zone does not include the actual difference between means (25). This is a general rule. If the 95% CI for the difference between means contains 0, then the P value (two-tailed) is greater than 0.05. If the 95% CI does not contain 0, then the P value is less than 0.05. If one end of the 95% CI is exactly 0, then the P value (two-tailed) is exactly 0.05. Similarly, if a 99% CI contains 0, then the P value is greater than 0.01, and if one end of a 99% CI is exactly 0, then the two-tailed P value is exactly 0.01.

Figure 23.4 shows what would have happened had the difference between means been equal to 20 (with the same sample size and SDs). The 95% CI has the same width as before, but is now shifted to the left and includes 0. The not significant zone hasn't changed but now includes the observed difference (20). For this altered example, the P value would be greater than 0.05 and not significant. The 95% CI starts at a negative number and goes to a positive number, crossing 0.

If you read a paper that reports a t test without a CI, it is usually easy to calculate it yourself, even without access to the raw data. Recall that the equation for the CI of a difference (Equation 7.3) requires you to know the difference between means, the critical value of t, and the SE of the difference. The difference is usually presented in the paper or tables, or it can be estimated from the graphs. The critical value of t comes from a table; it is not the same as the t ratio calculated from the data. The SE of the difference is rarely presented in published papers. However, the t ratio is usually presented, and it equals the difference between means (which you know) divided by the SE of the difference. Thus you can calculate the SE of the difference from the t ratio and the difference.

As an example, let's assume that the data for our example were presented incompletely as mean values (124 and 99), sample size (N = 5 in each group), t ratio (2.54), and P value (P < 0.05). How can we calculate the 95% CI of the difference? The t

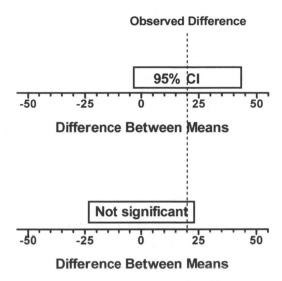

Figure 23.4. The relationship between CIs and statistical significance, continued. In this example, the 95% CI includes 0 and the range of not significant values includes the observed difference.

ratio (2.54) is equal to the difference (25) divided by the SE of the difference. Thus the SE of the difference must equal 25/2.54 or 9.84. The critical value of t for 8 df for a 95% CI is 2.306 (from Table A5.3 in the Appendix). Thus the 95% CI for the difference between the means is 25 ± 2.306 × 9.84, which equals 2.3 to 47.7.

Note the similarities and differences in calculating the P value and the CI. To calculate the P value, you start with the difference between means, calculate the t ratio, and end up with a probability. To calculate the 95% CI, you start with the probability (95%), find the critical value of t that corresponds to that probability, and end up with a range of differences (the confidence interval). It is easy to get confused by the different uses of t. When calculating a P value, you first calculate the t ratio for that particular set of data and determine the probability (P). When calculating a CI, you start with a probability (usually 95% confidence) and use a table to find the critical value of t for the appropriate number of df. When calculating a CI, it is a mistake to use the t ratio actually calculated for the data rather than the critical value of t needed for 95% confidence. If you mistakenly do this, one end of the CI will equal exactly 0. In this book we use the variable t to refer to a calculated ratio and t* to refer to a critical value you look up on a table. Other books do it differently.

CALCULATING THE POWER OF A t TEST*

You already know that statistical power is the answer to this question: If the difference between the means of two populations equals a hypothetical value Δ_H, then what is

---

*This section contains the equations you need to calculate statistics yourself. You may skip it without loss of continuity.

the probability that you will obtain a significant P value when comparing two random samples of the size actually used?

When you attempt to calculate the power of a study, you encounter two problems:

- You need to know the standard deviation of the populations, but only know the standard deviations of your samples.
- You need to know the distribution of t if the null hypothesis is *not* true. This is difficult to calculate, and even difficult to look up since the tables are enormous.

These problems make it difficult to calculate the power exactly. However, it is pretty easy to calculate the power approximately, and Equation 23.6 shows one way to do it.* This equation should only be used when df $\geq$ 10.

$$Z_{power} = \left(\frac{\Delta_H}{\text{SE of difference}} - t^* \right) \left[\frac{1}{\sqrt{1 + \dfrac{t^{*2}}{2 \cdot df}}} \right] \tag{23.6}$$

Let's look at the variables in the equation.

- t^* is the critical value of the t distribution. When calculating a t test, if $t = t^*$ then $P = \alpha$. Its value depends on the value you pick for α and on the number of df. Use Table A5.3 in the Appendix. Use the column for 95% confidence if you set α (two-tailed) equal to 0.05. Use the column for 99% confidence if you set $\alpha = 0.01$.
- Δ_H is the hypothetical difference between population means. Deciding on a value for Δ_H is not always easy. It is a scientific decision, not a statistical one. Δ_H is a hypothetical difference that you think would be worth detecting. Tiny differences are never worth detecting; huge differences always are. Given the context of the experiment, you need to decide on reasonable definitions for *tiny* and *huge* and to find a value in the middle that you think would be definitely important enough to detect. You can then calculate the power of the study to find a significant difference if the real difference was that large. You don't have to choose a single value for Δ_H. Instead you can calculate the power of a study for several values of Δ_H and make a graph or table of power versus Δ_H.
- If you are calculating the power of a published study, then you know the SE of the difference, or can calculate it from the SD (or SEM) and size of the samples using Equation 23.2 or 23.3. If you are calculating the power of a study you plan to perform, then you still use those equations, but estimate the value of the SDs from pilot studies or published articles.

To convert z_{power} to power, you need to find the probability that a random number chosen from a standard normal distribution (mean = 0, SD = 1) has a value less than z_{power}. The answers are tabulated in Table A5.6 in the Appendix. Half of the values in the standard normal distribution are negative and half are positive. So the power is 50% when $z_{power} = 0$, the power is greater than 50% when z_{power} is positive, and the power is less than 50% when z_{power} is negative. In equation 23.6, the term in brackets

---

*This equation is not well known. I adapted it from W. L. Hays, *Statistics,* 4th ed., Harcourt Brace, Orlando, 1988.

on the right side is always close to 1.0 and so affects the calculation of power by at most a few percentage points. Leaving out that term, an easier approximation is shown in Equation 23.7.

$$z_{power} \approx \frac{\Delta_H}{\text{SE of difference}} - t^* \qquad (23.7)$$

To make sense of this equation, let's see how it is affected by the variables.

- If you increase Δ_H, you will increase the value of z_{power}, and increase the power. A study has more power to detect a big difference than to detect a small one.
- If you lower α, you will increase t^*, which will decrease the value of z_{power}, and decrease the power. In other words, if you set a stricter threshold for statistical significance, the power of the study will decrease. If you want fewer Type I errors, you have to accept more Type II errors.
- If you increase the sample size, the SE of the difference will be smaller, so z_{power} will be larger and the power will be larger. Larger studies are more powerful than small studies.
- If the data have less scatter, the SE of the difference will be smaller, so z_{power} will be larger and the power will be higher. A study has more power to detect differences when the data have little scatter.

EXAMPLE 23.2

Motulsky et al. asked whether people with hypertension (high blood pressure) have altered numbers of α_2 adrenergic receptors on their platelets.* There has been a lot of speculation about the role of the autonomic nervous system in hypertension, so these investigators asked about possible changes in the number of receptors for epinephrine. They studied platelets because they are accessible in a blood sample and are thought to represent receptors in other organs. They studied 18 hypertensive men and 17 controls of the same age range. The platelets of hypertensives had an average of 257 ± 14 receptors per platelet (mean $\pm$ SEM) and the controls had an average of 263 ± 21 receptors per platelet. The authors concluded that receptor number was not significantly different. If the true difference between mean receptor number in controls and hypertensives was 50 receptors/cell, what was the power of this study to find a significant difference with $P < 0.05$ (two-tailed)?

First calculate the SE of the mean difference using equation 23.3 (after calculating the SDs from the SEMs). The SE of the mean difference is 25.0 receptors/cell. From Table A5.3, $t^* = 2.04$ for 33 df. From the question, $\Delta_H = 50$. Using Equation 23.7, $z_{power} = -0.04$. From Table A5.6, the power is about 48%. Using the longer equation 23.6 would not change the answer much. If the difference in mean number of receptors in the overall populations really was 50 sites/cell, there was only about a 48% chance that a study of this size would have found a significant difference with $P < 0.05$.

---

*HJ Motulsky, DT O'Connor, PA Insel. Platelet α_2 receptors in treated and untreated essential hypertension. *Clinical Science* 64:265–272, 1983.

PROBLEMS

1. Calculate the t test for Example 7.1. The authors compared stool output between treated and untreated babies. The results were 260 ± 254 (SD) for 84 control babies and 182 ± 197 for 85 treated babies.
2. Calculate the t test for Problem 2 in Chapter 7 (transdermal nicotine example).
3. A. Explain why power goes up as Δ gets larger.
   B. Explain why power goes down as SD gets larger.
   C. Explain why power goes up as N gets larger.
   D. Explain why power goes down as α gets smaller.
4. For the study in example 23.2, calculate the following:
   A. Calculate the t test and determine the t ratio and the P value.
   B. What was the power of that study to detect a mean difference of 25 sites/cell?
   C. What was the power of that study to detect a mean difference of 100 sites/cell?
   D. If the study were repeated with 50 subjects in each group, what would the power be to detect a difference of 50 sites/cell?

24

Comparing Two Means:
The Randomization and
Mann-Whitney Tests

INTRODUCTION TO NONPARAMETRIC TESTS

As you learned in the last chapter, the unpaired t test is based on the assumption that the data are sampled from two populations that follow a Gaussian distribution. Because the t test is based on an assumption about the distribution of values in the population, it is termed a *parametric* test. Most parametric tests, like the t test, assume that the populations are Gaussian.

Nonparametric tests make no rigid assumptions about the distribution of the populations. This chapter presents two nonparametric tests that can be used to compare two groups. The most commonly used nonparametric test is the *Mann-Whitney rank sum test*. Like many nonparametric tests, this test works by performing all calculations on the ranks of the values (rather than the actual data values). Because calculations are based on rank rather than values, this test is barely influenced if a value is especially high or low. These values are called *outliers*. Tests that are resilient to outliers are called *robust tests*. The Mann-Whitney test is a robust nonparametric test.

Other nonparametric tests analyze the actual data values. Before presenting the Mann-Whitney test, I present the *exact randomization test,* a nonparametric test that uses the actual data rather than ranks.

WHAT IS THE EXACT RANDOMIZATION TEST?

The randomization test, also called the *exact probability test* or the *permutation test,* is easier to understand than the t test. It is presented here as an aid to learning. If you understand the randomization test, then you really understand P values.

You don't have to think about populations and samples. Instead, you only have to think about the data you actually collected. If the null hypothesis is true, then each particular value could just as easily have come from one group as from the other. Thus, any shuffling of the values between the two groups (maintaining the sample size of each) would be just as likely as any other arrangement. To calculate the P value,

217

find out what fraction of all these rearrangements would result in means that are as far apart (or more so) than actually observed.

To perform the exact randomization test, do the following:

1. List all possible ways to shuffle the observed values between the two groups, without changing the number of observations in each group.
2. Calculate the difference between means for each of these possible arrangements.
3. Compute the fraction of the arrangements that lead to a difference between means that is as large or larger than the difference that was experimentally observed.

Conceptually this is quite easy. The hard part is keeping the details straight.

CALCULATING THE EXACT RANDOMIZATION TEST*

In the BP example, we have five measurements in each group. To calculate the exact randomization test, the first step is to list all possible ways to shuffle the data. How many ways can the 10 data points be divided up into two groups of five each? The answer turns out to be 252. You can either trust me (and skip the next paragraph), list all the possibilities yourself, or read the next paragraph.

We start with the 10 numbers, and want to select five for group A (MS1), which leaves the other five for group B (MS2). There are 10 possible values that can be selected to place in the first position of group A. Then there are nine remaining values that can be selected for the second position of group A. Then there are eight remaining values that can be selected as the third value of group A. ... Finally there are six possible values left to place in the fifth position group A. The remaining values become group B, and no choices are left. Thus there are $10 \times 9 \times 8 \times 7 \times 6$ permutations of values that can be selected for group A. However, this calculation is sensitive to the order in which the values were chosen for group A. Every set of five values is counted many times. Because we don't care about order that calculation substantially over counts the number of possibilities. To correct for this, we must divide by the number of ways each group of five numbers can be arranged within group A, and this is 5! ($5 \times 4 \times 3 \times 2 \times 1$). So the number of ways to divide the 10 numbers into two groups is:

$$\frac{10 \times 9 \times 8 \times 7 \times 6}{5 \times 4 \times 3 \times 2 \times 1} = 252 \tag{24.1}$$

If there were no difference in the distribution of BPs in the two populations, then each of these 252 ways of shuffling the data should have been equally probable. How many of these possibilities lead to group A having a mean value at least 25.0 units smaller than the mean of group B? We can list them, by putting the largest values in one group and the smallest in the other (Table 24.1).

Of the 252 different ways the data points can be shuffled between groups A and B (maintaining five values in each group), two arrangements (including the one actually observed) have a difference of 25, four arrangements have a larger difference, and the

---

*This section is more advanced than the rest. You may skip it without loss of continuity.

Table 24.1. Rearrangements Where the Mean Values of Group A Are $\geq$ 25 Units Smaller than Group B

Group	Values	Difference Between Means
A	80, 90, 95, 110, 120	25 (Actual data)
B	105, 115, 125, 130, 145	
A	80, 90, 95, 105, 120	27
B	110, 115, 125, 130, 145	
A	80, 90, 95, 105, 110	31
B	115, 120, 125, 130, 145	
A	80, 90, 95, 110, 115	27
B	105, 120, 125, 130, 145	
A	80, 90, 95, 105, 115	29
B	110, 120, 125, 130, 145	
A	80, 90, 95, 105, 125	25
B	110, 115, 120, 130, 145	

remaining 247 ways have smaller differences. So if we start with the null hypothesis that BP is unrelated to medical school class, we can state there is only a 2.38% (6/252) chance that the five individuals in group A would have a mean BP 25 or more units smaller than that of the five individuals in group B. The one-tailed P value is 0.0238. By symmetry, there is an equal chance (under the null hypothesis) that group A would have a mean BP 25 or more units higher than group B. The two-tailed P value is twice the one-tailed P value, or 0.0476.

This analysis is shown in Figure 24.1. It shows all 252 ways the data can be rearranged. The X axis shows possible differences between the mean blood pressures. The Y axis shows the number of rearrangements that give a particular difference between means. The area under the curve represents all 252 ways to rearrange the data. The shaded area on the right represent the six possibilities in which the mean difference equals or exceeds the observed value. The one-tailed P value, therefore, is that shaded area divided by the total area under the entire graph. The two-tailed P value is that area plus the matching area on the other side (or tail) of the distribution.

Unlike the t distribution, which is smooth, this distribution is chunky. There are a limited number of possible ways to rearrange the data and thus only a limited number of possible differences.

LARGE SAMPLES: THE APPROXIMATE RANDOMIZATION TEST

The exact randomization test requires the computer program to systematically examine every possible rearrangement of the data. With large data sets, the number of possible rearrangements becomes astronomical, and the exact randomization test is beyond the capability of even the fastest computers. Instead of examining all rearrangements of the data, you can just look at some rearrangements using the approximate randomization test:

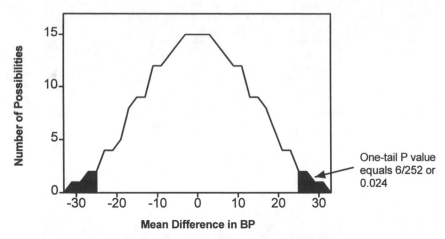

Figure 24.1. Exact randomization test. The figure shows all 252 ways that the ten data points can be shuffled with five data points in each group. The X axis shows the difference between means, and the Y axis shows the number of possible arrangements that would lead to that big a difference between means. The dark area on the right shows the six arrangements that lead to a difference between means as large or larger than observed in the example. The dark area on the left shows the six arrangements that lead to a difference between means as large or larger than observed in the example but in the other direction. The two-tailed P value is 12/252 or 0.048.

1. Randomly shuffle the values between the two groups.
2. Tabulate the answer to this question: Is the difference between the means of the two groups in the shuffled data larger or smaller than the difference between the means of the actual samples?
3. Repeat steps 1 and 2 many (perhaps several thousand) times.*
4. The P value is the fraction of the rearrangements in which the difference between means was greater than or equal to the difference you observed. This P value is an estimate of the true P value. The accuracy of the estimate depends on how many times steps 1 and 2 were repeated. Since the P value is a proportion, you may calculate a confidence interval as explained in Chapter 2.

Whereas the exact randomization test systematically examines all possible rearrangements, the approximate randomization test examines a random selection of the possibilities.

THE RELATIONSHIP BETWEEN THE RANDOMIZATION TEST AND THE t TEST

At first glance, the P value from the randomization test seems to answer a very different question than the P value from the t test.

To interpret the P value from the t test, you must imagine a Gaussian population from which you select many different samples. The P value is the fraction of those

---

*It is possible that you may randomly shuffle the values the same way on two different iterations. That's OK.

samples in which the means are as far apart (or more so) than you observed with the actual data. You don't actually have to generate all those samples, as it can all be done by mathematical calculations.

To interpret the P value from the randomization test, you don't have to think about sampling different data from the population. Instead, you only deal with the actual values you obtained in the experiment. Rather than hypothesizing about other data that might have been obtained from the theoretical population, you hypothesize about how these particular values could have been shuffled between the two groups. You (or your computer) actually have to shuffle the data; there is no mathematical shortcut. The P value is the fraction of the rearrangements that have means as far apart or more so than actually observed.

Even though it sounds like the P values from the two methods answer different questions, the two are usually similar, especially with large samples.

The randomization test is rarely used, and you may never encounter a paper that analyzes data using it. The randomization test (with more than a tiny sample) requires a computer, and computers weren't cheap until recently. And even with a computer, randomization tests can be slow. For these reasons, the t test is far more popular than the randomization test. If you find the logic of a randomization test to be easier to follow than the logic of the t test, you can consider the P value from a t test to be an approximation for the P value from the randomization test.

MANN-WHITNEY TEST

The Mann-Whitney test is similar to the randomization tests, except that all calculations are done on the ranks rather than the actual values. In calculating the Mann-Whitney test, the first step is to rank the data points without paying attention to which are in group A and which are in group B. In the example, all the blood pressure values are different (there are no ties), so ranking is straightforward. If there were ties, each of the tied values would receive the average of the two or more ranks for which they tie. In the BP example, the ranks look like Table 24.2.

Next, add up the ranks of the values in each group. Because the values in group B tend to be larger, these values have higher ranks, resulting in a larger sum of ranks. The sum of the ranks in group A is 18; the sum of the ranks of group B is 37.

We need to calculate the answer to the following question: If the distribution of ranks between groups A and B were distributed randomly, what is the probability that the difference between the sum of ranks would be so large? This question can be answered in a manner similar to the exact randomization test discussed in the previous section. We know that there are 252 ways to arrange 10 ranks between 2 groups (keeping 5 in each group). In how many of those arrangements is the sum of ranks in group A 18 or smaller and the sum of ranks in group B 37 or larger? Keeping track

Table 24.2. Ranks in Blood Pressure Example

Group A values	120	80	90	110	95
Group A ranks	7	1	2	5	3
Group B values	105	130	145	125	115
Group B ranks	4	9	10	8	6

Table 24.3. Arrangements Where the Sum of Ranks Are as Far Apart as Actually Observed

Group	Ranks	Sum of Ranks
A	1, 2, 3, 5, 7	18 (actual)
B	4, 6, 8, 9, 10	37
A	1, 2, 3, 5, 6	17
B	4, 7, 8, 9, 10	38
A	1, 2, 3, 4, 5	15
B	6, 7, 8, 9, 10	40
A	1, 2, 3, 4, 7	17
B	5, 6, 8, 9, 10	38
A	1, 2, 4, 5, 6	18
B	3, 7, 8, 9, 10	37
A	1, 2, 3, 4, 6	16
B	5, 7, 8, 9, 10	39
A	1, 2, 3, 4, 8	18
B	5, 6, 7, 9, 10	37

of the details is tedious, but it takes only a few minutes work to figure out that the answer is seven. The possibilities are shown in Table 24.3. Since the order of the values within each group is not important, we have arranged each group from low rank to high rank (Table 24.3).

There are three ways to get ranks sums of 18 and 37, and four ways to get the difference of rank sums to be even more different (17 and 38, 16 and 39, 15 and 40). There are 245 other ways to arrange the values so that the difference in ranks is less extreme or in the opposite direction. The one-tailed P value is, therefore, 7/252 = 0.028. The two-tailed P value, which includes the possibility that the rank difference is equally large but with group A, is twice the one-tailed value or 0.056.

Because the t test uses additional information (or rather an additional assumption) it is more powerful than the Mann-Whitney test when the assumptions of the t test are true. With large samples, the difference in power is trivial. With smaller samples, the difference is more pronounced. If you have seven or fewer data points (total in both groups), the Mann-Whitney test can never report a two-tailed P value less than 0.05 no matter how different the groups are.

PERFORMING THE MANN-WHITNEY TEST*

Like most statistical tests, the calculations are best left to computers. If you wish to calculate the test yourself, follow these steps. Although the logic is simple, there are two different ways to calculate the test. The method actually described by Mann and Whitney involves calculating a variable termed U. An alternative, but equivalent,

*This section contains the equations you need to calculate statistics yourself. You may skip it without loss of continuity.

method described by Wilcoxon involves calculating a variable termed T. Each method uses distinct set of tables. Both are presented here to accommodate whatever set of tables you find. Because equivalent tests were developed by Wilcoxon and Mann and Whitney, you will see this test referred to by either name. Wilcoxon also developed a nonparametric test for paired data that we will discuss in the next section; the tables needed for the two tests are quite different, and it is easy to mix them up.

1. Rank all values in both groups. When ranking, don't pay attention to which group a value is in. Give the smallest value in either value a rank of one, and give the largest value in either group a rank equal to the total number of data points in both groups. If two or more values are identical, assign each the average of the ranks for which they tie. Thus if two values tied for the fifth and sixth ranks, assign each a rank of 5.5. If three values tied for ranks 11, 12, and 13, assign each a rank of 12. Rank according to the actual values, not the absolute values. Thus negative numbers always have lower ranks than positive numbers.
2. Sum the ranks in each group. Call the sum of ranks T_a and T_b. Look in an appropriate table to find the P value. Although this table is usually labeled for the "Mann-Whitney" or "rank sum" tests, it is sometimes labeled for the "Wilcoxon rank sum test," which must be distinguished from the "Wilcoxon signed rank test" (used for paired data and described in the next chapter). You will need to find the appropriate spot in the table to account for the number of values in each group. This book does not include tables for the Mann-Whitney test.
3. Some tables give the P value for U instead of T. U is calculated as the smaller of these two terms (n_a and n_b are the numbers of data points in the two groups).

$$U = T_a - \frac{n_a(n_a + 1)}{2} \text{ or } U = T_b - \frac{n_b(n_b + 1)}{2}. \qquad (24.2)$$

Use an appropriate table to find the P value. You will need to find the appropriate spot in the table to account for the number of values in each group.
4. Tables are available for small numbers of subjects only. If the number of data points is large (more than about 10 per group), you may use the following method to approximate a P value. After calculating U, use Equation 24.3 to calculate z:

$$z = \frac{|U - n_a n_b/2|}{\sqrt{n_a n_b (n_a + n_b + 1)/12}}. \qquad (24.3)$$

Use Table A5.2 in the Appendix to determine a P value from z, where z is from a standard Gaussian distribution. This equation yields an approximate P value; the larger the sample size, the better this approximation. If there were many ties in the data, the denominator of this equation should be adjusted slightly. In this case, refer to a more advanced book or use a computer program.

ASSUMPTIONS OF THE MANN-WHITNEY TEST

The Mann-Whitney test is based on these assumptions:

• The samples are randomly selected from, or at least representative of, the larger populations.

- The two samples were obtained independently. If the subjects were matched, or if the two samples represent before and after measurements, then the Wilcoxon test should be used instead (see next chapter).
- The observations within each sample were obtained independently. Choosing any one member of the population does not alter the chance of sampling any one else.
- The two populations don't have to follow any particular distribution, but the two distributions must have the same shape.

WHEN TO USE NONPARAMETRIC TESTS

It is not always easy to decide when it is appropriate to use a nonparametric test. This topic is discussed at length in Chapter 37 (pages 297–300).

PROBLEMS

1. The data compare the number of beta-adrenergic receptors on lymphocytes of control subjects and those taking a drug.

Control	Drug
1162	892
1095	903
1327	1164
1261	1002
1103	961
1235	875

A. Calculate the Mann-Whitney test.
B. Calculate the t test.

25

Comparing Two Paired Groups: Paired t and Wilcoxon Tests

WHEN TO USE SPECIAL TESTS FOR PAIRED DATA

Often experiments are designed so that the same patients or experimental preparations are measured before and after an intervention. If you used the unpaired t test or the Mann-Whitney test with such data, you would ignore part of the experimental design. You would be throwing away information, which is never a good thing to do. The problem is that the unpaired tests do not distinguish variation among subjects from variation due to differences between groups. When subjects are matched or paired, you should use a special paired test instead.

Paired analyses are appropriate in these types of protocols:

- You measure a variable in each subject before and after an intervention.
- You recruit subjects as pairs, matched for variables such as age, postal code, or diagnosis. One of the pair receives an intervention; the other does not (or receives an alternative treatment).
- You measure a variable in twins or child/parent pairs.
- You run a laboratory experiment several times, each time with a control and treated preparation handled in parallel.

More generally, you should use a paired test whenever the value of one subject in the first group is expected to be more similar to a particular subject in the second group than to a random subject in the second group. When the experimental design incorporates pairing, you should use a statistical test that takes pairing into account. Of course the pairing has to be done as part of the protocol, before the results are collected. The decision about pairing is a question of experimental design and should be made long before the data are analyzed.

CALCULATING THE PAIRED t TEST

To calculate the paired t test, follow these steps:

1. For each subject, calculate the change in the variable (keeping track of sign). If you have matched pairs, rather than before and after measurements, then calculate the difference between the two.

2. Calculate the mean and SE of these differences.
3. Calculate t

$$t = \frac{\text{Mean differences}}{\text{SE of differences}}. \tag{25.1}$$

Note that the denominator is the standard error (SE) of the differences. To calculate this value, list all the differences (calculated for each pair) and calculate the standard error of the mean (SEM). The SE of the difference for the paired t test is *not* the same as the SE of the difference for the unpaired test. If the pairing is effective, this SE of the difference will be smaller in the paired test.
4. Determine the P value. The degrees of freedom equal the number of pairs minus one. A paired t test has half as many degrees of freedom as the corresponding unpaired test.
5. Calculate a 95% confidence interval (CI) for the mean differences between pairs:

$$\text{Mean difference} \pm t^* \cdot \text{SE of differences}. \tag{25.2}$$

When calculating the CI, remember to use the critical value of t* from a table. Don't use the value of t calculated for these data calculated in Step 3 above. The number of degrees of freedom equals the number of subjects (or number of pairs) minus 1.

To continue the blood pressure example, let's assume that we followed five students and measured their blood pressure in both the first and second years and collected the data shown in Table 25.1. The first column is the same as before; the second column is different.

Each row represents a single student whose blood pressure was measured twice, a year apart. The last column shows the increase in blood pressure for each student. The mean increase is 8, and the SEM of the increase is 3.74. The t ratio, therefore, equals 2.14. Because there are five subjects, the number of df is four. The two-tailed P value is 0.0993; the 95% CI for the change in blood pressure ranges from −2.3 to 18.4. With so few subjects, the CI is quite wide.

If these data had been analyzed by an unpaired two-sample t test, the two-tailed P value would have been 0.370. By making repeated measurements on the same subject and using this information in the analysis, the experiment is more powerful and generates a smaller P value and a narrower CI.

Table 25.1. Blood Pressure Data for Five Students

	MS1	MS2	Change
Student A	120	115	−5
Student B	80	95	15
Student C	90	105	15
Student D	110	120	10
Student E	95	100	5

ASSUMPTIONS OF PAIRED t TEST

- The pairs must be randomly selected from, or at least representative of, a larger population.
- The samples must be paired or matched. The matching between values in column A and those in column B must be based on experimental design and decided before the data are collected. It is not OK to ''pair'' data points after they have been collected.
- Each pair must be selected independently of the others.
- The distribution of differences in the population must approximate a Gaussian distribution.

"RATIO" t TESTS*

Paired t tests are usually calculated, as described above, by calculating the difference between pairs (by subtraction). This method tests an alternative hypothesis that the intervention always causes the same average *absolute* difference, regardless of the starting value. With many experimental protocols, the intervention will cause the same average *relative* difference. In these kinds of experiments, you would expect that the size of the absolute difference would vary with the starting value, but the relative change (as a ratio compared to the starting value) would be constant.

One thought would be to express all the data as a ratio, and then ask whether the mean ratio differs significantly from 1.0 (using a one-sample t test, as described in the next section). The problem with this method is that the ratio is intrinsically an asymmetrical measure. All possibilities in which the variable decreases are expressed in ratios between 0 and 1, and all possibilities where the variable increases are expressed in ratios greater than 1. The logarithm of the ratio is a more symmetrical measure, and the distribution of the logarithms is more likely to be Gaussian. The logarithm of the ratio is equal to the difference of logarithms:

$$\log\left(\frac{after}{before}\right) = \log(after) - \log(before). \tag{25.3}$$

Here is an easy way to analyze paired data in which you think that the relative difference is a more consistent measure than the absolute difference. Don't calculate a ratio. Simply convert all values (before values and after values) to logarithms (either natural logarithms or logarithms base 10), and then perform a paired t test on the logarithms. In addition to the P value, the paired test will also yield the 95% CI for the difference between mean logarithms. Take the antilogarithm of both confidence limits to express the CI in terms of ratios.

The decision as to whether to calculate the paired t test on actual values (test for absolute differences) or on the logarithm of the values (test for relative differences) is a scientific decision that should be based on the kinds of data you are analyzing. Ideally

---

*This section is more advanced than the rest. You may skip it without loss of continuity.

Table 25.2. Enzyme
Activity in Control
Versus Treated Cells

Control	Treated
24	52
6	11
16	28
5	8
2	4

this decision should be made before the data are collected. The decision can have a big impact, as shown in Table 25.2.

These data (which are not real) were collected to test whether treating cultured cells with a drug increases the activity of an enzyme. Five different clones of the cell were tested. With each clone, control and treated cells were tested side by side. In all five clones, the treatment increased the activity of the enzyme. An unpaired t test would be inappropriate, because the experiment was paired or matched. A paired t test yields a t of 2.07 and a P value (two tailed) of .107. Why is there such a high P value with such striking results? Recall that t is the ratio of the mean change divided by the SEM of the differences. In this example the SE is large, as the differences range from 2 to 28. It makes scientific sense in this example to think that the treatment will increase the enzyme activity by a certain percentage, rather than a certain amount. If the cells happened to have a high control enzyme activity one day, you'd expect the drug to have a bigger absolute effect but the same relative effect. Therefore, we take the logarithm of all values and then perform the paired t test on the logarithms. The t ratio is 11.81 and the P value is 0.0003. The 95% CI for the difference of logarithms (base 10) is 0.21 to 0.33. Expressed this way, the CI is hard to interpret. Convert both ends of the CI to antilogarithms, and the 95% CI for the ratio (treated/control) is 1.62 to 2.14. The doubling in activity with treatment is very unlikely to be a coincidence.

THE WILCOXON SIGNED RANK SUM TEST

The nonparametric method to compare two paired groups is termed the *Wilcoxon Signed Rank Sum test*. It is used in exactly the same situation as the paired t test discussed earlier.

As usual, the details are best left to a computer program but are presented here for reference:

1. For each pair, calculate the difference. Keep track of the sign. A decrease is a negative number, an increase is positive.
2. Temporarily ignoring the sign, rank the absolute value of the differences. If any differences are equal to 0, ignore them entirely. Rank the remaining differences (the smallest difference is number 1).
3. Add up the ranks of all positive differences and of all negative differences. These sums are labeled T.

4. Using an appropriate table, find the P value corresponding to the values of T. Most tables use the smaller value of T. When using the table you will need to use N, the number of pairs (excluding pairs whose difference equals 0). If the two values of T are far apart, the P value will be small.

For the blood pressure example, there are five data pairs, T = 1.5 and 13.5, and the two-tailed P value is 0.1250. If you have five or fewer pairs of data points, the Wilcoxon test can never report a two-tailed P value less than 0.05, no matter how different the groups.

The Wilcoxon Signed Rank Sum test depends on these assumptions:

- The pairs must be randomly selected from, or at least representative of, a larger population.
- The samples must be paired or matched. The matching between values in Column A and those in Column B must be based on experimental design and decided before the data are collected. It is not OK to "pair" data points after they have been collected.
- Each pair must be selected independently of the others.

PROBLEMS

1. These data are the same as those presented in Problem 1 from Chapter 24. The experimenters compared the number of beta-adrenergic receptors on lymphocytes of control subjects and those taking a drug. Assume that each subject was first measured as a control and then while taking the drug.

Control	Drug
1162	892
1095	903
1327	1164
1261	1002
1103	961
1235	875

A. Calculate the paired t test.
B. Calculate the Wilcoxon test.

26

Comparing Observed and Expected Counts

ANALYZING COUNTED DATA

Assume that an average of 10% of patients die during or immediately following a certain risky operation. But last month 16 of 75 patients died. You want to know whether the increase reflects a real change or whether it is just a coincidence. Statistical calculations cannot answer that question definitively, but they can answer a related one: If the probability of dying has remained at 10%, what is the probability of observing 16 or more deaths out of 75 patients? If the probability of dying remained at 10%, we would expect 10% × 75 = 7.5 deaths in an average sample of 75 patients. But in a particular sample of 75 patients, we might see more or less than the expected numbers. The data can be summarized as given in Table 26.1.

Note that this table is not a contingency table. In a contingency table, the columns must be alternative categories (i.e., male/female or alive/dead). Here the table compares observed and expected counts, which are not alternative outcomes.

The null hypothesis is that the observed data are sampled from populations with the expected frequencies. We need to combine together the discrepancies between observed and expected, and then calculate a P value answering this question: If the null hypothesis were true, what is the chance of randomly selecting subjects with this large a discrepancy between observed and expected counts?

We can combine the observed and expected counts into a variable χ^2 (pronounced ki square). Chi-square is calulated as

$$\chi^2 = \sum \frac{(\text{observed} - \text{expected})^2}{\text{expected}}. \tag{26.1}$$

In this example,

$$\chi^2 = \frac{(59 - 67.5)^2}{67.5} + \frac{(16 - 7.5)^2}{7.5} = 10.70. \tag{26.2}$$

Statisticians have derived the distribution of chi-square when the null hypothesis is true. As the discrepancies between observed and expected values increase, the value of chi-square becomes larger, and the resulting P value becomes smaller. The value of chi-square also tends to increase with the number of categories, so you must account for the number of categories when obtaining a P value from chi-square. This is expressed

Table 26.1. Data for Counted Data
Example

	# Observed	# Expected
Alive	59	67.5
Dead	16	7.5
Total	75	75

as the number of degrees of freedom. In this example there is only one degree of freedom. This makes sense. Once you know the total number of patients and the number who died, you automatically know the number who lived. The general rule is that the number of degrees of freedom equals the number of categories minus one.

Using a computer program or Table A5.7 we can determine that the P value is 0.0011. Since P is so low, we suspect that the null hypothesis is not true and that some factor (beyond chance) is responsible for the increase in death rate. Chi-square is also used to analyze contingency tables, as discussed in the next chapter.

THE YATES' CONTINUITY CORRECTION

Although the observed counts, by definition, must be integers, the expected counts in this case appropriately include a fractional component. Therefore, it is impossible for chi-square to exactly equal 0, which makes it impossible for the P value to exactly equal 1.0. In fact, the P value calculated from chi-square will always be a bit too small. This is part of a larger problem of trying to map the discrete probability distribution of counts to the continuous chi-square distributions. One solution is to avoid the continuous distributions and to calculate exact probabilities using more appropriate (and somewhat more difficult) methods based on the binomial distribution. Another solution is to try to correct the chi-square calculations to give a more appropriate P value.

The correction is termed the *continuity* or *Yates' correction*. With the correction, the formula becomes:

$$\chi^2 = \Sigma \frac{(|observed - expected| - 0.5)^2}{expected}. \tag{26.3}$$

In this example, the resulting chi-squared value is 9.48, which leads to a P value of 0.0021. Unfortunately, the revised equation overcorrects, so the value of chi-square is too low and the P value is too high. Because it is usually safer to overestimate the P value than to underestimate it, the correction is termed *conservative,* and many statisticians advise using the correction routinely. Others advise never using it. The continuity correction is only used when there are one or two categories.

The correction makes little difference when the numbers are reasonably large. In the example, there is little difference between P = 0.0011 and P = 0.0021. When the numbers are smaller, the correction makes more difference. When the numbers are very small (expected values less than five), the chi-square method does not work well at all, either with or without the correction, and you should use other methods (based on the binomial distribution, not detailed in this book) to calculate the P value.

WHERE DOES THE EQUATION COME FROM?*

To understand the basis of the chi-square equation, it helps to see it in this form:

$$\chi^2 = \Sigma \left(\frac{\text{observed} - \text{expected}}{\sqrt{\text{expected}}} \right)^2. \qquad (26.4)$$

The numerator is the discrepancy between the expected and observed counts. The denominator is the standard error of the expected number. As you'll learn in Chapter 28, counted data distribute according to the Poisson distribution, and the standard error of the number of expected counts approximately equals the square root of the expected number.

PROBLEMS

1. You hypothesize that a disease is inherited as an autosomal dominant trait. That means that you expect that, on average, half of the children of people with the disease will have the disease and half will not. As a preliminary test of this hypothesis, we obtain data from 40 children of patients and find that only 14 have the disease. Is this discrepancy enough to make us reject the hypothesis?

---

*This section is more advanced than the rest. You may skip it without loss of continuity.

Comparing Two Proportions

At the beginning of Chapter 8 you learned about the four kinds of experimental designs whose results can be expressed as two proportions. Chapters 8 and 9 showed you how to calculate the 95% confidence interval for the difference or ratio of those proportions. This chapter teaches you how to calculate a P value testing the null hypothesis that the two proportions are equal in the overall population. There are two tests you can use, Fisher's exact test or the Chi-square test.

FISHER'S EXACT TEST

Although the details are messy, the idea of Fisher's exact test is fairly simple and is similar to the exact randomization test you learned about in Chapter 24. Without changing the row or column totals in the contingency table, construct all possible tables. Fisher's test calculates the probability that each of these tables would have been obtained by chance. The two-sided P value is the sum of the probability of obtaining the observed table plus the probabilities of obtaining all other hypothetical tables that are even less likely.* The calculations are messy, and you shouldn't try to do them yourself. Tables are available, but are fairly tricky to use. Calculate Fisher's test with a computer program. You already learned how to interpret the P value in Chapter 10.

CHI-SQUARE TEST FOR 2 × 2 CONTINGENCY TABLES

Although the chi-square test is not as accurate as Fisher's test, it is far easier to calculate by hand and is easier to understand. Recall Example 8.1, which examined disease progression in subjects taking AZT or placebo. The results are shown again in Table 27.1.

First we must predict the expected data had the null hypothesis been exactly true (with no random variation). You can calculate these *expected* results solely from the data in the contingency table. One's expectations from previous data, theory, or hunch are not used when analyzing contingency tables.

---

*Because of the discrete and asymmetrical nature of the exact distribution, the one-sided P value calculated by Fisher's test does not always equal exactly half the two-sided P value, but is usually is close. In rare cases, the one- and two-sided P values can be identical.

Table 27.1. Example 8.1 Observed Data

	Disease Progressed	No Progression	Total
AZT	76	399	475
Placebo	129	332	461
Total	205	731	936

Calculating the expected values is quite easy. For example, how many of the AZT treated subjects would you expect to have disease progression if the null hypothesis were true? Combining both groups (looking at column totals), the disease progressed in 205/936 = 21.9% of the subjects. If the null hypothesis were true, we'd also expect to observe disease progression in 21.9% of the AZT treated subjects. That means we expect to see disease progression in 0.219 × 475 = 103.9 of the AZT treated subjects.

The rest of the table can be completed by similar logic or (easier) by subtraction (since you know the row and column totals). The expected data are given in Table 27.2.

Of course we did not observe the expected values exactly. One reason is that the expected values include fractions. That's OK, as these values are the average of what you'd expect to see if you repeated the experiment many times. The other reason we did not observe the expected values exactly is random sampling. You expect to see different results in different experiments, even if the null hypothesis is true.

To quantify how far our results were from the expected results, we need to pool together all the discrepancies between observed and expected values. As discussed in the previous chapter, the discrepancies between observed and expected values are pooled by calculating a single number, χ^2 chi-square. The larger the discrepancy between observed and expected values, the larger the value of χ^2. The details of the calculations are presented in the next section. For this example, $\chi^2 = 18.9$.

Statisticians have computed the probability distribution of chi-square under the null hypothesis. This distribution has been programmed into computers and tabulated, allowing you to figure out the P value. A large χ^2 value corresponds to a small P value. For this example, the P value is very low, less than 0.0001. If the null hypothesis were true, there is less than a 0.01% chance of observing such a large discrepancy between observed and expected counts.

HOW TO CALCULATE THE CHI-SQUARE TEST FOR A 2 × 2 CONTINGENCY TABLE*

1. Create a table of expected values. The expected probability that an observation will be in a certain cell is equal to the probability that an observation will be in the correct row times the probability that it will be in the correct column. Multiply the expected probability times the grand total to calculate the expected count. Turn those sentences into equations and simplify to Equation 27.1:

---

*This section contains the equations you need to calculate statistics yourself. You may skip it without loss of continuity.

Table 27.2. Example 8.1 Expected Data

	Disease Progressed	No Progression	Total
AZT	103.9	371.1	475
Placebo	101.1	359.9	461
Total	205	731	936

$$\text{Expected count} = \frac{\text{row total}}{\text{grand total}} \cdot \frac{\text{column total}}{\text{grand total}} \cdot \text{grand total} \qquad (27.1)$$
$$= \frac{\text{row total} \cdot \text{column total}}{\text{grand total}}.$$

2. Calculate chi-square using Equation 27.2:

$$\chi^2 = \sum_{\text{all cells}} \frac{(|\text{observed} - \text{expected}| - [0.5])^2}{\text{expected}}. \qquad (27.2)$$

Calculate the sum for all four cells in the contingency table.

The term in brackets [0.5] is the Yates' continuity correction, already discussed in the last chapter. The result of Equation 27.2 does not follow the chi-square distribution exactly, and the Yates' correction is an attempt to make it come closer. Some statisticians recommend including this correction (for 2×2 tables), but others suggest leaving it out. Either way, this test only calculates an approximate P value.

3. Look up the P value in Table A5.7 in the Appendix. The larger the value of chi-square, the smaller the P value. There is one degree of freedom. The resulting P value is two-sided. For a one-sided P value, halve the two-sided P value.

ASSUMPTIONS

Both chi-square and Fisher's test are based on these assumptions:

- *Random sampling.* The data must be randomly selected from, or at least representative of, a larger population.
- *The data must form a contingency table.* The values must be the *number* of subjects actually observed, not percentages or rates. The categories that define the rows must be mutually exclusive. The categories that define the columns must also be mutually exclusive. Many 2×2 tables are not contingency tables, and analyses of such data by chi-square or Fisher's test would not be meaningful.
- *For the chi-square test, the values must not be too small.* Don't use the chi-square test if the total number of subjects is less than 20 or any expected value is less than 5. Fisher's test can be used for any number of subjects, although most programs won't calculate Fisher's test with huge numbers (thousands) of subjects.
- *Independent observations.* Each subject must be independently selected from the population.
- *Independent samples.* No pairing or matching. If you have paired data, see the discussion of McNemar's test in Chapter 29.

CHOOSING BETWEEN CHI-SQUARE AND FISHER'S TEST

You may obtain a P value from a 2×2 contingency table using either the chi-square test or Fisher's exact test. Which should you use? The answer depends on the size of your sample.

Small samples. Use Fisher's test. How small is small? Any cutoff is arbitrary but a reasonable rule is this: The sample size is too small for the chi-square test if the total sample size is less than about 20 to 40, or the expected number of subjects in any cell is less than 5. Note that the criteria applies to the expected numbers, not the observed numbers.

Moderate-sized samples. If you have fewer than several hundred subjects, then you should select Fisher's test if you are using a computer program. If you don't have access to a program that performs Fisher's test, you can get by using the chi-square test instead. If you calculate the chi-square test, you must decide whether to use Yates' correction. Different statisticians give different recommendations, and there is no solid consensus. In either case (with or without the Yates' correction), the chi-square test gives an approximate P value. Rather than deciding which estimate is better for which purpose, you are better off using a program that calculates the exact P value with Fisher's test.

Large samples. If your sample size is huge (thousands of subjects), you should use the chi-square test. With such large samples, Fisher's test is slow to calculate, may lead to calculation problems (computer overflow), and is no more accurate than the chi-square test. With large samples, it doesn't matter whether or not you use Yates' correction, as the values of chi-square and the P value will be almost identical either way.

Chi-square and Fisher's test are alternative methods of calculation. The data going into the two tests are exactly the same (a 2×2 contingency table), and the resulting P value is intended to have the same meaning. Fisher's test always gives a theoretically correct P value, but calculating a Fisher's test without a computer is tedious, even with tables. In contrast, the chi-sauare test is easy to calculate, but the resulting P value is only an approximation. When the number of subjects is sufficiently large, the approximation is useful. When the number of subjects is small, the approximation is invalid and leads to misleading results.

When you read the literature, you may run across a third way to calculate a P value, using the *z test*. Although the z test looks quite different than the chi-square test, the two tests are mathematically equivalent and calculate identical P values. Like the chi-square test, you must decide whether or not you wish to include Yates' continuity correction when calculating the z test.

CALCULATING POWER*

When interpreting the results of a study that reaches a conclusion of not significant, it is useful to calculate the power of that study. Recall the definition of power. If you

---

*This section contains the equations you need to calculate statistics yourself. You may skip it without loss of continuity.

hypothesize that the true proportions in the two populations are p_1 and p_2, power is the chance that you will obtain a significant difference between the proportions when you compare random samples of a certain size. In a prospective or experimental study p_1 and p_2 are the incidence rates in the two populations; in a retrospective study p_1 and p_2 are the fraction of cases and controls exposed to the risk factor; in a cross-sectional study p_1 and p_2 are the prevalance rates in the two populations.

The power of a study depends on what proportions p_1 and p_2 you hypothesize for the populations. All studies have tiny power if p_1 and p_2 are close together and enormous power if p_1 and p_2 are far apart. You might imagine that power is a function of the difference between the two hypothesized population proportions $(p_1 - p_2)$, or perhaps their ratio (p_1/p_2). It turns out, however, that power depends not only on the difference or ratio of p1 and p2, but also on their particular values. The power to detect a difference between $p_1 = 0.10$ and $p_2 = 0.20$ is not the same as the power to detect a difference between $p_1 = 0.20$ and $p_2 = 0.30$ (consistent difference), or $p_1 = 0.20$ and $p_2 = 0.40$ (consistent ratio)

There are several methods to calculate power. One method would be to rearrange Equation 22.6 (used to calculate sample size) so that it calculates power. However, this leads to some messy calculations. Instead, what follows is adapted from a book by Cohen* that contains extensive tables for determining power in many situations. To calculate power approximately, first calculate H using Equation 27.3:

$$H = 2 \cdot |arcsine \sqrt{p_1} - arcsine \sqrt{p_2}|. \tag{27.3}$$

It seems surprising to see arcsines show up in statistical equations, but it turns out that power is a consistent function of the difference of the arcsines of the square roots of the proportions.† The power of a study will be the same for any pair of values for p_1 and p_2 that yields the same value of H.

To save you the trouble of calculating the arcsine of square roots, I've prepared Tables 27.3 and 27.4. The tables are symmetrical, so it doesn't matter which of the two proportions you label p_1 and which you label p_2 (you'll get the same answer if you switch the two). The first table covers all values of p_1 and p_2, the second table gives more detail for rare events (smaller values of p_1 and p_2).

Now that you've calculated H, you're ready to calculate power from Equation 27.4:

$$z_{power} = H \sqrt{\frac{N}{2}} - z^*. \tag{27.4}$$

Set z^* to the critical value of z that corresponds to the α you have chosen. If you set α to its conventional value of 0.05, then $z^* = 1.96$. If you set $\alpha = 0.10$, then $z^* = 1.65$. If you set $\alpha = 0.01$, then $z^* = 2.58$. All three of these values assume that you will be calculating two-tailed P values. N in the equation is the number of subjects in *each* group, so the total number of subjects is 2 * N.

If the two groups have different number of subjects, then calculate N as the harmonic mean of N_1 and N_2 (the numbers of subjects in each group). The harmonic

---

*Jacob Cohen. *Statistical Power Analysis for the Behavioral Sciences,* 2nd ed. Lawrence Erlbaum, Hillsdale, New Jersey, 1988.

†The arcsine is the inverse of the sin. Arcsin(x) is the angle (in radians) whose sine equals x.

Table 27.3. Values of H Used to Calculate Power

p_1										p_2									
	0.05	0.10	0.15	0.20	0.25	0.30	0.35	0.40	0.45	0.50	0.55	0.60	0.65	0.70	0.75	0.80	0.85	0.90	0.95
0.00	0.00	0.19	0.34	0.48	0.60	0.71	0.82	0.92	1.02	1.12	1.22	1.32	1.42	1.53	1.64	1.76	1.90	2.05	2.24
0.10	0.19	0.00	0.15	0.28	0.40	0.52	0.62	0.73	0.83	0.93	1.03	1.13	1.23	1.34	1.45	1.57	1.70	1.85	2.05
0.15	0.34	0.15	0.00	0.13	0.25	0.36	0.47	0.57	0.68	0.78	0.88	0.98	1.08	1.19	1.30	1.42	1.55	1.70	1.90
0.20	0.48	0.28	0.13	0.00	0.12	0.23	0.34	0.44	0.54	0.64	0.74	0.84	0.95	1.06	1.17	1.29	1.42	1.57	1.76
0.25	0.60	0.40	0.25	0.12	0.00	0.11	0.22	0.32	0.42	0.52	0.62	0.72	0.83	0.94	1.05	1.17	1.30	1.45	1.64
0.30	0.71	0.52	0.36	0.23	0.11	0.00	0.11	0.21	0.31	0.41	0.51	0.61	0.72	0.82	0.94	1.06	1.19	1.34	1.53
0.35	0.82	0.62	0.47	0.34	0.22	0.11	0.00	0.10	0.20	0.30	0.40	0.51	0.61	0.72	0.83	0.95	1.08	1.23	1.42
0.40	0.92	0.73	0.57	0.44	0.32	0.21	0.10	0.00	0.10	0.20	0.30	0.40	0.51	0.61	0.72	0.84	0.98	1.13	1.32
0.45	1.02	0.83	0.68	0.54	0.42	0.31	0.20	0.10	0.00	0.10	0.20	0.30	0.40	0.51	0.62	0.74	0.88	1.03	1.22
0.50	1.12	0.93	0.78	0.64	0.52	0.41	0.30	0.20	0.10	0.00	0.10	0.20	0.30	0.41	0.52	0.64	0.78	0.93	1.12
0.55	1.22	1.03	0.88	0.74	0.62	0.51	0.40	0.30	0.20	0.10	0.00	0.10	0.20	0.31	0.42	0.54	0.68	0.83	1.02
0.60	1.32	1.13	0.98	0.84	0.72	0.61	0.51	0.40	0.30	0.20	0.10	0.00	0.10	0.21	0.32	0.44	0.57	0.73	0.92
0.65	1.42	1.23	1.08	0.95	0.83	0.72	0.61	0.51	0.40	0.30	0.20	0.10	0.00	0.11	0.22	0.34	0.47	0.62	0.82
0.70	1.53	1.34	1.19	1.06	0.94	0.82	0.72	0.61	0.51	0.41	0.31	0.21	0.11	0.00	0.11	0.23	0.36	0.52	0.71
0.75	1.64	1.45	1.30	1.17	1.05	0.94	0.83	0.72	0.62	0.52	0.42	0.32	0.22	0.11	0.00	0.12	0.25	0.40	0.60
0.80	1.76	1.57	1.42	1.29	1.17	1.06	0.95	0.84	0.74	0.64	0.54	0.44	0.34	0.23	0.12	0.00	0.13	0.28	0.48
0.85	1.90	1.70	1.55	1.42	1.30	1.19	1.08	0.98	0.88	0.78	0.68	0.57	0.47	0.36	0.25	0.13	0.00	0.15	0.34
0.90	2.05	1.85	1.70	1.57	1.45	1.34	1.23	1.13	1.03	0.93	0.83	0.73	0.62	0.52	0.40	0.28	0.15	0.00	0.19
0.95	2.24	2.05	1.90	1.76	1.64	1.53	1.42	1.32	1.22	1.12	1.02	0.92	0.82	0.71	0.60	0.48	0.34	0.19	0.00

Table 27.4. Values of H Used to Calculate Power (small values of p_1 and p_2)

p_1	p_2											
	0.01	0.02	0.03	0.04	0.05	0.06	0.07	0.08	0.09	0.10	0.11	0.12
0.01	0.000	0.083	0.148	0.202	0.251	0.295	0.335	0.373	0.409	0.442	0.476	0.507
0.02	0.083	0.000	0.064	0.119	0.167	0.211	0.252	0.290	0.326	0.360	0.392	0.424
0.03	0.148	0.064	0.000	0.055	0.103	0.147	0.187	0.225	0.261	0.295	0.328	0.359
0.04	0.202	0.119	0.055	0.000	0.048	0.092	0.133	0.171	0.207	0.241	0.273	0.305
0.05	0.251	0.167	0.103	0.048	0.000	0.044	0.084	0.122	0.158	0.192	0.225	0.256
0.06	0.295	0.211	0.147	0.092	0.044	0.000	0.041	0.079	0.114	0.149	0.181	0.213
0.07	0.335	0.252	0.187	0.133	0.084	0.041	0.000	0.038	0.074	0.108	0.141	0.172
0.08	0.373	0.290	0.225	0.171	0.122	0.079	0.038	0.000	0.036	0.070	0.103	0.134
0.09	0.409	0.326	0.261	0.207	0.158	0.114	0.074	0.036	0.000	0.034	0.067	0.098
0.10	0.443	0.360	0.295	0.241	0.192	0.149	0.108	0.070	0.034	0.000	0.033	0.064
0.11	0.476	0.392	0.328	0.273	0.225	0.181	0.141	0.103	0.067	0.033	0.000	0.031
0.12	0.507	0.424	0.359	0.305	0.256	0.213	0.172	0.134	0.098	0.064	0.031	0.000
0.13	0.537	0.454	0.390	0.335	0.287	0.243	0.202	0.164	0.128	0.094	0.062	0.030
0.14	0.567	0.483	0.419	0.364	0.316	0.272	0.231	0.193	0.158	0.123	0.091	0.060
0.15	0.595	0.512	0.447	0.393	0.344	0.300	0.260	0.222	0.186	0.152	0.119	0.088
0.16	0.623	0.539	0.475	0.420	0.372	0.328	0.288	0.250	0.214	0.180	0.147	0.116
0.17	0.650	0.566	0.502	0.447	0.399	0.355	0.314	0.276	0.241	0.206	0.174	0.142
0.18	0.676	0.593	0.528	0.474	0.425	0.381	0.341	0.303	0.267	0.233	0.200	0.169
0.19	0.702	0.618	0.554	0.499	0.451	0.407	0.367	0.329	0.293	0.259	0.226	0.195

mean is the reciprocal of the mean of the reciprocals. Calculate N to use in Equation 27.4 by using Equation 27.5:

$$N = \frac{1}{\left(\dfrac{1}{2}\right)\left(\dfrac{1}{N_1} + \dfrac{1}{N_2}\right)} = \frac{2 \cdot N_1 \cdot N_2}{N_1 + N_2}. \tag{27.5}$$

To convert z_{power} to power, you need to find the probability that a random number chosen from a standard normal distribution (mean $= 0$, SD $= 1$) would have a value less than z_{power}. If z_{power} equals 0, the power is 50%, because half of the normal distribution has negative values of z. Table A5.6 in the Appendix converts z_{power} to power.

Rather than calculate z_{power}, you can estimate power directly from Figure 27.1. Find H on the X axis. Read up until you intersect the line corresponding to N, the number of subjects in each group. Then read over to the left to find the power. The figure assumes that you are using the conventional value of $\alpha = 0.05$.

Example 12.1 (Again)

This study investigated the usefulness of routine ultrasound to prevent neonatal deaths and complications. The authors found that use of routine ultrasound did not significantly reduce neonatal complications or deaths compared to the use of ultrasound only for specific reasons. The rate of complications and deaths is about 5%. What would be

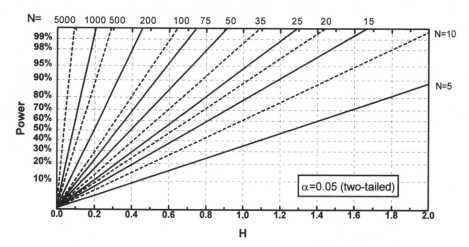

Figure 27.1. Determine the power of a study to detect differences between two proportions. First determine H from Table 27.3 or use this equation: $H = 2 * |arcsine(\sqrt{p_1}) - arcsine(\sqrt{p_2})|$. The variables p_1 and p_2 are the two proportions that you hypothesize exist in the overall populations. Find the value H on the X axis. Now read up to see where it intersects a line corresponding to the sample size. The variable N on this graph is the number of subjects in *each* group, assuming that the sample sizes are equal (or nearly so). Then read to the left to find the power. If the population proportions really are p_1 and p_2, and you perform a study with N subjects in each group, *power* is the probability that your study will find a statistically significant difference with $P < 0.05$.

the power of this study to detect a 20% reduction in the rate of complications and deaths (down to 4%)? Set $p_1 = 0.04$ and $p_2 = 0.05$. From the table, $H = 0.048$. Figure 27.1 is not useful with such a large sample size, so we'll use equation 27.4. Set $z^* = 1.96$, $H = 0.048$, and $N = 7600$. You can calculate that $z_{power} = 1.02$. From Table A5.6 in the Appendix, you can read off that the power is about 84%. If your hypothesis were true, there is an 84% chance that a study of this size will find a significant difference with $P < 0.05$, and a 16% chance that it will find no significant difference. Similarly, you can calculate the other values shown in Table 12.2.

Example 27.1

Among patients taking certain drug, 40% develop a certain side effect. You want to test a second drug that is supposed to prevent the side effects. If your experiment has 30 subjects in each group, what is its power to detect a relative reduction in the rate of side effects of 50% (down to 20%)? Your assumptions are that $p_1 = 0.40$ and $p_2 = 0.20$. From the table, $H = 0.44$. The graph does not have a line for $N = 30$ but has a line for $N = 25$ and $N = 35$. Read up Figure 27.1 at $H = 0.44$ and interpolate between the lines for $N = 25$ and $N = 35$. Read over to the Y axis. The power is about 40%. If your hypothesis is true, there is only about a 40% chance that your study will end up with a significant result, leaving a 60% chance that your study will incorrectly conclude that there is no significant difference.

Example 27.2

After a difficult operation, 10% of patients die. You read a paper that tested a new procedure in a randomized trial. The investigators compared mortality rates in 50 subjects given the standard operation and 50 given the new operation. The difference in mortality rates was not significant with $P > 0.05$. Before interpreting these negative results, you need to know the power of the study to detect a significant difference if the new procedure is really better. To calculate the power, you first need to decide how large a difference is worth detecting. Let's start by asking what power the study had to detect a reduction in mortality of 50%. Set $p_1 = 0.10$ (the mortality rate with the standard procedure) and $p_2 = 0.05$ (half that rate). Look up H in the table: $H = 0.192$. From the graph, you can see that the power is only about 17%. Even if the new procedure really does reduce mortality by 50%, a study with $N = 50$ only has only about a 17% chance of finding a statistically significant ($\alpha = 0.05$) difference in mortality rates. Since the study had poor power to detect an important reduction in mortality, you really can't conclude anything from the negative study. To increase the power will require more subjects.

PROBLEMS

1. Perform the chi-square test for the cat-scratch disease data for Example 9.1. State the null hypothesis, show the calculations, and interpret the results.

2. (Same as Problem 2 from Chapter 8.) Goran-Larsson et al. wondered whether hypermobile joints caused symptoms in musicians.* They sent questionnaires to many musicians and asked about hypermobility of the joints and about symptoms of pain and stiffness. They asked about all joints, but this problem concerns only the data they collected about the wrists. In contrast, 18% of musicians without hypermobility had such symptoms. Analyze these data as fully as possible.

3. Will a study with N = 50 have greater power to detect a difference between $p_1 = 0.10$ and $p_2 = 0.20$, or between $p_1 = 0.20$ and $p_2 = 0.40$?

4. Calculate a P value from the data in Problem 4 of Chapter 8. Interpret it.

5. In response to many case reports of connective tissue diseases after breast implants, the FDA called for a moratorium on breast implants in 1992. Gabriel and investigators did a prospective study to determine if there really was an association between breast implants and connective tissue (and other) diseases.† They studied 749 women who had received a breast implant and twice that many control subjects. They analyzed their data using survival analysis to account for different times between implant and disease and to correct for differences in age. You can analyze the key findings more simply. They found that 5 of the cases and 10 of the controls developed connective tissue disease.

A. What is the relative risk and P value? What is the null hypothesis?

B. What is the 95% CI for the relative risk?

C. If breast implants really doubled the risk of connective tissue disease, what is the power of a study this size to detect a statistically significant result with P < 0.05?

*L. Goran-Larsson, J Baum, GS Mudholkar, GD Lokkia. Benefits and disadvantages of joint hypermobility among musicians. *N Engl J Med* 329:1079–1082, 1993.

†SE Gabriel, WM O'Fallon, LT Kurland, CM Beard, JE Woods, LJ Melton. Risk of connective tissue diseases and other disorders after breast implantation. *N Engl J Med* 330:1697–1702, 1994.

VIII

INTRODUCTION TO
ADVANCED
STATISTICAL TESTS

With the information you've learned so far, you'll only be able to follow the statistical portions of about half of the articles in the medical literature. The rest of the articles use more advanced statistical tests. To learn these tests in depth, you will need to take an advanced statistics course or study an advanced text. The chapters in this section give you a general feel for how some of these tests are used, and this should be sufficient to interpret the results. You may read the chapters in any order without loss of continuity.

28

The Confidence Interval
of Counted Variables

So far we have discussed three kinds of variables: those expressed as measurements on a continuous scale (such as blood pressure or weight), those expressed as proportions (such as the proportion of patients that survive an operation or the proportion of children who are male), and survival times. This chapter deals with a fourth kind of outcome variable, those that are expressed as counts, such as the number of deliveries in an obstetrics ward each day, the number of eosinophils seen in one microscope field, or the number of radioactive disintegrations detected by a scintillation counter in a minute.

THE POISSON DISTRIBUTION

Variables that denote the number of occurrences of an event or object in a certain unit of time or space are distributed according to the *Poisson distribution,* named after a French mathematician. This distribution is based on the assumptions that the events occur randomly, independently of one another, and with an average rate that doesn't change over time. As usual, this equation works in the wrong direction, from population to sample. Given the average (population) number of occurrences in one unit of time (or space), the Poisson distribution predicts the frequency distribution for the number of events that will occur in each unit of time (or space).

Figure 28.1 shows an example. Assume that the time at which babies are born is entirely random, with no effect of season, phase of the moon, or time of day. In a certain very busy hospital an average of six babies are born per hour. That's our population. The X axis of Figure 28.1 shows various numbers of babies born per hour sampled, and the Y axis shows how often you would expect to see each possibility. It is most common to observe six babies born each hour, but there is a small chance that no babies will be born in a certain hour or that 15 babies will be born in that hour. Consult more advanced statistical books for equations and tables of the Poisson distribution.

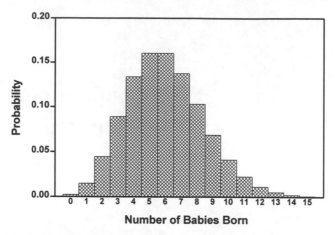

Figure 28.1. The Poisson distribution. If the average number of events in a time interval (for this example babies born per hour) equals 6, this figure shows the distribution you expect to see for number of events per time interval. From the first two bars, you can see that there is less than 1% chance of observing 0 events in a particular time interval, and about a 2% chance of observing a single event.

UNDERSTANDING THE CONFIDENCE INTERVAL OF A COUNT

When you analyze data, you have information about one sample, in this case the number of events per time or the number of objects per volume. By chance the number of events (or objects) that you counted may be higher or lower than the average number that occur in that unit time (or volume). You want to make inferences about the overall average number of events per time or the number of objects per volume.

Understanding the CI should be familiar to you by now. From the count in your sample, C, it is possible to calculate (Table 28.1) the 95% confidence interval (CI) for the average number of events per unit time (or objects per volume). You can be 95% sure that the overall average number of events per unit time (or objects per volume) lies within this CI.

You can obtain the 95% CI of a count from Table A5.9 in the Appendix (abridged in Table 28.1), or calculate an approximate CI as shown in the next section. For example, assume that you carefully dissect a bagel and find 10 raisins. Assuming that variability in raison count is entirely due to chance, you can be 95% sure that the average number of raisins per bagel is between 4.8 and 18.4. You can also say that you expect that about 95% of bagels will have between 5 and 18 raisins. Here is a more realistic example. You have counted 120 radioactive counts in one minute. The 95% CI for the average number of counts per minute is 99.5 to 143.5.

Your interpretation of the CI depends on the following assumptions:

• Events occur randomly.
• Each event occurs independent of other events.
• The average rate doesn't change over time.

Table 28.1. The Confidence
Interval of a Count

C	95% CI
0	0.00 to 3.69
1	0.03 to 5.57
2	0.24 to 7.22
3	0.62 to 8.77
4	1.09 to 10.24
5	1.62 to 11.67
6	2.20 to 13.06
7	2.81 to 14.42
8	3.45 to 15.76
9	4.12 to 17.08
10	4.40 to 18.39

CALCULATING THE CONFIDENCE INTERVAL OF A COUNT*

Look up the CI of a count in Table 28.1 or Table A5.9 in the Appendix. If C is the observed number of counts, then the table shows the 95% CI for the average number of counts in the same unit of time or space. When C is large (greater than 25 or so), you may calculate an approximate 95% CI using Equation 28.1. Note that you do not need to deal with the t distribution or the concept of degrees of freedom.

$$\text{Approximate 95\% CI of C: } (C - 1.96\sqrt{C}) \text{ to } (C + 1.96\sqrt{C}). \qquad (28.1)$$

If you count 120 radioactive counts, the approximate 95% CI is 98.5 to 141.5. The exact CI tabulated in the Appendix was calculated from a more elaborate equation derived from the Poisson distribution.

What if you have repeated the count several times? Assume that we have counted radioactivity in triplicate tubes for 1 minute and observe 175, 162, and 193 counts. How could you calculate the 95% CI for the average number of counts? You could use two approaches. One approach would be to treat the counts as measurements, and calculate the 95% CI using Equation 5.1, resulting in a 95% CI of 138.0 to 215.3. The problem with this approach is that it assumes that the counts follow a Gaussian distribution, which we know is not true. It is better to use an approach based on the Poisson distribution. To do this, first calculate the sum of all counts (530 counts in 3 minutes) and calculate the 95% CI of the sum from Equation 28.1. The result is a 95% CI of 484.9 to 575.1. This is the CI for the average number of counts in 3 minutes. Divide each confidence limit by 3 to obtain the 95% CI for the average number of counts in 1 minute: 161.6 to 191.7.

CHANGING SCALES

When calculating the CI of a count, it is essential that you base the calculations on the number of events (or objects) actually observed. Don't normalize to a more conve-

---

*This section is more advanced than the rest. You may skip it without loss of continuity.

nient time scale or volume until after you have calculated the CI. If you attempt to calculate a CI based on the normalized count, the results will be meaningless.

Example 28.1

You have placed a radioactive sample into a gamma counter and counted for 10 minutes. The counter tells you that there were 225 counts per minute. What is the 95% CI? Since you counted for 10 minutes, the instrument must have detected 2250 radioactive disintegrations. The 95% CI of this value extends from 2157 to 2343 counts per 10 minutes, so the 95% CI for the average number of counts per minute extends from 216 to 234.

Example 28.2

You read that exposure to an environmental toxin caused 1.6 deaths per 1000 person-years exposure. What is the 95% CI? To calculate the CI, you must know the exact number of deaths that were observed in the study. This study observed 16 deaths in observations of 10,000 person years (they might have studied 10,000 people for 1 year, or 500 people for 20 years). Using Table A5.9 in the Appendix to find that the 95% CI for the number of deaths ranges from 9.15 to 25.98. Dividing by the denominator, this means that the 95% CI for the death rate ranges from 0.92 to 2.6 deaths per 1000 person-years exposure.

OBJECTIVES

You should be able to identify variables that distribute according to the Poisson distribution. Using calculator and book, you should be able to calculate the 95% CI of the mean of a Poisson variable.

PROBLEMS

1. You use a hemocytometer to count white blood cells. When you look in the microscope you see lots of squares, and 25 of these enclose 0.1 microliter. You count the number of cells in nine squares and find 50 white blood cells. Can you calculate the 95% CI for the number of cells per microliter? If so, calculate the interval. If not, what information do you need? What assumptions must you make?
2. In 1988 a paper in *Nature* (333:816, 1988) caused a major stir in the popular and scientific press. The authors claimed that antibodies diluted with even 10^{-120} of the starting concentration stimulated basophils to degranulate. With so much dilution, the probability that even a single molecule of antibody remains in the tube is almost 0. The investigators hypothesized that the water somehow "remembered" that it had seen antibody. These results purported to give credence to homeopathy, the theory that extremely low concentrations of drugs are therapeutic.

    The assay is simple. Add a test solution to basophils and incubate. Then stain. Count the number of stained cells in a certain volume. Compare the number of cells in control and treated tubes. A low number indicates that many cells had degranulated, since cells that had degranulated won't take up the stain.

The authors present the "mean and standard error of basophil number actually counted in triplicate." In the first experiment, the three control tubes were reported as 81.3 ± 1.2, 81.6 ± 1.4, and 80.0 ± 1.5. A second experiment on another day gave very similar results. The tubes treated with the dilute antibody gave much lower numbers of stained cells, indicating that the dilute antibody had caused degranulation. For this problem, think only about the control tubes.

A. Why are these control values surprising?
B. How could these results have been obtained?

29

Further Analyses of Contingency Tables

McNEMAR'S CHI-SQUARE TEST FOR PAIRED OBSERVATIONS

In Chapter 9, you learned how to calculate an odds ratio (with its 95% confidence interval) from a matched case-control study. This section explains how to calculate a P value. We'll use the same example used at the end of Chapter 9. These data are shown in Table 29.1 as a contingency table. Analyzing this table with the chi-square test yields a P value of 0.015. This is not exactly invalid, but it is not the best way to analyze the data. Each control subject was matched to a particular case subject, but this matching was not considered in the analysis. Table 29.2 displays the same data in a way that emphasizes the matching.

Table 29.2 is not a contingency table, so the usual chi-square and Fisher's tests would be inappropriate. Each entry in the table represents a matched pair, which is why the total number of entries in this table is half that of Table 29.1. While it is possible to convert the paired table to the conventional table, it is not possible to convert a conventional table into a paired table without access to the original data.

Use *McNemar's test* to calculate a P value. This test uses only the numbers of discordant pairs, that is, the number of pairs for which the control has the risk factor but the case does not (in this case 4) and the number of pairs for which the case has the risk factor but the control does not (in this case 25). Call these two numbers R and S. The other pairs were either both exposed to the risk factor or both unexposed; these pairs contribute no information about the association between risk factor and disease.

To perform McNemar's test, calculate chi-square using Equation 29.1. (The -1 is the Yates' continuity correction, which may be omitted.)

$$\chi^2 = \frac{(|R - S| - 1)^2}{R + S}.$$ (29.1)

For our example, chi-square = 13.79 which has one degree of freedom. The two-tailed P value is 0.0002.

McNemar's test, like the chi-square test, is approximate and should not be used for small samples. If the sum of R + S is less than 10, you should use an exact test based on the binomial distribution (not discussed further in this book).

The results obtained with InStat are reported in Table 29.3. The program also reports the odds ratio, and its 95% CI as calculated using the method at the end of Chapter 9.

Table 29.1. Data for Paired Observations Expressed as a Contingency Table

	Risk Factor +	Risk Factor −	Total
Case	36	98	134
Control	19	115	134
Total	55	213	268

CHI-SQUARE TEST WITH LARGE TABLES (MORE THAN TWO ROWS OR COLUMNS)

Calculating chi-square with more than two rows or columns uses the same steps as used for a 2×2 table. Follow these steps:

1. Create a table of expected values. The expected probability that an observation will be in a certain cell is equal to the probability that an observation will be in the correct row times the probability that it will be in the correct column. Multiply the expected probability times the grand total to calculate the expected count. Turn those sentences into equations and simplify to Equation 29.2:

$$\text{Expected count} = \frac{\text{row total}}{\text{grand total}} \cdot \frac{\text{column total}}{\text{grand total}} \cdot \text{grand total} \qquad (29.2)$$

$$= \frac{\text{row total} \cdot \text{column total}}{\text{grand total}}.$$

Note that the expected values are calculated from the data, you shouldn't calculate the expected values from theory or hunch.

2. Calculate chi-square using Equation 29.3:

$$\chi^2 = \sum_{\text{all cells}} \frac{(\text{observed} - \text{expected})^2}{\text{expected}}. \qquad (29.3)$$

Calculate the sum for all cells in the contingency table. For a table with four rows and two columns, the sum will have eight components. Note that Yates' continuity correction is only used for tables with two columns and two rows. You should not apply Yates' correction to larger tables.

Table 29.2. Data for Matched-Pair Observations Displayed to Emphasize the Matching

	Cases		
	Risk Factor +	Risk Factor −	Total
Controls			
Risk Factor+	13	4	17
Risk Factor−	25	92	117
Total	38	96	134

Table 29.3. InStat Output for McNemar's Test

McNemar's Paired Test

McNemar chi-square (with Yates correction) = 13.793.
The two-sided P value is 0.0002, considered extremely significant.
There is a significant association between rows and columns.
McNemar odds ratio = 0.1600.
95% confidence interval for the odds ratio = 0.04724 to 0.4832.

---

3. Look up the P value in Table A5.8 in the Appendix. The larger the value of chi-square, the smaller the P value. The number of degrees of freedom equals $(R - 1) \times (K - 1)$, where R is the number of rows and K is the number of columns.

The chi-square test is based on these familiar assumptions:

- *Random sampling.* The data must be randomly selected from, or at least representative of, a larger population.
- *The data must form a contingency table.* The values must be the number of subjects. The categories that define the rows must be mutually exclusive. The categories that define the columns must also be mutually exclusive. Many data tables are not contingency tables and should not be analyzed with the chi-square test.
- The chi-square test is based on several simplifying assumptions that are *valid only with large data sets.* How large is large? Any rule is somewhat arbitrary, but here is a commonly used rule for deciding when it is OK to use the chi-square test. More than 80% of the expected values must be greater or equal to 5, and all the expected values must be greater or equal to 2 (greater than or equal to 1 if the table is large enough to have more than 30 df). Note that the rule involves expected values, not observed values. If the rule is violated, you can usually combine two or more rows or columns in a logical way. Otherwise you can use a program that calculates an exact test (analogous to Fisher's exact test).
- *Independent observations.* Each subject must be independently selected from the population.

For example, Table 29.4 is a contingency table for the blood pressure data for all four classes.

The results from InStat are shown in Table 29.5. The P value is 0.0908. The null hypothesis is that the prevalence of hypertension is identical in all four classes. If that

Table 29.4. Contingency Table for Blood Pressure Data

	Blood Pressure High	Blood Pressure Not High
I	5	25
II	11	19
III	12	19
IV	14	16

Table 29.5. InStat Output for Chi-Square Test

Chi-square: 6.472.   Degrees of freedom: 3.
Table size: 4 rows, 2 columns.
The P value is 0.0908.
The row and column variables are not significantly associated.

null hypothesis were true, there is a 9.1% chance of randomly selecting subjects with as much variation as observed here (or more).

CHI-SQUARE TEST FOR TREND*

The standard calculations for chi-square do not take into account any order among the rows or columns. If your experiment was designed so that the rows have a distinct order (i.e., time points, ages, or doses), this information is not used in the standard chi-square calculations.

In the blood pressure example, the rows are ordered, and you may wish to ask whether there is a linear trend, that is, whether the prevalence of hypertension changes linearly with class. The chi-square test for trend calculates a P value for the null hypothesis that there is no correlation between row (class) number and the proportion of subjects who are hypertensive (in left row). This test uses more information and tests a narrower set of alternative hypotheses than does the chi-square test for independence. Accordingly the P value is lower. For this example, P = 0.0178. If the null hypothesis were true, there is only a 1.8% chance that random sampling would lead to a linear trend this strong or stronger.

The details of the chi-square test for trend are not presented here, but the results of InStat are shown in Table 29.6.

MANTEL-HAENSZEL CHI-SQUARE TEST

Chi-square (or Fisher's) is appropriate when subjects are classified by two criteria: one for the columns and another for the rows. Sometimes subjects are classified by three criteria (for example, by presence or absence of disease, by the presence or absence of a possible risk factor, and by city). One is tempted to ignore the third criterion (city) when forming a contingency table, but this approach can be misleading. Pooling data into one contingency table violates one of the assumptions of chi-square (that the subjects are selected independently). Such a violation can lead you to miss a real association or to find an apparent association that doesn't really exist.

The proper analysis is the Mantel-Haenszel test, which essentially combines together a stack of contingency tables. It is used in two situations. One use is when you wish to combine results from several groups in one study. Many large studies are performed in several medical centers, and the results are pooled. The other situation

---

*This section is more advanced than the rest. You may skip it without loss of continuity.

Table 29.6. Chi-Square Test for Trend

Note: This analysis is useful only if the categories defining the rows are arranged in a natural order (i.e.,
       age groups, dose, or time), with equal spacing between rows.

Chi-square for trend = 5.614 (1 degree of freedom).
The P value is 0.0178.

There is a significant linear trend among the ordered categories defining the rows and the proportion of
subjects in the left column.

is when you want to combine several different studies. Combining several studies is
called *meta-analysis*.

Of course it only makes sense to combine several 2 × 2 tables when they are
similar. The Mantel-Haenszel method is based on the assumption that all tables show
data sampled from populations with the same odds ratio, even though the prevalence
or incidence rates may vary between populations. The Mantel-Haenszel calculations
generate the overall, or pooled, odds ratio and its CI. Additionally, it calculates a P
value testing the null hypothesis that the pooled odds ratio equals 1. In other words,
the null hypothesis is that there is no association between rows and columns, taking
into account the confounding effect of the various tables that are pooled together.

The details of the Mantel-Haenszel method are not presented here. The necessary
steps are not difficult, however, and can be performed by hand (if necessary) without
a lot of trouble. This test is a derivative of the chi-square test for 2 × 2 tables and
accordingly should not be used when the number of subjects is small.

30

Comparing Three or More Means: Analysis of Variance

In Chapters 23 and 24 you learned how to compare the mean of two groups. Comparing three or more means is more difficult.

WHAT'S WRONG WITH MULTIPLE t TESTS?

Your first thought might be to perform several tests. For example, to compare four groups you might perform six t tests (1 vs. 2, 1 vs. 3, 1 vs. 4, 2 vs. 3, 2 vs. 4, 3 vs. 4). You've already learned about the problem of multiple comparisons (Chapter 13), so you can see the problem with this approach. If all four populations have the same mean, there is a 5% chance that each particular t test would yield a "significant" P value. But you'd be making six comparisons. Even if all null hypotheses were true, the chance that at least one of these P values would be less than 0.05 by coincidence is far higher than 5%.

When you have collected data in more than two groups, you should not perform a separate t test for each pair of groups. Instead you should compare all the groups at once with a method termed *one-way analysis of variance* (ANOVA).

ONE-WAY ANOVA

One-way ANOVA tests the null hypothesis that all populations have identical means. It generates a P value that answers this question: If the null hypothesis is true, what is the probability that the means of randomly selected samples will vary as much (or more) than actually occurred?

As the name implies, ANOVA analyzes the variance among values. Recall that the first step in calculating the variance is to sum the squares of the differences between each value and the mean. This is called the *sum of squares*. The variance is the mean sum of squares, and the standard deviation (SD) is the square root of the variance.

When you combine data from several groups, the variance has two components. If the groups have different means, some of the variation comes from differences *among* the group means. The rest of the variation comes from differences among the subjects *within* each group. The latter can be quantified by summing the squares of the differences of each value from its group mean. This is called the *within-groups sum of squares*. The total variation can be quantified by summing the squares of the difference of each value from the *grand* mean. This is the total sum of squares. If the null hypothesis is true, the within-groups sum of squares will have a value close to the total sum of squares because each value will be nearly as close to the grand mean as to its group mean. If the null hypothesis is false, each value will tend to be closer to its group mean than to the grand mean, and you expect the within-group sum of squares to be quite a bit smaller than the total sum of squares. One-way ANOVA computes a P value answering this question: If the null hypothesis were true, what is the chance that an experiment of this size would have such a large a difference (or larger) between the total sum of squares and the within-groups sum of squares?

Entire books have been written on ANOVA, and the description here only gives you a glimpse of how the method works. The easiest way to understand the results of ANOVA is to follow an example.

AN EXAMPLE

Continuing Example 13.5 from page 120, the results are in Table 30.1.

The calculations are best left to computer programs, and the equations are not shown here. The output of InStat is shown in Table 30.2. Other programs would produce similar output, except that some programs use the word *model* instead of *between columns, error* instead of *residuals,* and *one factor* instead of *one-way* ANOVA. The first column shows the degrees of freedom. The degrees of freedom for treatments equals the number of groups minus 1. The degrees of freedom for the residuals equals the total number of subjects minus the number of groups.

The second column (sum of squares) shows how the total variability is divided into variation due to systematic differences between groups and residual variation within the groups. The key to ANOVA is that the within-group sum of squares and the treatment sum of squares sum to the total sum of squares. This is not an obvious relationship, but it can be proven with simple algebra. In this example $0.93/16.45 = 5.6\%$ of the variability is due to systematic differences between groups, and the remainder of the variability is due to differences within the groups. The third column shows *mean squares*. For each row, the mean square equals the sum-of-squares value divided by its degrees of freedom.

Table 30.1. LH Levels in Three Groups of Women

Group	log(LH) ± SEM	N
Nonrunners	0.52 ± 0.027	88
Recreational runners	0.38 ± 0.034	89
Elite runners	0.40 ± 0.049	28

Table 30.2. InStat Results for One-Way ANOVA

Source of Variation	Degrees of Freedom	Sum of Squares	Mean Square
Treatments (between groups)	2	0.9268	0.4634
Residuals (within groups)	202	16.450	0.0814
Total	204	17.377	

F = 5.690

The P value is 0.0039, considered very significant.
Variation among column means is significantly greater than expected by chance.

The ratio of the two mean squares is called F, after Fisher (who invented ANOVA). In this example, F = 5.690. If the null hypothesis is true, F is likely to have a value close to 1. If the null hypothesis is not true, F is likely to have a value greater than 1. This is not obvious, but can be proven without too much trouble. Statisticians have derived the probability distribution of F under the null hypothesis for various degrees of freedom and can calculate a P value that corresponds to any value of F. In other words, the P value also answers this question: If the null hypothesis is true, what is the chance that the F ratio (with a certain numbers of df for the numerator and denominator) would have a value as large or larger than calculated for this experiment? The larger the value of F, the smaller the P value.

The program calculated the P value from F and both degrees of freedom. For F = 5.690 with 2 and 202 degrees of freedom, P = 0.0039.* Interpreting this P value should be familiar to you by now. If the null hypothesis were true, there is a 0.4% chance that the means of three randomly selected samples (N = 5 each) would have sample means that differ from one another as much or more than actually occurred in this example. More simply, if the null hypothesis were true, there is only a 0.4% chance of obtaining a F ratio greater than 5.690 in an experiment of this size. Use Table A5.10 in the Appendix to determine P from F. Note the following:

- The alternative hypothesis is that one or more of the population means differs from the rest. Rejecting the null hypothesis does not indicate that every population mean differs from every other population mean.
- In our example, the three groups have a natural order. The ANOVA calculations do *not* take into account this order. ANOVA treats the groups as categories, and the calculations don't take into account how the groups are related to each other.
- Don't get misled by the term *variance*. The word *variance* refers to the statistical method being used, not the hypothesis being tested. ANOVA tests whether group means differ significantly from each other. ANOVA does not test whether the variances of the groups are different.
- With ANOVA, the concept of one- and two-tailed P values does not apply. The P value has many tails.

---

*To determine a P value from F, you have to know two different df values, one for the numerator and one for the denominator. Don't mix up the two df values, or you will get the wrong P value.

ASSUMPTIONS OF ANOVA

One-way ANOVA is based on the same assumptions as the t test:

- The samples are randomly selected from, or at least representative of, the larger populations.
- The samples were obtained independently. If the subjects were matched, or if the samples represent before, during, and after measurements, then repeated measures ANOVA should be used instead (see later in this chapter).
- The observations within each sample were obtained independently. The relationships between all the observations in a group should be the same. You don't have four independent measurements if you measure two animals twice, if the four animals are from two litters, or if two animals were assessed in one experiment in January and two others were assessed in an experiment in June.
- The data are sampled from populations that approximate a Gaussian distribution. The SD of all populations must be identical. This assumption is less important with large samples, and when the sample sizes are equal.

MATHEMATICAL MODEL OF ONE-WAY ANOVA*

One-way ANOVA fits data to this model:

$$Y_{i,j} = \text{grand mean} + \text{group effect}_i + \varepsilon_{i,j}. \qquad (30.1)$$

$Y_{i,j}$ is the value of the ith subject in the jth group. Each group effect is the difference between the mean of population i and the grand mean. Each $\varepsilon_{i,j}$ is a random value from a Gaussian population with a mean of 0. The null hypothesis is that all the group effects equal 0. In the null hypothesis is true, all the difference among values can be explained by Gaussian scatter around the grand mean, and none of the differences among values are due to group effects.

MULTIPLE COMPARISON POST TESTS

The P value for the example was less than 0.05, so we conclude that there is a statistically significant difference among the means of the three populations. Now you want to calculate confidence intervals (CIs) for the difference between group means and determine which groups are statistically significantly different from which other groups. There are three problems:

- *Multiple comparisons.* If the null hypothesis is true, then 5% of all comparisons will result in a significant result (unless you make special corrections). If you make many comparisons, the chance of making a Type I error is greater than 5%.
- *Pooled standard deviation.* When comparing two groups, you pool the SDs. With multiple comparisons, you can pool the SDs of all the groups, not just the two you

---

*This section is more advanced than the rest. You may skip it without loss of continuity.

are comparing. This gives you a more accurate measure of the population SD. Using this extra information gives you more statistical power. Pooling the SDs is only useful if you assume that the SD of all the populations is equal, and this is one of the assumptions of ANOVA.

- There is also a third, more abstract, problem in that the *comparisons are not entirely independent*. Once you have compared group 1 with group 2 and group 1 with group 3, you already know something about the comparison of group 2 with group 3.

Statisticians have developed numerous methods for calculating comparing pairs of groups after ANOVA. These methods, each named after the person who derived it, are called *multiple comparison post tests*. Most computer programs offer several different post tests and selecting one can be a bit confusing. My approach is to answer the following questions:

- Are the columns arranged in a natural order, perhaps time points or doses? If so, consider the post test for trend. Instead of testing pairs of means, the post test for trend asks whether there is a significant linear trend between group order and group mean. It is really a form of linear regression.
- Do you wish to make elaborate comparisons? For example, you might want to compare the mean of all treated groups with the mean of a control group. Or you might want to compare the mean of groups A and B with the mean of groups C, D, and E. Statisticians call such comparisons *contrasts*. To calculate contrasts, use Scheffe's method. To allow for the infinite variety of possible contrasts, Scheffe's method generates CIs that are quite wide, and it has less statistical power to detect differences than do the other post tests.
- Do you wish to compare one control group to all other groups (without comparing the other groups among themselves)? If so, select Dunnett's test.
- Based on experimental design, does it make sense to only compare selected pairs of means? If so, select the Bonferroni test for selected pairs of means, and select those pairs based on experimental design. It is not fair to first look at the data and then decide which pairs you want to compare. By looking at the data first, you have implicitly compared all columns, and you should do this explicitly.
- Do you want to compare all pairs of means? Three methods are commonly used: Bonferroni, Tukey, and Student-Newman-Keuls. The Bonferroni test is well known and relatively easy to understand, but it generates CIs that are too wide and its power is too low. With only three or four groups, the discrepancy is small. Don't use Bonferroni's method with five or more groups. The Tukey and Student-Newman-Keuls tests are related, and report identical results when comparing the largest mean with the smallest. With other comparisons, Tukey's method is more conservative but may miss real differences too often. The Student-Newman-Keuls method is more powerful but may mistakenly find significant differences too often. Statisticians do not agree about which one to use.

The results from the example were analyzed using the Tukey method, and the results from InStat are shown in Table 30.3. Setting α to 0.05 for the entire family of comparisons, nonrunners are statistically significantly different than recreational runners. No other differences are statistically significant.

Table 30.3. InStat Results for Tukey's Post Test

Comparison	Mean Difference	q	P Value
Nonrunners vs Recreational	0.1400	2.741	** P < 0.01
Nonrunners vs Elite	0.1200	2.741	ns P > 0.05
Recreational vs Elite	−0.02000	0.4574	ns P > 0.05

Difference	Mean Difference	Lower 95% CI	Upper 95% CI
Nonrunners — Recreational	0.1400	0.03823	0.2418
Nonrunners — Elite	0.1200	−0.02688	0.2669
Recreational — Elite	−0.02000	−0.1667	0.1267

REPEATED-MEASURES ANOVA

The difference between ordinary and repeated-measures ANOVA is the same as the difference between the unpaired and paired t tests. Use repeated-measures ANOVA to analyze data collected in three kinds of experiments:

- Measurements are made repeatedly in each subject, perhaps before, during, and after an intervention.
- Subjects are recruited as matched sets, matched for variables such as age, zip code, or diagnosis. Each subject in the set receives a different intervention (or placebo).
- A laboratory experiment is run several times, each time with several treatments (or a control and several treatments) handled in parallel. More generally, you should use a repeated-measures test whenever the value of one subject in the first group is expected to be closer to a particular value in the other group than to a random subject in the other group.

When the experimental design incorporates matching, use the repeated-measures test, as it is usually more powerful than ordinary ANOVA. Of course the matching has to be done as part of the protocol, before the results are collected. The decision about pairing is a question of experimental design and should be made long before the data are analyzed.

While the calculations for the repeated-measures ANOVA are different from those for ordinary ANOVA, the interpretation of the P value is the same. The same kind of post tests can be performed.

NONPARAMETRIC ANOVA

If you are not able to assume that your data are sampled from Gaussian populations, you can calculate the P value using a nonparametric test. The nonparametric test analogous to one-way ANOVA is called the *Kruskal-Wallis test*. The nonparametric test analogous to repeated-measures one-way ANOVA is called *Friedman's test*. Like the Mann-Whitney test you learned about in Chapter 24, these tests first rank the data from low to high and then analyze the distribution of the ranks among groups.

TWO-WAY ANOVA

The term *one-way* means that the subjects are categorized in one way. In the example, they were categorized by amount of running. In two-way ANOVA the subjects are simultaneously categorized in two ways. For example, you might study a response to three different drugs in both men and women. Each subject is categorized in two ways: by gender and by which drug he or she received.

Entire books have been written about ANOVA, and this section only touches the surface. Two-way ANOVA is complicated to understand, because it simultaneously asks three questions and computes three P values:

1. *Interaction.* The null hypothesis is that there is no interaction between the two factors. In our example, the null hypothesis is that the difference in effect between men and women is the same for all drugs. The P value answers this question: If the null hypothesis is true, what is the chance that randomly sampled subjects would show as much (or more) interaction than you have observed? If the interaction is statistically significant, you will usually find it quite difficult to interpret the individual P values for each factor.
2. *First factor.* The null hypothesis is that the population means are identical for each category of the first factor. In our example, the null hypothesis is that the mean response is the same for each drug in the overall population and that all observed differences are due to chance.
3. *Second factor.* The null hypothesis is that the population means are identical for each category of the second factor. In our example, the null hypothesis is that the mean response is the same for men and women in the overall population, and that all observed differences are due to chance. The P value answers this question: If the null hypothesis is true and you randomly select subjects, what is the chance of obtaining means as different (or more so) than you observed?

As with one-way ANOVA, special methods have been developed to calculate post tests focusing on particular differences. Methods have also been developed to deal with repeated measures.

PERSPECTIVE ON ANOVA

Since this chapter is only eight pages long and stuck towards the end, I may have set a record for writing the statistics book that places the least emphasis on ANOVA! I've written this book to help you interpret results in biology and medicine, and this doesn't require knowing too much about ANOVA. What's important is that you understand the issues of multiple comparisons (see Chapter 13) and the basic idea of one-way ANOVA (as explained in this chapter). In my experience, relatively few biomedical papers rely on fancier forms of ANOVA. In contrast, social scientists use ANOVA, in all its forms, quite frequently.

If you're not satisfied with interpreting results, but want to delve into the math and learn how statistical tests *really* work, then you'll need to learn far more about ANOVA. In various forms, ANOVA is central to much of statistical analysis. For example, the t test is really a simple form of ANOVA. Performing one-way ANOVA with two groups is equivalent to performing an unpaired t test. Additionally, linear regression can be explained in terms of ANOVA (partitioning variance into a portion explained by the linear model and a portion that is not). See Appendix 1 for suggested readings.

31

Multiple Regression

As you learned in Chapter 19, simple linear regression determines the best equation to predict Y from a single variable, X. Multiple linear regression finds the equation that best predicts Y from multiple independent variables.

THE USES OF MULTIPLE REGRESSION

Multiple regression can be used for three different purposes:

- To adjust data. Here you are mostly interested in the effect of one particular X variable but wish to adjust the data for differences in other X variables.
- Devise an equation to predict Y from several X variables for future subjects. Medical school admission committees might want to use multiple regression to obtain an equation that predicts performance in medical school from variables known at the time of application. Cardiologists might want to use multiple regression to obtain an equation that predicts cardiac output from the blood pressure, pulse rate, and weight of a patient.
- Explore relationships among multiple variables to find out which X variables influence Y. It can be fascinating to analyze huge data sets that contain lots of variables and see which ones matter.

Example 31.1

Staessen and colleagues investigated the relationship between lead exposure and kidney function.* It is known that heavy exposure to lead can damage kidneys. It is also known that kidney function decreases with age and that most people accumulate small amounts of lead as they get older. These investigators wanted to know whether accumulation of lead could explain some of the decrease in kidney function in aging. They studied over 1000 men† and measured the concentration of lead in the blood and assessed renal function (as creatinine clearance). The two variables correlated, but

---

*JA Staessen, RR Lauwerys, J-P Buchet, CJ Bulpitt, D Rondia, Y Vanrenterghem, A Amery. Impairment of renal function with increasing blood lead concentrations in the general population. *N Engl J Med* 327:151–156, 1992.

†They independently studied about 1000 women as well. The data for men and women were very similar.

this is not a useful finding because it is known that lead concentration increases with age and that creatinine clearance decreases with age. So the investigators used multiple regression to adjust for age and other factors.

The Y value was the creatinine clearance. The main X variable was the logarithm of lead concentrations. They used the logarithm of concentration, rather than the lead concentration itself, to make the data conform more closely to a Gaussian distribution. Other X variables were age, body mass, log gamma-glutamyl transpeptidase (a measure of liver function), and previous exposure to diuretics (coded as 0 or 1). The investigators asked this question: After adjusting for the other variables, is there a substantial linear relationship between the logarithm of lead concentration and creatinine clearance? You'll see the answer on the next page.

THE MULTIPLE REGRESSION MODEL

The multiple regression model is expressed by Equation 31.1:

$$Y = \alpha + \beta_1 \cdot X_1 + B_2 \cdot X_2 + \beta_3 \cdot X_3 \ldots + \varepsilon. \qquad (31.1)$$

Each X can be a measured variable, a transformation of a measured variable (i.e., age squared, or logarithm of gamma-glutamyl transpeptidase activity) or a discrete variable (i.e., gender, previous diuretic therapy). In the latter case, X is called a *dummy variable*. For example, X could be a dummy variable for sex where $X = 1$ is male and $X = 0$ is female. It is also possible to use dummy variables when there are more than two categories (for example, four medical school classes). Consult more advanced books if you need to do this, as it is far trickier than you would imagine. The final variable ε represents random variability (error). Like ordinary linear regression, multiple regression assumes that the random scatter follows a Gaussian distribution.

The multiple linear regression model does not distinguish between the X variable(s) you really care about and the other X variable(s) that you are adjusting for. You make that distinction when interpreting the results.

ASSUMPTIONS OF MULTIPLE REGRESSION

The assumptions of multiple regression are similar to those of linear regression:

- *Sample from population.* Your sample is randomly sampled from, or at least representative of, a larger population.
- *Linearity.* Increasing a X variable by one unit increases (or decreases) Y by the same amount at all values of X. In the example, this means that the creatinine clearance would decrease by a certain amount for any increase in log(lead), regardless of the lead level.
- *No interaction.* Increasing a X variable by one unit increases (or decreases) Y by a certain amount, regardless of the values of the other X variables. In the example,

this means that the creatinine clearance would decrease by a certain amount for any increase in log(lead), regardless of age.

- *Independent observations.* Knowing Y for any particular subject provides no information about Y in other subjects. In the example, this would preclude multiple measurement in one family.
- *Gaussian distribution.* For any set of X values, the random scatter of Y follows a Gaussian distribution, at least approximately.
- *Homoscedasticity.* This fancy word means that the SD of Y values is always the same, regardless of X variables. In the example, this means that the SD of creatinine clearance is the same for all ages and all lead levels.

INTERPRETING THE RESULTS OF MULTIPLE REGRESSION

When running a multiple regression program, you must give the program values of Y and each X variable for each subject. The multiple regression program finds the best-fit values of each β coefficient, along with a 95% CI for each. Here is how to think about the value: If X_i changes by one unit while all other X values are constant, then Y increases by β_i (Y decreases if β_i is negative). In the example, the β coefficient for the X variable "previous diuretic therapy" was -8.8 ml/min. This means that after adjusting for all the other variables, subjects who had taken diuretics previously had a mean creatinine clearance that was 8.8 ml/min lower than subjects who had not taken diuretics. The β coefficient for log(lead) was -9.5 ml/min. That means that after adjusting for all other variables, an increase in log(lead) of 1 unit (which means that the lead concentration increased 10-fold) is associated with a decrease in creatinine clearance of 9.5 ml/min. The 95% CI ranged from -18.1 to -0.9 ml/min.

Multiple regression programs also calculate R^2, which is the fraction of the overall variance in Y "explained" by variation in all the X variables.* The programs generate one value of R^2 for the entire regression, not one for each X. In the example, $R^2 = 0.27$, meaning that 27% of the variability in creatinine clearance was explained by variability in lead, age, and all the other variables. That leaves 73% of the variability that is not explained by variability in any of the measured variables. Many programs also report an adjusted R^2. The adjustment lowers the value based on the number of X variables. The adjusted R^2 value is a better measure of goodness of fit than the unadjusted value.

Multiple regression programs calculate several P value. One P value tests the overall null hypothesis that all β coefficients equal 0 in the overall population. In other words, the null hypothesis is that none of the X variables influence Y. If that P value is low, you can reject the overall null hypothesis that none of the X variables influence the outcome. In the example, the investigators did not report the overall P value. More interestingly, the programs report individual P values for each β_i coefficient, testing the null hypothesis that that particular β_i coefficient equals 0. In other words, each

---

*With multiple regression, the tradition is to use an upper case R. With linear regression, the tradition is to use a lower case r.

null hypothesis is that that particular X variable is not linearly related to Y, after adjusting for all other X variables. In the example, the authors state that the P value for lead was <0.05. If there really was no association between lead concentrations and creatinine clearance, you'd find such a large correlation in a study this size (after adjusting for age and the other variables) less than 5% of the time.

Multiple regression is a powerful technique, but it can be difficult to interpret the results. One problem is that it is difficult to visualize the results graphically. Here are some other problems that can complicate the interpretation of the results of multiple regression:

- *Outliers.* As with simple linear regression, a single outlying point can greatly distort the results of multiple regression. With simple linear regression, this is usually obvious if you look at a graph of the data. It is harder to visualize multiple regression data graphically and easier to be misled by a few outliers.
- *Too few subjects.* A general rule of thumb is that there should be 5 to 10 data points for each X variable. With fewer data points, it is too easy to be misled by spurious findings.
- *Inappropriate model.* Even though the multiple regression model may seem complicated to you, it is too simple for many kinds of data. The regression model assumes that each X variable independently contributes towards the value of Y. In the example, the model assumes that increasing log(lead) by a certain amount (equivalent to multiplying the lead concentration a certain factor) will decrease creatinine clearance by the same amount, regardless of whether the creatinine clearance started out high or low, and regardless of the age of the subject. Despite this simplification, multiple linear regression has proved to be useful for many kinds of biological data. Some investigators extend the multiple regression model to situations where the influence of one X variable depends on the value of another.
- *Unfocused studies.* If you give the computer enough variables, some relationships are bound to turn up by chance, and these may mean nothing. You've already learned about this problem of multiple comparisons in Chapter 13. Caution is needed to avoid being impressed by chance correlation.

CHOOSING WHICH X VARIABLES TO INCLUDE IN A MODEL

If you have collected many X variables, multiple regression programs can select a subset of those X variables that are most useful in predicting Y. There are several approaches, and this topic is more complicated than it seems.

The best approach is to fit the data to every possible model (each model includes some X variables and may exclude others). This approach is termed *all subsets* regression. With many variables and large data sets, the computer time required is prohibitive. To conserve computer time, other algorithms perform multiple regression in a stepwise manner. One approach is to start with a very simple model and add new X variables one at a time, always adding the X variable that most improves the model's ability to predict Y. Another approach is to start with the full model (include all X variables) and then sequentially eliminate the X variable that contributes the least to the model.

In either case, the program needs to make a decision as to when to stop adding or subtracting X variables from the model.

Automated methods for choosing variables to include in a multiple regression equation can be a useful way to explore data and generate hypotheses. It is important to realize that the final selection of X variables chosen by the program for the model may depend on the order you present the X variables to the program and on the quirks of the method that you select. When deciding which variables to include and which to reject, none of the methods use clinical or scientific judgment and none take into account biological plausibility. When looking at the relationship among many X variables, the program can't use common sense. That's your job.

Often several of the X variables may be correlated with each other (the statistical term is *multicolinearity*). This happens when the two variables contain redundant information. For example, you might want to predict blood pressure (Y) from age (X1), height (X2), and weight (X3). Height and weight are correlated with each other and are somewhat redundant ways of measuring the subject's size. When you feed all the data to a multiple regression program, it may decide to keep height in the final model and reject the weight. Or it may keep weight and reject height. You need to use common sense to realize that the two variables are interrelated.

THE TERM *MULTIVARIATE STATISTICS*

The word *multivariate* is used inconsistently. Its looser definition refers to any method that examines multiple variables at once. Under this definition, multiple regression is a multivariate method. A more strict definition of the word *multivariate* refers only to methods that simultaneously examine several outcomes. Since multiple regression is used to predict or model one outcome from multiple explanatory variables, it is not a "multivariate" method under the strict definition. The method for examining multiple outcome variables at one time is called multiple analysis of variance (MANOVA, not discussed in this book).

32

Logistic Regression

Note to basic scientists: The methods described in this chapter are rarely used outside of clinical studies. You may skip this chapter without loss of continuity.

INTRODUCTION TO LOGISTIC REGRESSION

Logistic regression quantifies the association between a risk factor (or treatment) and a disease (or any event), after adjusting for other variables.

Example 32.1

Patients with severe atherosclerosis of the aorta can be treated with surgery. Many of these patients also have severe atherosclerosis of the coronary arteries, which places them at high risk for a heart attack during surgery or soon thereafter. Since the operation is elective, it would be useful to predict which patients are especially prone to heart attacks before deciding whether to operate on their aorta. Baron and colleagues investigated whether patients with a low ejection fraction (which means that their heart is not pumping properly) were more likely to have heart attacks (or other cardiac complications) during (or soon after) the aortic surgery.*

They studied 457 patients who received aortic grafts, 86 of whom had a cardiac complication. First they analyzed the data using the simple techniques of Chapter 8. Subjects with a low ejection fraction were more likely to have a cardiac complication. The odds ratio was 2.1, with a confidence interval (CI) ranging from 1.2 to 3.7. The P value was less than 0.01. From this simple analysis, you'd conclude that patients with low ejection fractions are more likely to have a cardiac complication. Therefore it seems reasonable to think that measuring the ejection fraction would be helpful in identifying patients at high risk. But that is not a fair conclusion without considering the age of the subject and whether he or she had previous heart problems.

*J-F Baron, O Mundler, M Bertrand, E Vicaut, E Barre, G Godet, CM Samama, P Coriat, E Kieffer, P Viars. Dipyridamole-thallium scintigraphy and gated radionuclide angiography to assess cardiac risk before abdominal aortic surgery. *N Engl J Med* 330:663–669, 1994.

Older patients are more likely to have a cardiac complication and are also more likely to have a low ejection fraction. Additionally, patients with previous heart problems are more likely to have cardiac complications and more likely to have a low ejection fractions. In other words, the relationship between ejection fraction and cardiac complications is confounded by age and previous heart problems. To untangle these interrelationships, the investigators used logistic regression to determine the odds ratio for the association between low ejection fraction and cardiac complications, after adjusting for age and history of previous heart disease. You'll learn a bit about how this method works later in the chapter but can understand the results now. After adjusting for age and previous heart problems, the odds ratio was 1.7, and the 95% CI ranged from 0.9 to 3.2. The P value was greater than 0.05 (the authors did not give the actual P value). The authors concluded that their data did not demonstrate a statistically significant association between low ejection fraction and cardiac complications after adjusting for age and a history of previous heart problems.

Although logistic regression is a formidable technique, this example shows that you can understand the results easily. The results are displayed as odds ratios, which you already have learned to interpret. The only difference is that the odds ratios computed by logistic regression adjust for differences in other variables.

HOW LOGISTIC REGRESSION WORKS

As its name suggests, logistic regression is similar to linear regression (and multiple regression). Linear regression finds an equation that predicts an outcome variable Y (which must be a measured, a variable that can take on any value) from one (or more) X variables. Logistic regression finds an equation that best predicts an outcome variable (which must be binary, such as the presence or absence of disease) from one or more X variables. Some of the X variables may be quantitative variables (i.e., age or blood pressure); others may be binary (i.e., gender, treatment received, or history of previous heart disease).

Instead of working with probabilities (which range only from 0 to 1), logistic regression works with the logarithm of the odds (which can take on any value, negative or positive). The logistic regression equation can be expressed in two equivalent forms, shown as Equations 32.1 and 32.2 (ln is the abbreviation for natural logarithm):

$$\ln(\text{odds}) = \alpha + X_1 \cdot \beta_1 + X_2 \cdot \beta_2 \ldots \tag{32.1}$$

α is the log of the odds in the baseline group, defined as individuals for which every X value equals 0. Each β_i equals the natural logarithm of the odds ratio (OR) for the ith X variable. Equation 32.2 expresses the model more conveniently. It defines the log of the overall odds ratio for any individual from the individual odds ratios for each X variable:

$$\ln(\text{OR}) = X_1 \cdot \ln(\text{OR}_1) + X_2 \cdot \ln(\text{OR}_2) \ldots \tag{32.2}$$

To run a logistic regression program, you must first enter the data for each subject. Code the outcome as $Y = 0$ if the outcome did not occur and $Y = 1$ if it did occur. Also enter all the X variables for each subject. The program finds the best-fit values for α and all the βs (Equation 32.1) or each individual adjusted odds ratio (Equation 32.2). Most programs also report the 95% CI for the odds ratios or the β coefficients.

The computations of logistic regression are more difficult than linear regression and will not be explained here. The method of least squares does not apply. Instead logistic regression programs use an iterative maximum likelihood method. The details are far beyond the scope of this book, but the idea of maximum likelihood was explained in Chapter 19.

Extensions of logistic regression are used to analyze retrospective (case-control) studies, studies with matched-pair data, and studies in which the outcome may have more than two possibilities. As with multiple linear regression (see previous chapter), logistic regression programs can be given a number of X variables and select those that are most useful in predicting the outcome.

ASSUMPTIONS OF LOGISTIC REGRESSION

Logistic regression depends on these assumptions:

- The subjects are randomly selected from, or representative of, a larger population.
- Each subject was selected independently of the others. Knowing the outcome of any one subject does not help you predict the outcome of any other subject.
- No interaction. The influence of any particular X variable is the same for all values of the other X variables. In the example, the model assumes that the low ejection fraction would increase the risk of old and young subjects equally.

INTERPRETING RESULTS FROM LOGISTIC REGRESSION

The logistic regression program either reports the best-fit values for each of the odds ratios (Equation 32.2) or reports each β coefficient (Equation 32.1). If you read a paper (or computer output) that used Equation 32.1, convert the β coefficients to odds ratios as follows: the odds ratio for the ith variable equals $e^{\beta i}$.

When X_i is a binary variable (i.e., X = 0 male, X = 1 female), the interpretation of the odds ratio is familiar. Holding all other X values constant, the odds ratio is the odds that the outcome will occur in someone who has X = 1 divided by the odds that the outcome will occur in someone who has X = 0. In the example, the authors coded one of the X variables as X = 0 if the patient did not have a low ejection fraction and as X = 1 if they did have a low ejection fraction. The odds ratio is the odds of having cardiac complications among patients with low ejection fractions divided by the odds of having cardiac complications among patients with normal ejection fractions. This is an adjusted odds ratio, as it corrects for the influence of all other X variables, such as age.

When X_i is a measurement, the interpretation of the odds ratio is less familiar. The odds ratio is the relative increase in odds when X_i increases by 1.0. Holding all other X values constant, the odds ratio is the odds that the outcome will occur in someone who has X_i = Z (where Z is any value) divided by the odds that the outcome will occur in someone who has $X_i = Z - 1$. The method assumes that the odds ratio is the same for all values of Z.

Note that the logistic regression programs do not distinguish between the variables you really care about (in the example, decreased ejection fraction) and the variables you are adjusting for (in the example, age and history of previous heart disease). Logistic regression finds the odds ratio for each X variable after adjusting for all others. In the example, the program would also give an adjusted odds ratio for the increased risk of cardiac complication with increased age and with previous history of heart disease. It is only when you interpret the data that you distinguish variables you care about from variables you are adjusting for.

Logistic regression programs report several P values. One P value tests the null hypothesis that the overall odds ratio in the population equals 1.0. In other words, this null hypothesis states that none of the X variables are associated with the outcome. If that P value is low, you can then look at individual P values for each odds ratio. Each P value tests the null hypothesis that the adjusted odds ratio for that X variable equals 1.0 in the overall population. In other words, the null hypothesis is that there is no association between that X variable and the presence or absence of disease, after adjusting for all other X variables.

Whenever you review the results of logistic regression, ask yourself these questions:

- Are there enough subjects? A general rule of thumb is that there should be 5 to 10 events for each X variable. Don't count the subjects, count the number of the least likely outcomes. With fewer events, it is too easy to be misled by spurious findings. In the example, there were 86 cardiac complications, so the authors could study 8 to 16 X variables.
- Is the model appropriate? Remember that the logistic regression model is simple, probably too simple to account for many kinds of data. The regression model assumes that each X variable contributes independently to the odds ratio. It is possible to extend the model to account for interactions.
- Do the results make sense? This is an important question when the computer chose the X variables to include in the model. If you give the computer enough variables, some relationships are bound to turn up by chance, and these may mean nothing.
- Is the study focused? If the investigators are exploring relationships among many variables with no clear hypothesis, distinguish between studies that generate hypotheses from studies that test hypotheses. If you study enough variables, some relationships are bound to turn up by chance. These kinds of unfocused studies are useful for generating hypotheses, but the hypotheses should be tested with different data.

33

Comparing Survival Curves

COMPARING TWO SURVIVAL CURVES

You've already learned (Chapter 6) how to interpret survival curves. It is common to compare two survival curves to compare two treatments. Compare two survival curves using the log-rank test. This test calculates a P value testing the null hypothesis that the survival curves are identical in the two populations. If that assumption is true, the P value is the probability of randomly selecting subjects whose survival curves are as different (or more so) than was actually observed.*

Example 33.1

Rosman and colleagues investigated whether diazepam would prevent febrile seizures in children.† They recruited about 400 children who had had at least one febrile seizure. Their parents were instructed to give medication to the children whenever they had a fever. Half were given diazepam and half were given placebo.‡ They analyzed the data in several ways, including survival analysis. Here the term *survival* is a bit misleading, as they compared time until the first seizure, not time until death. When they compared the times to first seizure with the log rank test, the placebo treated subjects tended to have seizures earlier and the P value was 0.06. The difference in survival curves was small. If diazepam was really no more effective than placebo, you'd expect that 6% of experiments this size would find a difference this large or larger. The authors did not reach a conclusion from this analysis because they analyzed the data in a fancier way, which we will discuss later in the chapter.

---

*You will sometimes see survival curves compared with the Mantel-Haenszel test. This test is essentially identical to the log-rank test. It is based on the same assumptions and produces virtually identical results.

†NP Rosman, T Colton, J Labazzo, PL Gilbert, NB Gardella, EM Kaye, C Van Bennekom, MR Winter. A controlled trial of diazepam administered during febrile illnesses to prevent recurrence of febrile seizures. *N Engl J Med* 329:79–84, 1993.

‡The experimental therapy (diazepam) was compared with placebo because there is no standard therapy known to be effective. Phenobarbital was previously used routinely to prevent febrile seizures, but recent evidence has shown that it is not effective.

ASSUMPTIONS OF THE LOG-RANK TEST

The log-rank test depends on these assumptions:

- The subjects are randomly sampled from, or at least are representative of, larger populations.
- The subjects were chosen independently.
- Consistent criteria. If patients are enrolled in the study over a period of years, it is important that the entry criteria don't change over time and that the definition of *survival* is consistent. This is obvious if the end point is death, but survival methods are often used for other end points.
- Baseline survival rate is not changing over time. If the subjects are enrolled in the study over a period of years, you must assume that the survival of the control patients enrolled early in the study would be the same (on average) as the survival of those enrolled later (otherwise you have to do a fancier analysis to adjust for the difference).
- The survival of the censored subjects would be the same, on average, as the survival of the remaining subjects.

The calculations of the log-rank test are tedious and best left to computer. The idea is pretty simple. For each time interval, compare the observed number of deaths in each group with the expected number of deaths if the null hypothesis were true. Combine all the observed and expected values into one chi square statistic and determine the P value from that.

A POTENTIAL TRAP: COMPARING SURVIVAL OF RESPONDERS VERSUS NONRESPONDERS

This approach sounds reasonable but is invalid.

> I treated a number of cancer patients with chemotherapy. The treatment seemed to work with some patients because the tumor became smaller. The tumor did not change size in other patients. I plotted separate survival curves for the responders and nonresponders, and compared them with the log-rank test. The two differ significantly, so I conclude that the treatment prolongs survival.

This analysis is not valid, because you only have one group of patients, not two. Dividing the patients into two groups based on response to treatment is not valid for two reasons:

- A patient cannot be defined to be a ''responder'' unless he or she survived long enough for you to measure the tumor. Any patient who died early in the study was defined to be a nonresponder. In other words, survival influenced which group the patient was assigned to. Therefore you can't learn anything by comparing survival in the two groups.
- The cancers may be heterogeneous. The patients who responded may have a different form of the cancer than those who didn't respond. The responders may have survived longer even if they hadn't been treated.

The general rule is clear: You must define the groups you are comparing (and measure the variables you plan to adjust for) before starting the experimental phase of the study. Be very wary of studies that use data collected during the experimental phase of the study to divide patients into groups or to adjust the data.

WILL ROGERS' PHENOMENON

Assume that you are tabulating survival for patients with a certain type of tumor. You separately track survival of patients whose cancer has metastasized and survival of patients whose cancer remains localized. As you would expect, average survival is longer for the patients without metastases. Now a fancier scanner becomes available, making it possible to detect metastases earlier. What happens to the survival of patients in the two groups?

The group of patients without metastases is now smaller. The patients who are removed from the group are those with small metastases that could not have been detected without the new technology. These patients tend to die sooner than the patients without detectable metastases. By taking away these patients, the average survival of the patients remaining in the "no metastases" group will improve.

What about the other group? The group of patients with metastases is now larger. The additional patients, however, are those with small metastases. These patients tend to live longer than patients with larger metastases. Thus the average survival of all patients in the "with-metastases" group will improve.

Changing the diagnostic method paradoxically increased the average survival of *both* groups! Feinstein termed this paradox the *Will Rogers' phenomenon* from a quote from the humorist Will Rogers ("When the Okies left California and went to Oklahoma, they raised the average intelligence in both states.")*

MULTIPLE REGRESSION WITH SURVIVAL DATA: PROPORTIONAL HAZARDS REGRESSION

Proportional hazards regression applies regression methodology to survival data. This method lets you compare survival in two or more groups after adjusting for other variables.

Example 33.1 Continued (Diazepam and Febrile Seizures)

The investigators performed proportional hazards regression to adjust for differences in age, number of previous febrile seizures, and several other variables. After those adjustments, they found that the relative risk was 0.61 with a 95% CI ranging from 0.39 to 0.94. Compared with subjects treated with placebo, subjects treated with diazepam had only 61% of the risk of having a febrile seizure. This reduction was statistically

---

*AR Feinstein, DA Sosin, CK Wells. Will Rogers phenomenon. Stage migration and new diagnostic techniques as a source of misleading statistics for survival in cancer. *New Engl J Med* 312:1604–1608, 1985.

significant with a P value of 0.027. If diazepam was ineffective, there is only a 2.7% chance of seeing such a low relative risk in a study of this size. This example shows that the results of proportional hazards regression are easy to interpret, even though the details of the analysis are complicated.

HOW PROPORTIONAL HAZARDS REGRESSION WORKS

A survival curve plots cumulative survival as a function of time. The slope or derivative of the survival curve is the rate of dying in a short time interval. This is termed the *hazard*. For example, if 20% of patients with a certain kind of cancer are expected to die this year, then the hazard is 20% per year. When comparing two groups, investigators often assume that the ratio of hazard functions is constant over time. For example, the hazard among treated patients might be one half the hazard in control patients. The death rates change over the course of the study, but at any particular time the treated patients' risk of dying is one half the risk of the control patients. Another way to say this is that the two hazard functions are proportional to one another. This is a reasonable assumption for many clinical situations.

The ratio of hazards is essentially a relative risk. If the ratio is 0.5, then the relative risk of dying in one group is half the risk of dying in the other group. Proportional hazards regression, also called *Cox regression* after the person who developed the method, uses regression methods to predict the relative risk based on one or more X variables.

The assumption of proportional hazards is not always reasonable. You would not expect the hazard functions of medical and surgical therapy for cancer to be proportional. You might expect that the surgical therapy to have the higher hazard at early times (because of deaths during the operation or soon thereafter) and medical therapy to have the higher hazard at longer times. In such situations, proportional hazards regression should be avoided or used only over restricted time intervals for which the assumption is reasonable.

Having accepted the proportional hazards assumption, we want to know how the hazard ratio is influenced by treatment or other variables. One thought might be to place the hazard ratio on the left side of a regression equation. It turns out that the results are cleaner when we take the natural logarithm first.* So the logarithm of the hazard ratio can be placed on the left side of the multiple regression equation to generate Equation 33.1:

$$\ln(\text{hazard ratio}) = \ln(\text{relative risk}) = X_1 \cdot \ln(RR_1) + X_2 \cdot \ln(RR_2). \quad (33.1)$$

The hazard ratio must be defined relative to a baseline group. The baseline group is subjects in which every X variable equals 0. You can see in Equation 33.1 that the

---

*The problem is that the relative risk is not symmetrical with respect to decreased or increased risk. All possible decreases in risk are expressed by relative risks between 0 and 1, while all possible increases in risk are expressed by a relative risk between 1 and infinity. By taking the logarithm of the relative risk, any negative value indicates a decreased risk, and any positive value indicates an increased risk. For example, a doubling of risk gives a $\ln(RR)$ of 0.69; a halving of risk gives a $\ln(RR)$ of -0.69.

logarithm of the hazard ratio in the baseline group equals 0, so the hazard ratio (by definition) equals 1.0 (antilogarithm of 0).

INTERPRETING THE RESULTS OF PROPORTIONAL HAZARDS REGRESSION

To run a proportional hazards regression program, you must first enter the data for each subject. Enter the survival time for each subject, along with a code indicating whether the subject died at that time* or was censored at that time (see Chapter 6 for a definition of censoring). Also enter all the X variables for each subject.

Programs that calculate proportional hazards compute the best-fit values for each of the relative risks (hazard ratios), along with their 95% CI. If you encounter a program (or publication) that reports β coefficients instead of odds ratios, it is easy to convert. The relative risk for variable X_i equals e^{β_i}.

Programs that calculate proportional hazards regression report several P values. One P value tests the overall null hypothesis that in the overall population all relative risks equal 1.0. In other words, the overall null hypothesis is that none of the X variables influence survival. If that P value is low, you can reject the overall null hypothesis that none of the X variables influence survival. You can then look at individual P values for each X variable, testing the null hypothesis that that particular relative risk equals 1.0.

Whenever you review the results of proportional hazards regression, ask yourself these questions:

• Are there enough subjects? A general rule of thumb is that there should be 5 to 10 deaths for each X variable. Don't count the subjects, count the number of deaths. With fewer events, it is too easy to be misled by spurious findings. For example, a study with 1000 patients, 25 of whom die, provides enough data to study the influence of at most 2 to 4 explanatory X variables.
• Is the model appropriate? The proportional hazards regression model is simple, probably too simple to account for many kinds of data, as it assumes that each X variable contributes independently to the relative risk. That means that a change in one X value has the same impact on the relative risk, regardless of the values of the other X values. Moreover, proportional hazards regression assumes that the hazard ratio is constant at all times. If these assumptions are not true, the analysis will be misleading.
• Do results make sense? This is an important question when the computer chose the X variables to include in the model. If you give the computer enough variables, some relationships are bound to turn up by chance, and these may mean nothing.
• Distinguish between studies that generate hypotheses and studies that test hypotheses. If you study enough variables, some relationships are bound to turn up, and these may be just a coincidence. It is OK to generate hypotheses with this kind of exploratory research, but you need to test the hypothesis with different data.

---

*The term *survival* is used generally. The event does not have to be death. Proportional hazards regression can be used with any outcome that happens at most once to each subject.

34

Using Nonlinear Regression
to Fit Curves

Linear regression calculations are easy to calculate and easy to understand. However, many relationships of biological interest are not linear, so it is often necessary to fit a curve that best predicts Y from X.

THE GOALS OF CURVE FITTING

You might have three reasons to fit a curve through data:

- *Artistic.* You just want the graph to look nice. In this case, you can draw the curve by hand using a flexible ruler or French curve. Or you can use any computerized curve-fitting method, without much thought about what the results mean, since all you want is an attractive graph.
- *Interpolation.* You want to draw a curve through standard points, so you can read off unknown values. You can use many different methods for obtaining the curve and don't need to worry about what the curve means.
- *Model fitting.* You want to fit a curve through your data, and you want the curve to follow an equation that has biological or chemical meaning. If this is your goal, read the rest of the chapter.

AN EXAMPLE

Pharmacologists perform radioligand binding experiments to figure out the number of receptors (B_{max}) in a tissue and their affinity for the drug (K_D).* The results of a typical experiment are shown in Figure 34.1. Various concentrations of a radioactively labeled drug (radioligand) were added to a suspension of membrane fragments. The concentrations are shown on the X axis, measured in nanomolar (nM). After waiting for the

---

*B_{max} is the maximum amount of binding at very high concentrations of radioligand. It is the receptor number, often expressed as receptors per cell, or femtomoles (fmol) of receptors per milligram of membrane protein. The K_d is the equilibrium dissociation constant. It is the concentration of radioligand needed to bind half the receptors at equilibrium.

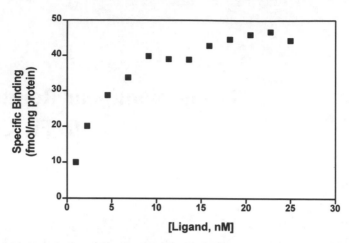

Figure 34.1. Typical radioligand binding experiment. The X axis shows the various concentrations of the radioactively labeled ligand. The Y axis shows the specific binding of the ligand to receptors. These data are analyzed in later figures.

binding to equilibrate, the specific binding was measured. The measurements were of the number of radioactive counts per minute. However, the experimenter knew the specific radioactivity of the radioligand and the amount of protein in the membrane suspension, so transformed the results to show femtomoles of receptor binding per milligram of membrane protein. That is shown on the Y axis.

The results make sense. At low concentrations of the radioligand, there is little binding. As the concentration increases, the binding increases. As the concentration increases further, the binding begins to plateau as you approach 100% saturation of receptors with ligand.

For many kinds of receptors, the radioligand binds reversibly to the receptors:

$$\text{Receptor} + \text{Ligand} \overset{K_d}{\rightleftharpoons} \text{Receptor} \cdot \text{Ligand}. \qquad (34.1)$$

From the chemical definition of equilibrium and the law of mass action, the relationship between the concentration of radioligand and the amount of binding at equilibrium follows Equation 34.2:

$$\text{Specific binding} = \frac{B_{max} \cdot [\text{Ligand}]}{K_d + [\text{Ligand}]}. \qquad (34.2)$$

$$Y = \frac{B_{max} \cdot X}{K_d + X}.$$

The relationship between X (concentration) and Y (binding) is not linear, so the methods of linear regression cannot be used. We need to fit a curve through the data points to find the best-fit values of B_{max} and K_d.

WHAT'S WRONG WITH TRANSFORMING CURVED DATA INTO STRAIGHT LINES?

Before the age of cheap microcomputers, nonlinear regression was not readily available to most scientists. Instead, scientists transformed their data to make the graph linear. They then used linear regression to analyze the transformed data. For the particular kind of data shown in the example, the Scatchard transformation linearizes the data, as shown in Figure 34.2.

Other commonly used linear transformations are the double reciprocal or Lineweaver-Burke plot used in enzymology and the log transform used in kinetic studies. All these methods are outdated and should rarely be used to analyze data. The problem is that linear regression assumes that the scatter of data around the line follows a Gaussian distribution with a standard deviation (SD) that does not depend on the value of X. This assumption is rarely true with transformed data. Moreover, linear regression assumes that X and Y are measured independently. With some transforms (for example, the Scatchard transform), the X and Y values are intertwined by the transform, invalidating linear regression.

Figure 34.3 shows the problem of transforming data. The left panel shows data from an experiment similar to the previous example but with more scatter. The solid curve was fit by nonlinear regression. The right panel plots the same data after the Scatchard transformation—it is a Scatchard plot. The Scatchard transformation amplifies and distorts the scatter.* The solid line was created by applying the Scatchard transform to the curve generated by nonlinear regression. The dotted line was created by calculating the best-fit linear regression line to the transformed values. The two lines are quite different, and correspond to different values for K_d and B_{max}. The values generated from nonlinear regression are likely to be closer to the true values.

Analyzing data by performing linear regression with transformed data is like using pH paper to measure acidity. Scientists rarely use pH paper any more, because it is easier and more accurate to measure pH with a pH meter. But pH paper is still useful when you need a crude measure of pH and a meter is not available. Similarly, fitting curves with computerized nonlinear regression is easier than first transforming the data and then calculating linear regression. Microcomputers cost less than pH meters, and curve-fitting software is readily available.

Although it is usually inappropriate to *analyze* transformed data, it is often helpful to *graph* transformed data. It is easier to visually interpret transformed data because the human eye and brain evolved to detect edges (lines), not to detect rectangular hyperbolas or exponential decay curves.

USING A NONLINEAR REGRESSION PROGRAM

Nonlinear regression fits an equation to data and finds values of the variables that make the equation best fit the data. To use nonlinear regression, follow these steps:

---

*There are some cases for which transforming the data makes sense. You should transform the data when the transformation makes the experimental scatter more uniform and more Gaussian. In most cases, however, the transformation will make the scatter less uniform and less Gaussian.

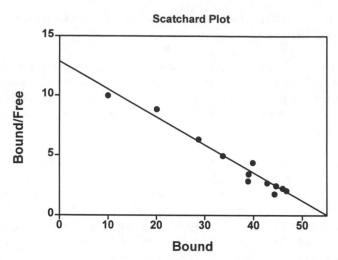

Figure 34.2. Scatchard transformation of the data in Figure 34.1. The X axis in the Scatchard plot is the same as the Y axis sin Figure 34.1. The Y axis of the Scatchard plot equals the Y axis from Figure 34.1 divided by the X axis of that figure. This transformation converts the curve into a straight line. The slope of the line equals the reciprocal of the K_d. The X intercept of the line equals the B_{max}.

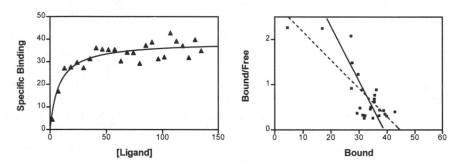

Figure 34.3. Why linearizing transforms should be avoided. The left panel shows data similar to those in Figure 34.1 but with more scatter. The curve was generated by nonlinear regression. The right panel shows the Scatchard plot of the same data. If you did not use nonlinear regression to fit a curve to the raw data, you would perform linear regression of the transformed data to generate the dotted line. You could then derive values for B_{max} and K_d from the slope and intercept of that line. But these values would be far from the values determined by nonlinear regression. To see this, look at the solid line that was generated by applying the Scatchard transform to the nonlinear regression curve. This represents the correct fit of the data, displayed on a Scatchard plot. The values of the B_{max} and K_d determined from the linear regression line (dotted) would be far from their correct values (solid line).

Table 34.1. Example Results Calculated by Nonlinear Regression

Variables	
B_{max}	53.3
K_d	3.86
Standard error	
B_{max}	1.33
K_d	0.378
95% Confidence intervals	
B_{max}	50.3 to 56.2
K_d	3.02 to 4.70
Goodness of fit	
Degrees of freedom	10
R^2	0.984
Absolute sum of squares	21.4

1. You must choose an equation that defines Y as a function of X and one or more variables. You should choose an equation that expresses a chemical or biological (or physical or genetic, etc.) model. Nonlinear regression programs rarely help you choose an equation, as the choice depends on the theory behind the kind of experiment you have done. The goal of nonlinear regression is to find the best values for the variables in the equation so that the curve comes closest to the points.

2. Provide an initial value for each variable in the equation.* Providing initial values is rarely a problem if you understand the equation in terms of biology or chemistry, and have looked at a graph of the data. For the example, we can estimate that B_{max} is a bit higher than the largest Y value, perhaps about 50. Since K_d is the concentration (X value) at which Y equals half its maximal value, a glance at Figure 34.1 tells you that the K_d is a bit less than 5. Many nonlinear regression programs automatically provide initial values if you select a built-in equation.

3. The program fits the curve. First it generates the curve defined by the initial values and calculates the sum of squares of the vertical distances of the points from the curve. Then it adjusts the variables to make the curve come closer to the data points. Several algorithms are available for doing this, but the one devised by Marquardt is used most often. The program repeatedly adjusts the variables until no adjustment is able to reduce the sum of the square of the vertical distance of the curve from the points.

Nonlinear regression is based on the same assumptions as linear regression (see Chapter 19).

THE RESULTS OF NONLINEAR REGRESSION

For the example, the results calculated by GraphPad Prism are shown in Table 34.1 (see also Figure 34.4).

---

*Variable refers to the parameters you wish to fit. In the example, K_d and B_{max} are variables. The equation also includes the variables X and Y, of course, but you don't have to estimate values of X and Y.

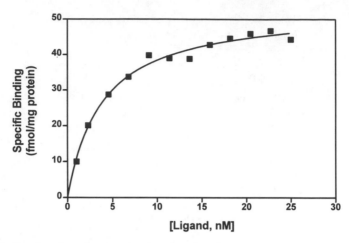

Figure 34.4. Results of nonlinear regression. The data from Figure 34.1 were fit with nonlinear regression to generate the curve.

The whole point of nonlinear regression is to find values for the variables, and these are reported first. If the equation expresses a biological or chemical model, these values can be interpreted in terms of biology or chemistry. For example, B_{max}, the number of receptors, equals 53.3 (it is measured in the units of the Y variable, in this case femtomoles of receptor per milligram of membrane protein). Next comes the standard error and the approximate 95% confidence interval (CI) of each variable. The 95% CIs reported by nonlinear regression programs are considered to be approximations, because they are always symmetrical, even though the real CI may be asymmetrical. They are optimistic approximations, as the true CIs may be wider.

Goodness of fit is quantified by R^2, a fraction between 0 and 1. When $R^2 = 0.0$, the best-fit curve fits the data no better than a horizontal line at the mean of all Y values. When $R^2 = 1.0$, the best-fit curve fits the data perfectly, going through every point. You can think of R^2 as the fraction of the total variance of Y that is explained by the model (equation).

Linear regression always finds unique results. Put the data in and the answers come out. There is no chance of ambiguity. Nonlinear regression is more tricky, and the program can report the wrong answers. This happens because nonlinear regression programs always adjust the variables in small steps to improve the goodness of fit. The program is done when making a small change to any variable would make the fit worse. Usually, this process finds the right answer. In some cases, however, this process does not find the best curve. To find the best-fit curve would require making a large change in one or more variables, but nonlinear regression programs always make small changes. This problem is termed finding a *local minimum*. You will rarely encounter a local minimum if your data have little scatter, you collected data over an appropriate range of X values, you have chosen an appropriate equation and have provided reasonable initial values.

POLYNOMIAL REGRESSION

Nonlinear regression techniques can fit data to any equation that defines Y as a function of X and other variables. One particular equation, the polynomial equation (Equation 34.3), deserves special emphasis:

$$Y = \alpha + \beta_1 \cdot X + \beta_2 \cdot X^2 + \beta_3 \cdot X^3 + \beta_4 \cdot X^4 \ldots \qquad (34.3)$$

The polynomial equation is special because it is not strictly a "nonlinear" equation, even though curves generated by this equation are curved.* Writing a program to fit data to the polynomial equation is far easier than writing a program to fit data to a nonlinear equation. For this reason, more programs offer polynomial regression than nonlinear regression. Polynomial regression programs are easy to use, as you only have to make one choice, how many terms to include in the equation. If the equation ends with the β_2 term above, it is called a *second-order equation*. If you stop after the β_3 term above, it is called a *third-order equation*. Unlike nonlinear regression, you don't have to enter initial estimated values of the variables, and you don't have to worry about the possibility that the results are a local minimum.

In addition to drawing the curve, such programs will yield an R^2 value, as well as values for all the parameters (α, β_1, β_2 ...) and a 95% CI for each. Polynomial regression can produce attractive curves, and these curves are often used as standard curves for interpolating unknown values. However, few biological or chemical processes follow models described by polynomial equations, so the parameters can rarely be interpreted in terms of chemistry or biology.

---

*Y is *not* a linear function of X, so the XY graph is curved. Y *is* a linear function of the parameters β_1, β_2, β_3, etc. so the equation is not "nonlinear."

35

Combining Probabilities

PROPAGATION OF ERRORS

Variables of biological or clinical interest often are not measured directly but instead are calculated from two or more measured variables. This section explains how to determine the confidence interval (CI) of the calculated variable.

Confidence Interval of a Difference

It is common to analyze a difference (D) determined by subtracting a baseline or non-specific measurement (B) from a total measurement (A). In other words, $D = A - B$. Both A and B are Gaussian variables, and you want to know the 95% CI for the difference between the means (D). Assuming that the observations are not paired and the two samples sizes are equal, you can calculate the SE of the difference between means from this equation:

$$SE_D^2 = \sqrt{SEM_A^2 + SEM_B^2}. \qquad (35.1)$$

Three notes:

- The SE of the difference is larger than both the SE of A and the SE of B.
- Equation 35.1 (identical to Equation 7.1) assumes that both groups have the same number of observations. If the sample sizes are different, use Equation 7.2 instead.
- Both A and B must be measured in the same units, and D is measured in the same units.

The 95% CI for the difference D is calculated using this equation, where t* is the critical value of the t distribution for 95% confidence. The number of degrees of freedom (df) equals the total number of observations in groups A and B minus 2.

$$D - t^* \cdot SE_D \text{ to } D + t^* \cdot SE_D$$
$$df = N_A + N_B. \qquad (35.2)$$

This equation assumes that the two variables A and B are not paired or matched, and are distributed in a Gaussian manner. If the two variables are paired (and correlated), then more complicated equations are needed that also include the correlation coefficient between the two variables.

Confidence Interval of a Sum

Calculating the CI of a sum uses identical equations to the CI of a difference. If a sum is defined by $S = A + B$, the 95% CI of S is given by Equations 35.1 and 35.2. Simply change D (for difference) to S (for sum).

Confidence Interval of a Ratio of Two Means

Another common situation is that the variable of interest is a quotient (Q) determined by dividing one measurement by another ($Q = A/B$). A and B are Gaussian variables, and you wish to know the 95% CI for the mean quotient Q. Since A and B can be measured in different units, there is no way to directly combine the SEs (which may also be measured in different units). Instead it is possible to combine the ratios of the SE of A or B divided by the means. If the SE of the denominator B is small compared to B, then you can approximate the CI of the quotient using Equation 35.3:

$$SE_Q = Q\sqrt{\frac{SEM_A^2}{A^2} + \frac{SEM_B^2}{B^2}}$$

(35.3)

95% CI: $Q - t^* \cdot SE_Q$ to $Q + t^* \cdot SE_Q$.

Since SEM_A is expressed in the same units as A, the ratio is unitless. The ratios of SEM_B/B and SE_Q/Q are also unitless.

If the SEM of the denominator is not small compared to the denominator, then you need to use a more complicated equation, as the CI of the quotient is not symmetrical around the quotient. The derivation of the equations (derived by Fieller) is not intuitive and will not be presented here.

First calculate the intermediate variable g using Equation 35.4:

$$g = \left(t^* \cdot \frac{SEM_B}{B}\right)^2.$$

(35.4)

If g has a value greater than or equal to 1, you cannot calculate the CI of the quotient. The value will only be this large when the CI for the denominator (B) includes 0. If the CI of the denominator includes 0, then it is impossible to calculate the CI of the quotient. If g has a value less than 1.0, calculate the CI for the quotient from Equation 35.5:

$$SE_Q = \frac{Q}{(1 - g)} \cdot \sqrt{(1 - g)\frac{SEM_A^2}{A^2} + \frac{SEM_B^2}{B^2}}.$$

(35.5)

95% CI: $\frac{Q}{(1 - g)} - t^* \cdot SE_Q$ to $\frac{Q}{(1 - g)} + t^* \cdot SE_Q$.

Calculate g using Equation 35.4 and look up t* for 95% confidence with the degrees of freedom equal to the total number of values in numerator and denominator minus two.

If the SEM of B is small, then g is very small, the CI is nearly symmetrical around A/B and Equation 35.3 is a reasonable approximation. If the SEM of B is moderately large, then the value of g will be a large fraction, the CI will not be centered around Q and you need Equation 35.5. If the SEM of B is large, then g will be greater than 1.0 and the CI cannot be calculated.

These equations are based on the assumption that the two variables A and B are not paired or matched, and are distributed in a Gaussian manner. If the two variables are paired (and thus correlated) then more complicated equations are needed that also include the correlation coefficient between the two variables.

EXAMPLE OF ERROR PROPAGATION

Adenylyl cyclase activity was measured under basal conditions and after stimulation by a hormone. Under basal conditions, the activity was 11 ± 3 pmol/min/mg protein. After stimulation, the activity was 105 ± 10 pmol/min/mg protein. Both values are expressed as the mean and SE of four replicate determinations. We will analyze these data as the difference between stimulated and basal levels and as the ratio of the two.

The difference between means is $105 - 11$ or 94. Using Equation 35.1, the SE of the difference is 10.4, and the 95% CI is 68 to 120 pmol/min/mg. We can be 95% sure that the true difference lies within this range.

Now let's look at the ratio of stimulated to basal activities. These data showed a 105/11 or 9.5-fold increase in enzyme activity. To calculate the 95% CI, first look up $t^* = 2.45$, then calculate g, which is 0.445. The 95% CI of the quotient is from 5.3 to 29.1. In other words, we can be 95% sure that, overall, the average stimulation is between 5-fold and 29-fold.

DECISION ANALYSIS

Making decisions can be difficult because one has to integrate various probabilities and various outcomes to decide what to do. Decision analysis is a procedure for explic-itly mapping out the probabilities and outcomes, in an effort to focus one's thinking about decisions and help reach the right one.

Decision analysis is useful because people often find it difficult to think about multiple possibilities and probabilities. As an example I will use a problem that has been widely discussed in diverse forums including *Parade* magazine (a supplement to many Sunday newspapers), the *New York Times,* and "Car Talk" (a talk show on National Public Radio that usually discusses car mechanics), as well as mathematical and statistical journals.

Here is the problem: You are on a game show and are presented with three doors. Behind one is a fancy new car worth US$10,000 (which gives a clue as to how many years this problem has been circulating). Behind the others is a booby prize. As the problem was initially presented, this booby prize was a goat. However, to avoid entanglement with animal activists, we'll just leave those two doors empty. You must choose one door and get to keep whatever is behind it. All doors have identical appearances, and the moderator (who knows which door has the car behind it) gives no clues. You pick a door. At this point, the moderator opens up one of the other two doors and shows you that there is no car behind it. He now offers you the chance to change your mind and choose the other door. Should you switch?

Before reading on, you should think about the problem and decide whether you should switch. There are no tricks or traps. Exactly one door has the prize; all doors appear identical; the moderator (who knows which door leads to the new car) has a perfect poker face and gives you no clues. There is never a car behind the door the moderator chooses to open. Most people find it difficult to think through the possibilities, and most people reach the wrong conclusion. Don't cheat. Think it through yourself before continuing.

A decision tree helps focus one's thinking (Figure 35.1). The first branch of the tree presents the choice you are confronted with: Switch or don't switch. The next branch presents the two possibilities: Either you picked the correct door the first time or you didn't. The outcomes, in this case, the dollar value of the prize behind the door, are shown at the end of the tree. In this tree, there is only one set of probabilities: What is the chance that your first choice was correct? Given the rules of the contest, and barring extrasensory perception, the probability that your initial guess was correct is one third, which leaves a two-thirds chance that the initial choice was wrong.

The tree presents the outcomes and the probabilities. Now sum the product of each outcome by the corresponding probability. If you stick with your initial choice (don't switch doors), the expected outcome is 1/3 * $10,000 + 2/3 * $0, which equals $3333. If you switch doors, the expected outcome is 2/3 * $10,000 + 1/3 * $0, which equals $6666. Thus your best choice is to switch! Of course, you can't be absolutely sure that switching doors will help. In fact, one third of the time you will be switching

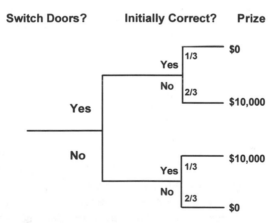

Figure 35.1. A decision tree. To determine the best choice, multiply the probability by the prize, and sum the possibilities for each decision. If you choose to switch doors, your expected prize (the average if you repeated the game many times) is $0 * (1/3) + $10,000 * (2/3), which equals $6,666. If you choose not to switch doors, your expected prize is $10,000 * (1/3) + $0 * (2/3), which is only $3,333. Over the long run, you'll certainly do better by switching. If you can only play the game once, you don't know which decision will be correct, but you have a greater chance of winning by switching than by not switching. This conclusion surprises many people.

away from the prize. But the other two thirds of the time you will be switching to the prize. If you repeat the game many times, you will maximize your winnings by switching doors every time. If you only get to play once, the outcome is a matter of chance. It is clear, however, that you have twice the chance of winning by switching doors. Many, indeed most, people intuitively reach the wrong decision. It is very hard to simultaneously think through two (or more) parallel tracks, as is needed.

Creating decision trees for medical decisions presents two additional complexities. First, the various outcomes (monetary cost, disability, pain, and death) are not directly comparable. Somehow these have to be normalized to one scale, called *utility*. This means explicitly putting a dollar value on death and disability. People cringe at making this decision explicit, although such decisions are made implicitly all the time. The other problem is that the various probabilities are not always known exactly. However, it is often possible to estimate the probabilities reasonably.

Most decision trees for medical applications have many more branches than the simple example. For example, a decision tree for whether or not to operate on a certain condition must include a branch for operative death and acquiring an infection during the surgery. That branch must have sub-branches to deal with the possibility of developing an allergic reaction or other severe reaction to antibiotics.

Once the tree is complete, the probabilities and outcomes can be propagated down the tree to find which decision is best. Before using such a tree, however, you want to know how robust it is. What if one of the utility values were changed just a bit? What if we doubled the probability of antibiotic reaction? When such questions are pursued systematically, the results are called *sensitivity analyses*.

Decision analysis should not be used as a way to avoid taking responsibility for decisions, nor for relegating decisions to computers. However, decision analysis is a very useful way to force one to think clearly about decisions and to explicitly state the assumptions (probabilities and relative utilities) upon which they are based.

META-ANALYSIS

If you read the medical literature pertaining to a clinical problem, you will often find several similar studies that reach somewhat different conclusions. How can one best pool the results of the studies?

There are two distinct problems. The first is the scientific problem of deciding which studies are best. Various investigators will have used different study designs and different patient populations, and have measured different end points. It can be difficult to decide which studies to emphasize and which to ignore. This is a fundamental challenge of evaluating clinical research, and there are no shortcuts. The second problem is more manageable. Once you have selected comparable studies (and eliminated seriously flawed studies), how can their results be pooled?

An obvious, but seriously flawed, solution is *majority rules*. Simply tabulate how many studies show a significant difference and how many don't, and go with the larger number. This is not a good way to pool data. The problem is that significance depends on both the size of the effect and on sample size. A better approach is to tabulate the CIs from each study. Sometimes it requires a bit of recalculating to create comparable

CIs for each study. These individual CIs can then be graphed and pooled (using methods not discussed here) to yield an overall CI.

Combining together the results of many studies is called *meta-analysis*. It takes a lot of work to do a good meta-analysis, and the pooled results can be far more useful than the individual studies. Medical journals often publish meta-analyses. In addition to tabulating the results, a good meta-analysis also discusses the strengths, weaknesses, and potential biases of each study. Even if each individual study is inconclusive, a meta-analysis can lead to clear conclusions.

When reading a meta-analysis, here are some questions to ask: How did the authors select papers to study? How many of the papers did they include in the formal analysis? Why did they exclude the others? Who were the patients in each study? How similar were the diagnoses and treatments? How were the results pooled? Did they look at various patient subgroups separately? Who paid for the study? How similar were the results in the studies?

IX

OVERVIEWS

36

Adjusting for Confounding Variables

This chapter reviews topics that were introduced in many different chapters.

WHAT IS A CONFOUNDING VARIABLE?

When comparing means, proportions, or survival times between two groups, you ordinarily must assume that the only difference between the two groups is the experimental treatment. What if the groups differ in other ways? What if one group was older than the other? What if the placebo-treated group were more sick to start with? In these situations, there are several possible explanations for differences between groups. It is difficult to know whether the difference is due to the treatment or due to the other differences between the groups.

When results can be explained in more than one way, the results are said to be *confounded*. The confounding variable is associated both with the outcome you care about and the treatment group or risk factor. The presence of confounding variables complicates the analysis of many studies.

Here is a more formal way to define confounding variables. In any experiment, there are three kinds of variables: independent, dependent, and extraneous. Variables that the experimenter manipulates (or used to select subjects) are called *independent variables*. The response that the experimenter measures is called the *dependent variable*. The whole point of the study is to see if the independent variable influences the dependent variable. Variables that affect the dependent variable but are not part of the experimental design are called *extraneous variables*. When an extraneous variable is systematically related to the independent variable, the two variables are said to be *confounded*, and it is difficult to tell whether differences in the dependent variable are due to the independent variable or to the extraneous variable.

DESIGNING STUDIES TO AVOID CONFOUNDING VARIABLES

A good study is designed to reduce the influence of extraneous variables. This is done using these techniques (most of which were already discussed in Chapter 20):

- *Random allocation of treatments.* Treatments should be assigned randomly. Any other scheme may potentially introduce confounding variables.

- *Blinding.* Physicians should not know which treatment the patient receives, and the patient should not know what treatment they are getting.
- *Matching.* Some studies compare pairs of subjects, matched for age, disease stage, ethnic group, etc.
- *Stratified randomization.* Some studies divide the subjects into groups (strata) based on age or disease stage. Subjects in each group are randomly divided between the two treatments. This ensures that the grouping variable (age) cannot confound the results.
- *Compulsive controls.* The two treatment groups must be treated identically, except for the experimental treatment.
- *Crossover design.* In some studies, each subject serves as his or her own control.

THE NEED TO CORRECT FOR CONFOUNDING VARIABLES

Example 36.1

Tables 36.1 and 36.2 present hypothetical results from two experiments that compare two operations. In both studies the relative risk was 1.00, so you conclude the overall success rate is identical for the two operations. There is no reason to prefer one over the other. These are fake data. With real data, even if there is no difference in the operations overall, you'd expect to see some differences in the sample you study.

What happens if you combine the two studies and analyze the pooled data? The results, shown in Table 36.3, are surprising.

Overall, operation I failed in $140/320 = 43.75\%$ of the subjects, and operation II failed in only $80/240 = 33.3\%$ of the subjects. The relative risk is 1.31. If you test the null hypothesis of no difference with Fisher's test, the P value is 0.014. Operation 2 is significantly better. If you take this analysis on face value (you shouldn't), you would conclude the difference between the two operations is statistically significant and that operation II is 31% better.

How is this possible? In each study, the two operations were indistinguishable. Why are the combined data different? The pooled analysis violates a key assumption of chi-square (or Fisher's) test, that every observation is independently sampled from one population. Here the subjects are sampled from two populations that differ substantially. Overall (pooling the two operations), 50% of the operations in the first study failed, while only 25% of the operations in the second study failed. Pooling the data into one contingency table is not valid and leads to an incorrect conclusion.

Table 36.1. Study 1: Results

| | Outcome | | | Analysis |
	Failed	Worked	Total	P = 1.00 RR = 1.00
Operation I	120	120	240	
Operation II	40	40	80	
Total	160	160	320	

Table 36.2. Study 2: Results

	Outcome			Analysis
				P = 1.00
	Failed	Worked	Total	RR = 1.00
Operation I	20	60	80	
Operation II	40	120	160	
Total	60	180	240	

To analyze these data properly, you need to use a fancier test to test this null hypothesis: After adjusting for differences between the failure rate of the two populations, there is no difference between the failure rates of the two operations. Several statistical tests can calculate a P value to test this null hypothesis. One is the Mantel-Haenszel test, which was briefly mentioned in Chapter 29, and the other is logistic regression, which was described in Chapter 32. Either way, the resulting P value is 1.00. You can conclude that there is no difference between the two operations after adjusting for differences between the populations used in the two studies.

In this example, the presence of a confounding variable created the illusion of a major difference between treatments. Adjusting for the confounding variable raised the P value. In other examples, a confounding variable could hide the presence of a real difference. In these cases, adjusting for the confounding variable would decrease P values.

STATISTICAL TESTS THAT CORRECT FOR CONFOUNDING VARIABLES

Many tests have been developed to test for differences between groups after adjusting for one or more confounding variables. Table 36.4 lists some tests that you will encounter in the medical literature.

While these tests are rarely used by basic scientists, many of the articles you read in clinical journals use one or more of these tests. Three regression methods are used most often, as regression techniques can analyze the same kinds of data analyzed by the Mantel-Haenszel test and by two-way ANOVA.

Table 36.3. Combined Results of Studies 1 and 2

	Outcome			Analysis
				P = 0.014
	Failed	Worked	Total	RR = 1.31
Operation I	140	180	320	
Operation II	80	160	240	
Total	220	340	560	

Table 36.4. Statistical Texts Correcting for Confounding Variables

| Test | Outcome Variable | What kinds of variables can it adjust for? | | |
		Binomial Variable (discrete categories)	Measurements (Continuous scale)	Several Variables
Two-way ANOVA	Continuous (measurement)	YES	NO*	NO
Mantel-Haenszel	Survival time or binomial (fraction)	YES	NO*	NO
Multiple regression	Continuous (measurement)	YES	YES	YES
Logistic regression	Binomial (fraction)	YES	YES	YES
Cox proportional hazards regression	Survival time	YES	YES	YES

*No, unless you divide the measurements into two or more categories. For example, you could divide cholesterol levels into low, medium and high.

INTERPRETING RESULTS

If you need to analyze data using these tests, you'll need to consult with a statistician or read more advanced books. You'll frequently encounter results of these tests in the biomedical literature and shouldn't have much trouble understanding the results. Here are some questions to ask yourself:

• Has the study been designed well? Fancy statistical tests cannot make up for poor experimental design.
• Can you see the main findings on graphs or tables? First try to understand the data. Then look at the P values.
• When you see a P value, are you sure you know what null hypothesis is being tested? These tests commonly generate more than one P value, and it is easy to get mixed up as to what they mean. Unless you can clearly state the null hypothesis, you shouldn't try to interpret the P value.
• How did the investigators decide which variables to adjust for? Sometimes the investigators collect many variables and let the computer decide which of these are confounding the results and worth adjusting for. This is a powerful technique, but one that can be abused.
• Are the investigators testing a focused hypothesis, or are they fishing for P values? Beware of unfocused studies in which the authors adjust for many combinations of variables. Even if all null hypotheses are true, you expect 5% of P values to be "significant." If the investigators gave the computer enough potentially confounding variables, it is easy to generate dozens or hundreds of P values. You may be misled if they only show you the significant P values without telling you the number of P values that are not significant. See the discussion of multiple comparisons in Chapter 13.

37

Choosing a Test

REVIEW OF AVAILABLE STATISTICAL TESTS

This book has discussed many different statistical tests. To select the right test, ask yourself two questions: What kind of data have you collected? What is your goal? Then refer to Table 37.1.

All tests are described in this book and are performed by InStat, except for tests marked with asterisks. Tests labeled with a single asterisk are briefly mentioned in this book, and tests labeled with two asterisks are not mentioned at all.

REVIEW OF NONPARAMETRIC TESTS

Choosing the right test to compare measurements is a bit tricky, as you must choose between two families of tests: parametric and nonparametric. Many statistical tests are based upon the assumption that the data are sampled from a Gaussian distribution. These tests are referred to as *parametric tests*. Commonly used parametric tests are listed in the first column of the table and include the t test and analysis of variance.

Tests that do not make assumptions about the population distribution are referred to as *nonparametric tests*. You've already learned a bit about nonparametric tests in previous chapters. All commonly used nonparametric tests rank the outcome variable from low to high and then analyze the ranks. These tests are listed in the second column of the table and include the Wilcoxon, Mann-Whitney test, and Krusak-Wallis tests. These tests are also called *distribution-free* tests.

CHOOSING BETWEEN PARAMETRIC AND NONPARAMETRIC TESTS: THE EASY CASES

Choosing between parametric and nonparametric tests is sometimes easy. You should definitely choose a parametric test if you are sure that your data are sampled from a population that follows a Gaussian distribution (at least approximately). You should definitely select a nonparametric test in three situations:

• The outcome is a rank or a score and the population is clearly not Gaussian. Examples include class ranking of students, the Apgar score for the health of newborn babies

Table 37.1. Selecting a Statistical Test

	Type of Data			
Goal	Measurement (from Gaussian Population)	Rank, Score, or Measurement (from Non-Gaussian Population)	Binomial (Two Possible Outcomes)	Survival Time
Describe one group	Mean, SD	Median, interquartile range	Proportion	Kaplan Meier survival curve
Compare one group to a hypothetical value	One-sample t test	Wilcoxon test	Chi-square or Binomial test**	—
Compare two unpaired groups	Unpaired t test	Mann-Whitney test	Fisher's test (chi-square for large samples)	Log-rank test or Mantel-Haenszel*
Compare two paired groups	Paired t test	Wilcoxon test	McNemar's test	Conditional proportional hazards regression**
Compare three or more unmatched groups	One-way ANOVA	Kruskal-Wallis test	Chi-square test	Cox proportional hazard regression*
Compare three or more matched groups	Repeated-measures ANOVA	Friedman test	Cochrane Q**	Conditional proportional hazards regression**
Quantify association between two variables	Pearson correlation	Spearman correlation	Contingency coefficients**	
Predict value from another measured variable	Simple linear regression or Nonlinear regression	Nonparametric regression**	Simple logistic regression*	Cox proportional hazard regression*
Predict value from several measured or binomial variables	Multiple linear regression* or Multiple nonlinear regression**		Multiple logistic regression*	Cox proportional hazard regression*

*Only briefly mentioned in this book.
**Not discussed in this book.

(measured on a scale of 0 to 10 and where all scores are integers), the visual analogue score for pain (measured on a continuous scale where 0 is no pain and 10 is unbearable pain), and the star scale commonly used by movie and restaurant critics (* is OK, ***** is fantastic).

• Some values are "off the scale," that is, too high or too low to measure. Even if the population is Gaussian, it is impossible to analyze such data with a parametric test since you don't know all of the values. Using a nonparametric test with these data is simple. Assign values too low to measure an arbitrary very low value and assign values too high to measure an arbitrary very high value. Then perform a nonparametric test. Since the nonparametric test only knows about the relative ranks of the values, it won't matter that you didn't know all the values exactly.

• The data are measurements, and you are sure that the population is not distributed in a Gaussian manner. If the data are not sampled from a Gaussian distribution, consider whether you can transform the values to make the distribution become Gaussian. For example, you might take the logarithm or reciprocal of all values. There are often biological or chemical reasons (as well as statistical ones) for performing a particular transform.

CHOOSING BETWEEN PARAMETRIC AND NONPARAMETRIC TESTS: THE HARD CASES

It is not always easy to decide whether a sample comes from a Gaussian population. Consider these points:

• If you collect many data points (over a hundred or so), you can look at the distribution of data and it will be fairly obvious whether the distribution is approximately bell shaped. A formal statistical test* can be used to test whether the distribution of the data differs significantly from a Gaussian distribution. With few data points, it is difficult to tell whether the data are Gaussian by inspection, and the formal test has little power to discriminate between Gaussian and non-Gaussian distributions.

• You should look at previous data as well. Remember, what matters is the distribution of the overall population, not the distribution of your sample. In deciding whether a population is Gaussian, look at all available data, not just data in the current experiment.

• Consider the source of scatter. When the scatter comes from numerous sources (with no one source contributing most of the scatter), you expect to find a roughly Gaussian distribution.

When in doubt, some people choose a parametric test (because they aren't sure the Gaussian assumption is violated), and others choose a nonparametric test (because they aren't sure the Gaussian assumption is met).

---

*Komogorov-Smirnoff test, not explained in this book.

CHOOSING BETWEEN PARAMETRIC AND NONPARAMETRIC TESTS: DOES IT MATTER?

Does it matter whether you choose a parametric or nonparametric test? The answer depends on sample size. There are four cases to think about:

- Large sample. What happens when you use a parametric test with data from a non-Gaussian population? The central limit theorem (discussed in Chapter 5) ensures that parametric tests work well with large samples even if the population is non-Gaussian. In other words, parametric tests are robust to deviations from Gaussian distributions, so long as the samples are large. The snag is that it is impossible to say how large is large enough, as it depends on the nature of the particular non-Gaussian distribution. Unless the population distribution is really weird, you are probably safe choosing a parametric test when there are at least two dozen data points in each group.
- Large sample. What happens when you use a nonparametric test with data from a Gaussian population? Nonparametric tests work well with large samples from Gaussian populations. The P values tend to be a bit too large, but the discrepancy is small. In other words, nonparametric tests are only slightly less powerful than parametric tests with large samples.
- Small samples. What happens when you use a parametric test with data from non-Gaussian populations? You can't rely on the central limit theorem, so the P value may be inaccurate.
- Small samples. When you use a nonparametric test with data from a Gaussian population, the P values tend to be too high. The nonparametric tests lack statistical power with small samples.

Thus, large data sets present no problems. It is usually easy to tell if the data come from a Gaussian population, but it doesn't really matter because the nonparametric tests are so powerful and the parametric tests are so robust. Small data sets present a dilemma. It is difficult to tell if the data come from a Gaussian population, but it matters a lot. The nonparametric tests are not powerful and the parametric tests are not robust.

ONE- OR TWO-SIDED P VALUE?

With many tests, you must choose whether you wish to calculate a one- or two-sided P value (same as one- or two-tailed P value). The difference between one- and two-sided P values was discussed in Chapter 10. Let's review the difference in the context of a t test. The P value is calculated for the null hypothesis that the two population means are equal, and any discrepancy between the two sample means is due to chance. If this null hypothesis is true, the one-sided P value is the probability that two sample means would differ as much as was observed (or further) in the direction specified by the hypothesis just by chance, even though the means of the overall populations are actually equal. The two-sided P value also includes the probability that the sample means would differ that much in the opposite direction (i.e., the other group has the larger mean). The two-sided P value is twice the one-sided P value.

A one-sided P value is appropriate when you can state with certainty (and before collecting any data) that there either will be no difference between the means or that the difference will go in a direction you can specify in advance (i.e., you have specified which group will have the larger mean). If you cannot specify the direction of any difference before collecting data, then a two-sided P value is more appropriate. If in doubt, select a two-sided P value.

If you select a one-sided test, you should do so before collecting any data and you need to state the direction of your experimental hypothesis. If the data go the other way, you must be willing to attribute that difference (or association or correlation) to chance, no matter how striking the data. If you would be intrigued, even a little, by data that goes in the "wrong" direction, then you should use a two-sided P value. For reasons discussed in Chapter 10, I recommend that you always calculate a two-sided P value.

PAIRED OR UNPAIRED TEST?

When comparing two groups, you need to decide whether to use a paired test. When comparing three or more groups, the term *paired* is not apt and the term *repeated measures* is used instead.

Use an unpaired test to compare groups when the individual values are not paired or matched with one another. Select a paired or repeated-measures test when values represent repeated measurements on one subject (before and after an intervention) or measurements on matched subjects. The paired or repeated-measures tests are also appropriate for repeated laboratory experiments run at different times, each with its own control.

You should select a paired test when values in one group are more closely correlated with a specific value in the other group than with random values in the other group. It is only appropriate to select a paired test when the subjects were matched or paired before the data were collected. You cannot base the pairing on the data you are analyzing.

FISHER'S TEST OR THE CHI-SQUARE TEST?

When analyzing contingency tables with two rows and two columns, you can use either Fisher's exact test or the chi-square test. The Fisher's test is the best choice as it always gives the exact P value. The chi-square test is simpler to calculate but yields only an approximate P value. If a computer is doing the calculations, you should choose Fisher's test unless you prefer the familiarity of the chi-square test. You should definitely avoid the chi-square test when the numbers in the contingency table are very small (any number less than about six). When the numbers are larger, the P values reported by the chi-square and Fisher's test will be very similar.

The chi-square test calculates approximate P values, and the Yates' continuity correction is designed to make the approximation better. Without the Yates' correction, the P values are too low. However, the correction goes too far, and the resulting P value is too high. Statisticians give different recommendations regarding Yates' correction. With large sample sizes, the Yates' correction makes little difference. If you select Fisher's test, the P value is exact and Yates' correction is not needed and is not available.

REGRESSION OR CORRELATION?

Linear regression and correlation are similar and easily confused. In some situations it makes sense to perform both calculations. Calculate linear correlation if you measured both X and Y in each subject and wish to quantify how well they are associated. Select the Pearson (parametric) correlation coefficient if you can assume that both X and Y are sampled from Gaussian populations. Otherwise choose the Spearman nonparametric correlation coefficient. Don't calculate the correlation coefficient (or its confidence interval) if you manipulated the X variable.

Calculate linear regressions only if one of the variables (X) is likely to precede or cause the other variable (Y). Definitely choose linear regression if you manipulated the X variable. It makes a big difference which variable is called X and which is called Y, as linear regression calculations are not symmetrical with respect to X and Y. If you swap the two variables, you will obtain a different regression line. In contrast, linear correlation calculations are symmetrical with respect to X and Y. If you swap the labels X and Y, you will still get the same correlation coefficient.

38

The Big Picture

When reading research papers that include statistical analyses, keep these points in mind.

LOOK AT THE DATA!

Statistical tests are useful because they objectively quantify uncertainty and help reduce data to a few values. While such calculations are helpful, inspection of statistical calculations should never replace looking at the data. The data are important; statistics merely summarize.

BEWARE OF VERY LARGE AND VERY SMALL SAMPLES

The problem with very large sample sizes is that tiny differences will be statistically significant, even if scientifically or clinically trivial. Make sure you look at the size of the differences, not just their statistical significance.

   The problem with small studies is that they have very little power. With small sample sizes, large and important differences can be insignificant. Before accepting the conclusion from a study that yielded a P value greater than 0.05 (and thus not significant), you should look at the confidence interval and calculate the power of the study to detect the smallest difference that you think would be clinically or scientifically important.

BEWARE OF MULTIPLE COMPARISONS

When analyzing random data, on average 1 out of 20 comparisons will be statistically significant by chance. Beware of large studies that make dozens or hundreds of comparisons, as you are likely to encounter spurious "significant" results. When reading papers, ask yourself how many hypotheses the investigators tested.

DON'T FOCUS ON AVERAGES: OUTLIERS MAY BE IMPORTANT

Statistical tests (t tests, ANOVA) compare averages. Variability in biological or clinical studies is not always primarily due to measurement uncertainties. Instead, the variability in the data reflects real biological diversity. Appreciate this diversity! Don't get mesmerized by averages; the extreme values can be more interesting. Nobel prizes have been won from studies of individuals whose values were far from the mean.

NON-GAUSSIAN DISTRIBUTIONS ARE NORMAL

Many statistical tests depend on the assumption that the data points come from a Gaussian distribution, and scientists often seem to think that nature is kind enough to see to it that all interesting variables are so distributed. This isn't true. Many interesting variables are not scattered according to a Gaussian distribution.

GARBAGE IN, GARBAGE OUT

Fancy statistics are not useful if the data have not been collected properly. Statistics are the easy part of interpreting data. The hard part is evaluating experimental methodology. Statistical tests cannot tell you whether the study was conducted properly. Many studies are designed badly, and statistical tests from such studies can give results that are misleading. If the data were not collected properly (or as statisticians say if the study was biased), statistical tests are hard to interpret. Thinking about experimental design is the challenging part of science. Perfect studies are rare, so we need to draw conclusions from imperfect studies. This requires judgment and intuition, and the statistical considerations are often minor.

CONFIDENCE LIMITS ARE AS INFORMATIVE AS P VALUES (MAYBE MORE SO)

The P value is the likelihood that the difference you observed would occur if the null hypothesis were true. The confidence interval tells you how certain you know the overall difference. The two are complementary, and the best policy is to report both. Many published papers show P values without confidence intervals. Usually it is not too difficult to construct a confidence interval from the data presented.

STATISTICALLY SIGNIFICANT DOES NOT MEAN SCIENTIFICALLY IMPORTANT

With large sample sizes, small differences can be statistically significant but may not be important. P values (and statistical significance) are purely the result of arithmetic; decisions about importance require judgment.

P < 0.05 IS NOT SACRED

There really is not much difference between p = 0.045 and p = 0.055! By convention, the first is statistically significant and the second is not, but this is completely arbitrary. A rigid cutoff for significance is useful in some situations but is not always needed. Don't be satisfied when someone tells you whether the P value is greater or less than some arbitrary cutoff; ask for its exact value. More important, ask for the size of the difference or association, and for a confidence interval.

DON'T OVERINTERPRET NOT SIGNIFICANT RESULTS

If a difference is not statistically significant, that means that the observed results are not inconsistent with the null hypothesis. It does *not* mean that the null hypothesis is true. You often read that "there is no evidence that A causes B." That does not necessarily mean that A doesn't cause B. It may mean that there are no data, or that all the studies have used few subjects and don't have adequate power to find a difference.

DON'T IGNORE PAIRING

Paired (or repeated-measures) experiments are very powerful. The use of matched pairs of subjects (or before and after measurements) controls for many sources of variability. Special methods are available for analyzing such data, and they should be used when appropriate.

CORRELATION OR ASSOCIATION DOES NOT IMPLY CAUSATION

A significant correlation or association between two variables may indicate that one variable causes the other. But it may just mean that both are related to a third variable that influences both. Or it may be a coincidence.

DISTINGUISH BETWEEN STUDIES DESIGNED TO GENERATE A HYPOTHESIS AND STUDIES DESIGNED TO TEST ONE

If you look at enough variables, or subdivide subjects enough ways, some "significant" relationships are bound to turn up by chance alone. Such an approach can be a useful way to generate hypotheses. But such a hypothesis then needs to be tested with new data.

DISTINGUISH BETWEEN STUDIES THAT MEASURE AN IMPORTANT OUTCOME AND STUDIES THAT MEASURE A PROXY OR SURROGATE OUTCOME

Measuring important outcomes (i.e., survival) can be time consuming and expensive. It is often far more practical to measure proxy or surrogate variables (i.e., white cell count). But, although the relationship between the proxy variable and the real variable may be "obvious," it may not be true. Treatments that improve results of lab tests may not improve health or survival.

PUBLISHED P VALUES ARE OPTIMISTIC (TOO LOW)

By the time you read a paper a great deal of selection has occurred. When experiments are successful, scientists continue the project. Lots of other projects get abandoned. And when the project is done, scientists are more likely to write up projects that lead to significant results. And journals are more likely to publish "positive" studies. If the null hypothesis were true, you would get a significant result 5% of the time. But those 5% of studies are more likely to get published than the other 95%.

CONFIDENCE INTERVALS ARE OPTIMISTIC (TOO NARROW)

The "real" confidence interval (CI) is nearly always wider than the CI calculated from a particular experiment. The strict interpretation of a CI assumes that your samples are randomly sampled from the population of interest. This is often not the case. Even if the sample were randomly selected, the population of interest (all patients) is larger and more heterogeneous than the population of patients studied (limited to certain ages, without other disease, living in a certain area, and receiving medical care from a certain medical center). Thus the true 95% CI (which you can't calculate) is often wider than the interval you calculate.

X

APPENDICES

Appendix I

References

Table A1.1 lists the statistics books that I have found to be most useful. For each topic I indicate up to three excellent references. In many cases, it was difficult to limit the recommendations to three books.

1. DG Altman. *Practical Statistics for Medical Research.* New York, Chapman and Hall, 1991, 596 pages. Compared to most statistics books, this one is much more oriented towards explaining what the tests mean. Lots of clinical examples.
2. JC Bailar III, F Mosteller. *Medical Uses of Statistics,* 2nd ed. Boston, New England Journal of Medicine Books, 1992, 449 pages. Uneven collection of articles from the *New England Journal of Medicine.* Very strong in study design and interpretation of results. Includes some surveys of how statistics are really used in the medical literature.
3. WW Daniel. *Applied Nonparametric Statistics,* 2nd ed. Boston, PWS-Kent, 1990, 635 pages. Comprehensive cookbook for nonparametric tests.
4. AR Feinstein. *Clinical Biostatistics.* St. Louis, MO, CV Mosby, 1977, 468 pages. Collection of essays about statistics. Very opinionated and thought provoking.
5. LD Fisher, G vanBelle. *Statistics. A Methodology for the Health Sciences.* New York, Wiley-Interscience, 1993, 991 pages. This book has more depth than most others, with many examples, explanations, and cautions.
6. JL Fleiss. *Statistical Methods for Rates and Proportions,* 2nd ed. New York, John Wiley and Sons, 1981, 321 pages. Thorough reference for analyses of rates and proportions. Oriented towards epidemiology.
7. SA Glantz. *Primer of Biostatistics,* 3rd ed. New York, McGraw Hill, 1992, 440 pages. Wonderful primer that really explains how basic statistical tests work. The explanations of ANOVA and power are wonderful. Nothing about analysis of survival curves, multiple regression, or logistic regression.
8. SA Glantz, BK Slinker. *Primer of Applied Regression and Analysis of Variance.* New York, McGraw-Hill, 1990, 777 pages. Very clear explanation of multiple and logistic regression, one- and two-way analysis of variance, and nonlinear regression.
9. GR Norman, DL Streiner. *PDQ Statistics.* New York, BC Decker, 1986, 172 pages. Short, funny, clear introduction to the basics with very brief explanations of many advanced tests.
10. WH Press, SA Teukolsky, WT Vetterling, BP Flannery. *Numerical Recipes in C. The Art of Scientific Computing,* 2nd ed. Cambridge, Cambridge University Press, 1992, 994 pages. Only a small part of this book deals with statistics. Look here if you plan to write computer programs that calculate P values, or if you want a mathematical explanation of regression.
11. WL Hays. *Statistics,* 4th ed. Orlando, Harcourt Brace Jovanovich, 1988, 1029 pages. Wonderfully lucid explanation of fundamental statistical principles. Contains a fair amount of math,

Table A1.1. Reference Books on Statistics

Topic	Reference (see list below)												
	1	2	3	4	5	6	7	8	9	10	11	12	13
Intuitive introduction to statistical principles	*						*		*				
Mathematical explanation of statistical principles					*		*				*		
Critique of statistical practice			*		*								
One-way ANOVA							*	*				*	
Two-way ANOVA								*			*	*	
Nonparametric tests				*								*	
Multiple and logistic regression						*		*					*
Nonlinear regression								*	*				
Survival curves	*	*			*								
Rates and proportions					*	*							
Sample size and power	*					*						*	
Computer algorithms										*			
Designing clinical studies	*	*	*										
Comprehensive reference	*				*							*	
Interpreting results	*			*	*								
Epidemiology					*	*							*

but with text that really explains what is going on. Little about nonparametric tests. It was written for social scientists, so has nothing about survival curves and has no biological or clinical examples.

12. JH Zar. *Biostatistical Analysis,* 2nd ed. Englewood Cliffs, NJ, Prentice-Hall, 1984, 696 pages. Good comprehensive reference for all aspects of biostatistics except for analyses of survival curves. If you need to find an equation or table, it will be here. Does not cover survival curves or logistic regression.

13. HA Kahn, CT Sempos. *Statistical Methods in Epidemiology.* New York, Oxford University Press, 1989, 291 pages. Good explanation of multivariate methods (logistic regression and proportional hazards regression).

Appendix 2

GraphPad InStat and GraphPad Prism

WHAT INSTAT DOES

GraphPad Instat is a simple and inexpensive statistical program available for both DOS and Macintosh computers.* It is a terrific companion to this book. Unlike heavy-duty programs designed for statisticians, InStat is designed for students, clinicians, and scientists. InStat explains choices clearly and presents results without unnecessary statistical jargon. Built-in help screens are available to help you choose a test and interpret the results. InStat is extremely easy to learn, and most users never need to read the manual.

InStat calculates the following tests, all with confidence intervals and exact P values:

Group comparisons: Paired and unpaired t tests; Mann-Whitney and Wilcoxon nonparametric tests; ordinary and repeated-measures ANOVA followed by Bonferroni, Tukey, Student-Newman-Keuls, or Dunnett post tests; Kruskal-Wallis or Friedman nonparametric tests, followed by Dunn post tests.

Contingency tables: Chi-square test (with or without Yates' correction); Fisher's exact test. McNemar's test; calculates 95% confidence interval for the difference of two proportions, relative risk or odds ratio; chi-square test for trend.

Linear regression and correlation: Linear regression, optionally forcing the line through a defined point; determines new points along the standard curve; Pearson linear correlation and Spearman nonparametric correlation.

Calculate sample size: Calculates sample size needed to compare one mean or proportion with a theoretical value, or to compare two means or two proportions; enters either the minimum difference to be detected as statistically significant (with specified values of α and β) or the desired span of the confidence interval.

Miscellaneous calculations: Calculates P from z, t, r, F, or chi-square; calculates the 95% confidence interval of a proportion or count; compares observed and expected counts using chi-square.

---

*By the time you read this, a special version for Windows may also be available.

WHAT INSTAT DOES NOT DO

InStat was designed to instantly analyze relatively small data sets. InStat is not able to select a subset of cases from a large file or crosstabulate data to form a contingency table. InStat is an ideal tool for analyzing experimental data and data from small clinical studies but is not designed to analyze large clinical studies. You may enter at most 500 rows and 26 columns of data.

InStat can perform most basic statistical calculations but cannot perform advanced analyses. Specifically, InStat cannot perform survival curve analyses, multiple regression, logistic regression, or two-way ANOVA (but note that InStat performs one-way repeated-measures ANOVA, which is the same as two-way ANOVA without replication). Although InStat cannot perform polynomial or nonlinear regression, these analyses can be performed by the companion program, GraphPad Prism (Windows).

GRAPHPAD PRISM

Prism creates publication-quality graphs (all graphs in this book were created with Prism). Unlike InStat, this program fits curves with nonlinear regression, creates and compares survival curves, and calculates two-way ANOVA.

Version 2, which may be available by the time you read this, will also perform basic statistical tests.

System Requirements

InStat is available for DOS and Macintosh computers. The DOS version requires DOS version 3.0 or later, approximately 440K of free RAM, and 1.2 MB hard disk space. The Macintosh version requires System 6 or later, 700K RAM, and 1 MB hard disk space. Prism requires Windows 3.1 or later and 3 MB hard disk space.

Availability

For more information, contact:

> GraphPad Software
> 10855 Sorrento Valley Road #203
> San Diego, CA 92121
> Phone: 619-457-3909 or 800-388-4723
> FAX: 619-457-8141
> Email: Sales@graphpad.com
> Internet: http://www.graphpad.com

As of 1995, InStat cost only $95, and Prism costs $495. Substantial discounts are available for students and purchases of multiple copies. Both programs are backed by a 90-day money-back guarantee.

Appendix 3

Analyzing Data With a Spreadsheet or Statistics Program

SPREADSHEET PROGRAM FUNCTIONS

Although designed for business use, spreadsheet programs are also very useful for analyzing scientific data and performing statistical calculations. The most popular spreadsheet is Microsoft Excel. Its statistical functions are powerful, but poorly documented. The table below lists some of the functions available. I use version 5 for Windows, and the Macintosh version is supposed to be identical. Other spreadsheet programs may have similar functions. Note that all the tables in Appendix 5 were created using Excel, and the formulae used to calculate the values are documented at the beginning of that Appendix.

To calculate	Use this function	Comments
The mean of a range of values.	AVERAGE (range)	
The median of a range of values.	MEDIAN (range)	
The SD of a range.	STDEV (range)	This function correctly uses $N - 1$ in the denominator, so calculates the "sample SD," which is what you usually want.
The "population SD" of a range.	STDEVP (range)	This function uses N in the denominator, so calculates the "population SD," which is only appropriate when you have collected data for the entire population.
The SEM of a range of values.	STDEV (range)/SQRT (COUNT (range)).	There is no function for the standard error, so you need to calculate the SEM as the SD divided by the square root of sample size.
The value of t* for DF degrees of freedom for P% confidence.	TINV $(1 - 0.01 * P, DF)$.	This finds the critical value of the t distribution to use in calculating P% confidence intervals.
A P value from z	$2* (1.0 - NORMSDIST (z))$	Two-tailed P value.

(continued)

To calculate	Use this function	Comments
A P value from t with DF degrees of freedom	TDIST (t, DF, 2)	Two-tailed P value.
A P value from χ^2 with DF degrees of freedom	CHIDIST (χ^2, DF).	If there is only DF, then the P value is two sided. If there is more than one DF, there is no distinction between one and two tails.
A P value from F with DFn degrees of freedom in the numerator and DFd degrees of freedom in the denominator	FDIST (F, DFn, DFd)	Don't mix up DFn and DFd!

ANALYZING DATA WITH SPREADSHEETS

Using the functions listed in the previous section, you can instruct Excel to do many statistical calculations. Beyond that, Excel can perform some statistical tests for you without you having to write any equations. Pull down the Tools menu and choose Analysis. (If Analysis does not appear on the menu, run the Excel Setup program again and include the Analysis ToolPak.) Then choose the analysis from the menu.

Here is a partial list of the analyses that Excel can perform from the analysis menu: One-way analysis of variance, two-way analysis of variance (no missing values allowed), descriptive statistics by column (mean, median, SD, etc.), frequency distributions, random number generation, linear regression, and t tests (paired and unpaired).

USING A STATISTICS PROGRAM

If you use a statistics program other than InStat and Excel, you may be confused by data entry. Table A3.1 shows how you enter data from two groups that you wish to compare with an unpaired t test using InStat or Excel. The two columns represent the two groups you are comparing.

Most other statistics programs expect you to enter the data differently, as shown in Table A3.2. This is sometimes called an *indexed format*. All of the values for both

Table A3.1.
Arrangement of Data
for InStat or Excel

Group 1	Group 2
2.4	3.7
4.3	5.4
3.4	4.5
2.9	3.9

Table A3.2.
Arrangement of Data
for Most Other
Programs

Group	Value
1	2.4
1	4.3
1	3.4
1	2.9
2	3.7
2	5.4
2	4.5
2	3.9

groups go into one column. Another column (the first in this example) tells you which group the value belongs to. In this format, each row represents one subject, and each column represents one variable. Groups are indicated by a variable. Although this format seems far more confusing when you simply want to compare two groups, it makes things easier when you want to perform more complicated analyses.

Appendix 4

Answers to Problems

CHAPTER 2

1. Of the first 100 patients to undergo a new operation, 6 die. Can you calculate the 95% CI for the probability of dying with this procedure? If so, calculate the interval. If not, what information do you need? What assumptions must you make?

Yes, it is possible to calculate a CI. This is a straightforward calculation of the 95% CI from a proportion. The proportion is 6/100. The denominator is too big to use the table in the Appendix, so we need to calculate the CI using Equation 2.1:

$$\text{Approximate 95\% CI of proportion: } p - 1.96\sqrt{\frac{p(1-p)}{N}} \text{ to } p + 1.96\sqrt{\frac{p(1-p)}{N}}$$

$$0.06 - 1.96\sqrt{\frac{0.06(1-0.06)}{100}} \text{ to } 0.06 + 1.96\sqrt{\frac{0.06(1-0.06)}{100}}$$

$$0.06 - 1.96 \cdot 0.024 \text{ to } 0.06 + 1.96 \cdot 0.024$$

$$0.013 \text{ to } 0.107$$

You can calculate the answer more exactly using a computer program. Using InStat, the 95% CI is 0.02 to 0.13.

To interpret these results, you need to define the population. If these were real data, you'd want to see how the investigators selected the patients. Are they particularly sick? Do they represent all patients with a certain condition, or only a small subset? If you can't figure out how the investigators selected their patients, then the data are not very useful. If you can define the population they selected from, you must assume that the 100 patients are representative of these patients and that each was selected independently. If you accept these assumptions, then you can be 95% sure that the overall death rate in the population lies somewhere within the 95% CI.

2. A new drug is tested in 100 patients and lowers blood pressure by an average of 6%. Can you calculate the 95% CI for the fractional lowering of blood pressure by this drug? If so, calculate the interval. If not, what information do you need? What

is the CI for the fractional lowering of blood pressure? What assumptions must you make?

Although this looks similar to the previous question, it is quite different. In the previous question, the percentage was really a proportion. In this question, the percentage is really change in a measured value. You can't use Equation 2.1. You will learn how to calculate CIs for these kinds of data in later chapters, but you need to know something about the scatter of the data (the standard deviation or standard error).

Beware of percentages. Percentages can express a proportion, a relative difference in proportions, or a relative difference in a measured value, or many other things. Use the methods of Chapter 2 only with proportions.

3. You use a hemocytometer to determine the viability of cells stained with Trypan Blue. You count 94 unstained cells (viable) and 6 stained cells (indicating that they are not viable). Can you calculate the 95% CI for the fraction of stained (dead) cells? If so, calculate the interval. If not, what information do you need? What assumptions must you make?

This is essentially the same problem as #1 above. The population is the entire tube of cells, and you have assessed the viability of a sample of these cells. Assuming that you mixed the tube well, it is reasonable to assume that your sample is randomly selected from the population. You can calculate a 95% CI for the proportion of viable cells in the entire population. Using Equation 2.1, you can say that you are 95% sure that the fraction of stained (dead) cells is between 0.013 and 0.107.

4. In 1989, 20 out of 125 second-year medical students in San Diego failed to pass the biostatistics course (until they came back for an oral exam). Can you calculate the 95% CI for the probability of passing the course? If so, calculate the interval. If not, what information do you need? What assumptions must you make?

It is easy to plug the numbers into Equation 2.1 (or into a computer program) to calculate a CI. But what would it mean? There were only 125 students in 1989, so we don't really have a sample. It would make sense to calculate a CI only if you imagined the 125 students in 1989 to be representative of students in other years (or other cities), and that there is no change in course content, student preparation, exam difficulty, or passing threshold. You would also have to assume that students in future years haven't read this question, as it would give them an incentive to study harder.*

Because those assumptions are unlikely to be true, I don't think it would be meaningful to calculate a CI. If you are persuaded that the assumptions are met, you can calculate that the 95% CI for passing ranges from 77% to 90% (using Equation 2.1) or 76% to 90% (using InStat).

5. Ross Perot won 19% of the vote in the 1992 Presidential Election in the United States. Can you calculate the 95% CI for the fraction of voters who voted for him? If so, calculate the interval. If not, what information do you need? What assumptions must you make?

---

*The students in 1990 read this question. They got the message that the course required work, and they studied hard. All passed.

It wouldn't be meaningful to calculate a CI. We know the exact fraction of voters who voted for each candidate. This is not a poll before the election—these are results from the election. Since we have not sampled from a population, there is no need for a CI.

6. In your city (population = 1 million) last year a rare disease had an incidence of 25/10,000 population. Can you calculate the 95% CI for the incidence rate? If so, calculate the interval. If not, what information do you need? What assumptions must you make?

First, ask yourself whether it makes sense to calculate a CI. You know the incidence of the disease in the entire population last year, so there is no need for a CI for the incidence of the disease last year. If you assume that the overall incidence of the disease is not changing, it may be reasonable to consider last year's population to be a sample from the populations of the next few years. So you can calculate a CI for the overall incidence rate for the next few years.

The problem is stated as number of disease cases per 10,000 population, but 10,000 is not really the denominator, as the data were collected from a population of 1,000,000. So the sample proportion is 2500 out of a population of 1,000,000. Use Equation 2.1 to determine that the CI ranges from 0.24% to 0.26%. Assuming that the incidence of the disease is not changing, you can be 95% sure that the true average incidence rate for the next few years will be somewhere within that range.

7. Is it possible for the lower end of a CI for a proportion to be negative? To be zero?

It is impossible for the end of the CI to be negative. The CI is a range of proportions, and the proportion cannot be negative. The lower confidence limit equals 0 only when the numerator of the proportion is 0.

8. Is it possible to calculate a 100% CI?

If you want to be 100% sure that an interval includes the population proportion, that interval must range from 0 to 100%. That is a 100% CI, but it is not the least bit useful.

CHAPTER 3

1. Estimate the SD of the age of the students in a typical medical school class (just a rough estimate will suffice).

Without data, you can't determine the SD exactly. But you can estimate it. The mean age of first-year medical students is probably about 23. Lots of students are 1 year older or younger. A few might be a lot younger or a lot older. If you assume that the age distribution is approximately Gaussian, you will include two thirds of the students in a range of ages that extends one SD either side of the mean. Therefore the SD is probably about 1 year. Obviously it will vary from class to class.

The point of this question is to remind you that the SD is not just a number that falls out of an equation. Rather, the SD describes the scatter of data. If you have a rough idea how the values are scattered, you can estimate the SD without any calculations.

2. Estimate the SD of the number of hours slept each night by adults.

Without data, you can't know the SD exactly. But you can estimate it. On average, adults sleep about 8 hours. Some sleep more, some less. If you go one SD from the mean, you'll include two thirds of the population, assuming that sleep time is roughly Gaussian. I'd estimate that the SD is about 1 hour. If this were true, then about two thirds of adults would sleep between 6 and 9 hours per night.

Again, the point of this question is to remind you that the SD is not just a number that falls out of an equation. Rather, the SD describes the scatter of data. If you have a rough idea how the values are scattered, you can estimate the SD without any calculations.

3. The following cholesterol levels were measured in 10 people (mg/dl):

260, 150, 160, 200, 210, 240, 220, 225, 210, 240

Calculate the mean, median, variance, standard deviation, and coefficient of variation. Make sure you include units.

$$\text{Mean} = 211.5 \text{ mg/dl}$$

$$\text{Median} = 215 \text{ mg/dl}$$

$$\text{Variance} = 1200.28 \text{ (mg/dl)}^2$$

$$\text{SD} = 34.645 \text{ mg/dl}$$

$$\text{CV} = 34.645/211.5 = 16.4\%$$

4. Add an 11th value (931) to the numbers in Question 3 and recalculate all values.

$$\text{Mean} = 276.9 \text{ mg/dl}$$

$$\text{Median} = 220 \text{ mg/dl}$$

$$\text{Variance} = 48,140 \text{ (mg/dl)}^2$$

$$\text{SD} = 219.4 \text{ mg/dl}$$

$$\text{CV} = 219.4/276.9 = 79.23\%$$

One value far from the rest increases the SD and variance considerably. It has little impact on the median.

5. Can a SD be 0? Can it be negative?

The SD will be 0 only if all values are exactly the same. The SD describes scatter above and below the mean, but it is always a positive number. Mathematically, the SD is the square root of the variance. By convention, the square root is always defined to be positive.

CHAPTER 4

1. The level of an enzyme in blood was measured in 20 patients, and the results were reported as 79.6 ± 7.3 units/ml (mean ± SD). No other information is given about

the distribution of the results. Estimate what the probability distribution might have looked like.

With no other information to go on, it is reasonable to speculate that the enzyme activity would be roughly Gaussian. Since we know the mean and SD, we can sketch this distribution (Figure A4.1).

2. The level of an enzyme in blood was measured in 20 patients and the results were reported as 9.6 ± 7.3 units/ml (mean $\pm$ SD). No other information is given about the distribution of the results. Estimate what the probability distribution might have looked like.

Again, you might start with the assumption that the distribution is roughly Gaussian. If that were true, then 95% of the population would be within 2 SDs of the mean. But to go down 2 SDs from the mean, you need to include negative numbers. But enzyme activity cannot be negative. Therefore, the distribution cannot possibly be Gaussian with the stated mean and SD. The distribution must either have a "tail" to the right, or it might be bimodal. Two possibilities are illustrated in Figure A4.2.

3. The values of serum sodium in healthy adults approximates a Gaussian distribution with a mean of 141 mM and a SD of 3 mM. Assuming this distribution, what fraction of adults has sodium levels less than 137 mM? What fraction has sodium levels between 137 and 145?

What fraction has values below 137 mM? $Z = (141 - 137)/3 = 1.33$. Interpolating from the third column of Table A5.2 in the Appendix, the answer is about 9.2%.

What fraction are between 137 and 145? To answer this question, calculate the fraction with values below 137 and the fraction above 145. Subtract these fractions from 100% to get the answer. You've already calculated the fraction below 137. What fraction have values above 145 mM? The value of $z = (145 - 137)/3 = 2.67$. From the second column of Table A5.2 in the Appendix, the answer is just a bit under 0.40%. The fraction with values between 137 and 145 is $100\% - 9.2\% - 0.4\% = 90.4\%$.

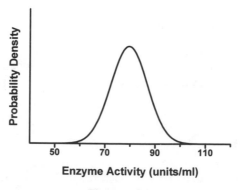

Figure A4.1.

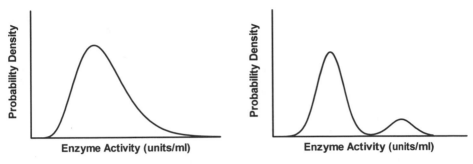

Figure A4.2.

4. You measure plasma glucose levels in five patients and the values are 146, 157, 165, 131, 157 mg/100 ml. Calculate the mean and SD. What is the 95% prediction interval? What does it mean?

The mean is 151.2 mg/100 ml. The SD is 13.2 mg/100 ml. With a sample of $N = 5$, the prediction interval extends 3.04 standard deviations on either side of the mean. So the 95% prediction interval goes from 111.1 to 191.3 mg/100 ml.

5. The Weschler IQ scale was created so that the mean is 100 and the standard deviation is 15. What fraction of the population has an IQ greater than 135?

The value of $z = (135 - 100)/15 = 2.33$. The problem asks what fraction of a standard Gaussian distribution has a z score greater than 2.33. From Table A5.2 in the Appendix, you can interpolate that the value is about 1%. If IQ scores are distributed in a Gaussian manner, 1% of scores are greater than 135.

CHAPTER 5

1. Figure 5.4 shows the probability distribution for the time that an ion channel stays open. Most channels stay open for a very short time, and some stay open longer. Imagine that you measure the open time of 10 channels in each experiment and calculate the average time. You then repeat the experiment many times. Sketch the expected shape of the distribution of the mean open times.

The probability distribution is an exponential curve. Lots of channels stay open a short period of time, a few stay open longer. If you repeatedly calculated the mean open time of 10 channels, the distribution of the means would follow a roughly Gaussian distribution. This is a consequence of the central limit theorem.

2. An enzyme level was measured in cultured cells. The experiment was repeated on 3 days; on each day the measurement was made in triplicate. The experimental conditions and cell cultures were identical on each day; the only purpose of the repeated experiments was to determine the value more precisely. The results are shown as enzyme activity in units per minute per milligram of membrane protein. Summarize these data as you would for publication. The readers have no interest

in the individual results for each day; just one overall mean with an indication of the scatter. Give the results as mean, error value, and N. Justify your decisions.

	Replicate 1	Replicate 2	Replicate 3
Monday	234	220	229
Tuesday	269	967	275
Wednesday	254	249	246

There are two difficulties with this problem. First, you need to decide what to do with Tuesday's second replicate. It is wildly out of whack with the others. Deciding what to do with outlying values is a difficult problem, and different scientists have different policies. In this case, I think it is clear that the value can't possibly be correct. Not only is it far far away from the other two replicates, but it is far far away from the six replicates measured on other days. Since the experiment was performed with cultured cells, there is no biological variability from day to day. The only contribution to scatter are experimental errors. It seems plain to me that the 967 value is clearly incorrect. First, I would look back at the lab notebook and see if the experimenter noticed anything funny about that tube during the experiment. If there was a known problem with that replicate, then I would feel very comfortable throwing it out. Even otherwise, I would delete it from the analysis, while recording in a lab notebook exactly what I had done. Including that value would throw off the calculations, so the result wouldn't be helpful.

If the rows represented different patients (rather than different experiments on the same cell cultures), then my answer would be different. With different patients, biological variability plays a role. Maybe the large number is correct and the two small numbers are wrong. Maybe the second patient is really different than the first and third. If possible, I'd do the experiment with patient #2 a second time to find out.

The second problem is how to pool the eight remaining values. It is not really fair to consider these to be eight independent measurements. The replicates in one day are closer to one another than they are to measurements made on other days. This is expected, since there are more contributions to variability between days than between replicates. So calculating the mean and SD or SEM of all eight values is not valid.

Instead, I would calculate the mean value for each day, and then calculate the mean and SEM of the three means and then calculate a 95% CI. This makes sense if you think of the population as the means of all possible experiments with N = 3. We can be 95% sure that the true mean lies somewhere within the CI.*

The calculations are given in the following table.

	Replicate 1	Replicate 2	Replicate 3	Mean
Monday	234	220	229	227.67
Tuesday	269	~~967~~	275	272.00
Wednesday	254	249	246	249.67
Grand Mean				249.78

*It would be better to use a technique that also takes into account the variability between replicates. But this is difficult to calculate, confusing to explain, and doesn't change the results very much.

The SEM of the three means is 12.8. The 95% CI for the grand mean is from 194.7 to 304.9. If you were to repeat a similar experiment many times, the 95% CI would contain the true mean in 95% of experiments.

"The experiment was performed three times in triplicate. After excluding one wildly high replicate, the mean for each experiment was calculated. The grand mean is 249.8 with a SEM of 12.8 (n = 3). The 95% CI ranges from 194.7 to 304.9."

3. Is the width of a 99% CI wider or narrower than the width of a 90% CI?

To be more confident, you need to span a wider range of values. The 99% CI is wider than the 90% CI.

4. The serum levels of a hormone (the Y factor) was measured to be 93 ± 1.5 (mean ± SEM) in 100 nonpregnant women and 110 ± 2.3 (mean ± SEM) in 100 women in the first trimester of pregnancy.

   A. Calculate the 95% CI for the mean value in each group.

With 100 subjects in each group, df = 99, and t* = 1.984 (from Table A5.3 in the Appendix). The 95% CIs are the mean plus or minus 1.984 times the SEM. For the nonpregnant women, the CI ranges from 90.0 to 96.0. For the pregnant women, the CI ranges from 105.4 to 114.6.

   B. Assuming that the measurement of Y is easy, cheap, and accurate, is it a useful assay to diagnose pregnancy?

NO! The SEM values only tell you about how well you know the means. The 95% CIs for the nonpregnant and pregnant women don't overlap (and don't even come close), so it is pretty clear that the mean values of the two populations are really different. In other words, it is very clear that the mean level of the Y factor increases with pregnancy. But that does not mean that measuring the level of Y will be helpful in diagnosing pregnancy. To assess its usefulness as a diagnostic test, you need to look at how much the two distributions overlap.

The 95% CI is the range of values that you can be 95% sure contains the population mean. To think about the diagnostic usefulness of this test, we need to think about the prediction interval, the range of values that you expect to contain 95% of the values in the population. First, you need to calculate the SDs. This is done by multiplying the SEM by the square root of sample size. The SD of the nonpregnant women is 15; the SD of the pregnant women is 23. For N = 100, the prediction interval extends 1.99 standard deviations either side of the mean (Table 4.2). For the nonpregnant women, the prediction interval extends from 63.2 to 122.9. For pregnant women, the prediction interval extends from 64.2 to 155.8. The two distributions overlap considerably, so the test would be completely useless as a diagnostic test.

The CI is only useful when you wish to make inferences about the *mean*. You must calculate the much wider prediction interval when you wish to make inferences about the actual distribution of the populations.

5. A paper reports that cell membranes have 1203 ± 64 (mean ± SEM) fmol of receptor per milligram of membrane protein. These data come from nine experiments.

   A. Calculate the 95% CI. Explain what it means in plain language.

With N = 9, df = 8 and t* = 2.306. So the 95% CI ranges from 1055.4 to 1350.6 fmol/mg protein. Assuming that each of the nine experiments was conducted independently, you can be 95% sure that the true mean (if you were to perform an infinite number of such experiments) lies within this range.

B. Calculate the 95% prediction interval. Explain what it means in plain language.

To calculate the prediction interval, you first have to calculate the SD. The SD equals the SEM times the square root of the sample size, or 192. For N = 9, the prediction interval extends 2.43 SDs on each side of the mean (Table 4.2). So the 95% prediction interval ranges from 736.4 to 1669.6 fmol/mg. The prediction interval is much wider than the CI.

C. Is it possible to sketch the distribution of the original data? If so, do it.

If you are willing to assume that population is distributed in a Gaussian manner, you can draw the distribution. This is a reasonable assumption, because the experimental scatter is due to many factors acting independently. From the mean and SD, we can draw this distribution in Figure A4.3.

D. Calculate the 90% CI.

With df = 8, t* for 90% CI is 1.86 (Table A5.3 in the Appendix). So the CI extends 1.86 SEMs on each side of the mean. The 90% CI ranges from 1084.0 to 1322.0.

E. Calculate the coefficient of variation.

The coefficient of variation (CV) is the SD divided by the mean, 192/1203 or 16.0%.

6. You measure blood pressure in 10 randomly selected subjects and calculate that the mean is 125 and the SD is 7.5 mmHg. Calculate the SEM and 95% CI. Then you measure blood pressure in 100 subjects randomly selected from the same population. What values do you expect to find for the SD and SEM?

The SEM is the SD divided by the square root of sample size, so it equals $7.5/\sqrt{10}$ or 2.37. With N = 10, df = 9, and t* = 2.26. The 95% CI extends 2.26 SEMs either side of the mean: 119.6 to 130.4 mmHg.

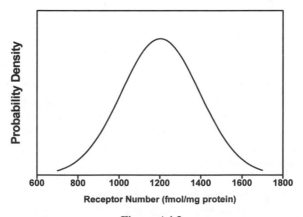

Figure A4.3.

When you increase the sample size, you don't expect the SD to change. The SD measures the scatter of the data, and the scatter doesn't change by sampling more subjects. You'll know the SD more accurately, so the value will probably change. But the SD is just as likely to increase as to decrease when you add more subjects. Your best guess is that the SD will still equal 7.5 mmHg.

The SEM quantifies how well you have determined the population mean. When you increase the sample size, your sample mean will probably be closer to the population mean. So the SEM gets smaller as the sample size gets larger. With N = 100, your best guess for the SEM is 7.5/$\sqrt{100}$, or 0.76 mmHg.

Many students mistakenly think that the SD ought to get smaller as the sample size gets larger. Not so. The SD quantifies variability or scatter, and the scatter is real. The values in the population arc not all the same. Collecting a larger sample helps you know the SD better, but the size of the SD does not change predictably as sample size increases. The SD quantifies an attribute of the population, and that attribute (scatter) doesn't change when you sample more subjects.

If you are still confused, think of how the mean will change when you sample more subjects. If you collect a larger sample, you'll know the population mean better and your sample mean from N = 100 won't exactly equal the sample mean from N = 10. But it is just as likely that the mean will increase as decrease. The sample mean quantifies an attribute of the population (its average), and the population doesn't change by sampling more subjects.

The standard error is different. It does not quantify an attribute of the population. It quantifies sampling error, and its value is determined (in part) by the sample size. Collecting bigger samples reduces the value of the standard error of the mean.

7. Why doesn't it ever make sense to calculate a CI from a population SD (as opposed to the sample SD)?

If you know the population SD, then you must know all the values in the population. If this is true, then you know the population mean exactly. So there is no point in calculating a CI. Confidence intervals are only useful when you are unsure about the true population mean, so want to calculate a CI to quantify your uncertainty.

8. Data were measured in 16 subjects, and the 95% CI was 97 to 132.

  A. Predict the distribution of the raw data.

Since the CI is symmetrical around the mean, the mean must equal the average of 97 and 132, or 114.5. With 16 subjects, the 95% CI extends about 2 SEMs on either side of the mean, so the SEM must be one quarter of the difference between 132 and 97, or about 8.75. The SD equals the SEM times the square root of sample size, or 35. With nothing else to go on, we assume that the population is roughly Gaussian. The prediction interval extends 2.20 SDs either side of the mean, from 37.5 to 191.5.

  B. Calculate the 99% CI.

For df = 15, t* = 2.95 (Appendix Table A5.3). The 99% CI extends 2.95 SEMs either side of the mean: from 88.7 to 140.3. The 99% CI is wider than the 95% CI. To be more sure that the interval includes the population mean, you have to make it wider.

9. Which is larger, the SD or the SEM? Are there any exceptions?

The SD is larger, without exception. The SD quantifies scatter; the SEM quantifies how well you know the population mean. The SEM must be smaller. If you look at Equation 5.2, you can see mathematically why the SEM is smaller. With that equation, the SEM would equal the SD if $N = 1$. But if $N = 1$, it wouldn't make any sense to think about scatter and the SD is not defined.

CHAPTER 6

1. Why are the CIs in Figure 6.3 asymmetrical?

Survival can't possibly be greater than 100% or less than 0%. At the early time points when the survival is close to 100% the upper confidence limit is constrained at 100% while the lower confidence limit is not constrained.

2. Why is it possible to calculate the median survival time when some of the subjects are still alive?

If you listed all the survival times in order, the median would be the middle time. If subjects are still alive, then you don't know their survival time. But you do know that their survival time will be longer than any subject who already died. So you can still find the median time, even if subjects are still alive. Median survival is the time at which half the subjects are still alive and half are dead.

3. Why are survival curves drawn as staircase curves?

The curves plot the survival of the subjects in your sample. Whenever a subject dies, the percent survival drops. The percent survival then remains at a steady plateau until the next patient dies.

You could imagine drawing the curve differently if your goal were to draw your best guess of survival in the population. You might then draw a sloping line from one patient's death to the next. Then the curve would not look like a staircase. But the convention is to draw survival curves showing the survival of your sample, so staircase curves are used.

4. A survival curve includes many subjects who were censored because they dropped out of the study because they felt too ill to come to the clinic appointments. Is the survival curve likely to overestimate or underestimate survival of the population?

Since the censored patients were sicker than the average subjects, on average they probably died sooner than the noncensored patients. So the remaining patients probably die later than the overall average, and so the data on the remaining patients overestimates the survival of the entire population.

A survival curve can only be interpreted when you assume that the survival of the censored subjects would, on average, equal the survival of the uncensored subjects. If many subjects are censored, this is an important assumption. If the study is large and only a few subjects were censored, then it doesn't matter so much.

5. A study began on January 1, 1991 and ended on December 31, 1994. How will each of these subjects appear on the graph?

A. Entered March 1, 1991. Died March 31, 1992.

This subject is shown as a downward step at time = 13 months.

B. Entered April 1, 1991. Left study March 1, 1992.

This subject was censored at 11 months. He or she will appear as a blip at time = 11 months.

C. Entered January 1, 1991. Still alive when study ended.

The subject was alive for the entire 48 months of the study. He or she will appear as a blip at time = 48 months. Since the graph ends at 48 months, the blip may not be visible.

D. Entered January 1, 1992. Still alive when study ended.

This subject was alive for 36 months after starting the trial. So he or she was censored at 36 months and will appear as a blip at time = 36 months.

E. Entered January 1, 1993. Still alive when study ended.

This subject was alive for 24 months at the time the study ended. So he or she appears as a censored subject at time = 24 months.

F. Entered January 1, 1994. Died December 30, 1994.

This subject died 12 months after starting the trial. So he or she will appear as a downward step at time = 12 months.

G. Entered July 1, 1992. Died of car crash March 1, 1993.

This is ambiguous. After 8 months the subject died of a car crash. This might count as a death or as a censored subject, depending on the decision of the investigator. Ideally, the experimental protocol should have specified how to handle this situation before the data were collected. Some studies only count deaths from the disease being studied, but most studies count deaths from all causes. Some studies analyze the data both ways. Why would you include a death from a car crash in a study of cancer? The problem is that it is impossible to know for sure that a death is really unrelated to the cancer (and its therapy). Maybe the car crash in this example would have been avoided if the study medication hadn't slowed the reaction time or if the cancer hadn't limited peripheral vision.

6. A survival study started with 100 subjects. Before the end of the fifth year, no patients had been censored and 25 had died. Five patients were censored in the sixth year, and five died in the seventh year. What is the percent survival at the end of the seventh year? What is the 95% CI for that fraction?

This book did not give the equations for calculating a survival curve, but it is pretty easy to figure out. At the end of the fifth year, 25 patients had died and 75 are still alive. So the fraction survival is 75%. During the sixth year, five subjects are censored.

This does not change percent survival, which remains at 75%. Now five subjects die in the next year. At the start of that year, 70 subjects were alive (100 − 25 − 5). At the end of the seventh year, 65 subjects are alive. So for subjects alive at the start of the seventh year, the probability of surviving that year is 65/70 or 92.86%. The probability of surviving up to the beginning of the seventh year, as we already calculated, is 75%. So the probability of surviving from time zero to the end of the seventh year is the product of those two probabilities, 75% × 92.86% = 69.6%.

To calculate the CI, you have to know the number of subjects still being followed at the end of the seventh year. This equals 100 − 25 − 5 − 5 = 65. Use Equation 6.1 with p = .696 and N = 65. The 95% CI is from 0.696 − 0.093 to 0.696 + 0.093, or 0.603 to 0.789, or 60.3% to 78.9%.

CHAPTER 7

1. (Same data as in Problem 4 in Chapter 5.) The serum levels of a hormone (the Y factor) was measured to be 93 ± 1.5 units/ml (mean ± SEM) in 100 nonpregnant women and 110 ± 2.3 units/ml in 100 women in the first trimester of pregnancy.

   A. What is the 95% CI for the difference between mean levels of the Y factor?

The SE of the difference is the square root of the sum of the square of the two SEM values (Equation 7.1), or 2.75 units/ml. With 100 subjects in each group, there are 100 + 100 − 2 = 198 degrees of freedom. With so many degrees of freedom t* = 1.97. The difference between the means is 110 − 93 = 17. The 95% CI extends 1.97 SEs on each side, or 11.6 to 22.4 units/ml. If we assume that both samples are representative of the overall populations, we can be 95% sure that the mean level of the Y factor in pregnant women is between 11.6 and 22.4 units/ml higher than in nonpregnant women.

   B. What assumptions do you have to make to answer that question?

We assume that the two samples are representative of the overall populations of pregnant and nonpregnant women of the appropriate age range. Since the samples are so large, we don't need to assume that the distributions are Gaussian. We do have to assume that the two populations have similar standard deviations, and that the subjects have each been selected independently.

2. Pullan et al. investigated the use of transdermal nicotine for treating ulcerative colitis. The plasma nicotine level at baseline was 0.5 ± 1.1 ng/ml (mean ± SD; N = 35). After 6 weeks of treatment, the plasma level was 8.2 ± 7.1 ng/ml (N = 30). Calculate a 95% CI for the increase in plasma nicotine level.

Using Equations 7.2 and 7.3, the 95% CI for the difference extends from 5.3 to 10.1 ng/ml.

What assumptions are you making? Are these valid?

• The subjects are representative of all patients with ulcerative colitis. This seems reasonable but to really know you'd have to investigate how the investigators recruited subjects.

- The populations follow a Gaussian distribution. Since the samples are fairly large, this isn't too important. From the data it is obvious that the populations cannot be Gaussian. If the populations were Gaussian, 95% of the values in the population would be within 2 SDs of the mean and the distribution would be symmetrical around the mean. With the values given in this problem, this would be possible only if nicotine levels could be negative.
- Each subject was selected independently of the others. Without knowing the details of the experimental design, it is impossible to evaluate this assumption. It seems reasonable.
- The values in the two groups are not paired or matched. Clearly this assumption is violated. Each subject was measured twice.
- The two populations have equal SDS. This assumption has been violated.

If you had access to all the data, how might you analyze the data?

There are two problems with the analysis. First, the data are paired, so the analysis should take into account the pairing. Second, and less important, the data are not sampled from a Gaussian distribution.

    If you had access to the raw data, you could calculate the change in nicotine for each subject and then calculate the mean change and the 95% CI of the mean change. This analysis does not require that the data be sampled from Gaussian distributions but does assume that the changes are distributed in a Gaussian manner. You could also express the results as the logarithm of concentration, which may result in a more Gaussian distribution.

3. You measure receptor number in cultured cells in the presence and absence of a hormone. Each experiment has its own control. Experiments were performed several months apart, so that the cells were of different passage numbers and were grown in different lots of serum. The results are reproduced below. Each value is receptor number in fmol/mg (femtomole of receptor per milligram of membrane protein). How would you summarize and graph these results? What CI would you calculate? The results are given in the following table.

Experiment No.	Control	Hormone	Change
Experiment 1	123	209	86
Experiment 2	64	103	39
Experiment 3	189	343	154
Experiment 4	265	485	220

The mean of the changes is 124.8 fmol/mg, with a SE of 39.6 fmol/mg. The 95% CI for the overall average change is from -1.1 to 250.6 fmol/mg. This is a very wide range, ranging from a slight decrease to an enormous increase.

    That approach is based on the assumption that the most consistent measure of change is the difference between the control and hormone values. In many experiments it is more consistent to express the change as a fold difference. For this example, you'd calculate the ratio of the hormone/control. You can see that this is a more consistent way to quantify the change, as the receptor number almost doubles in all four experiments. If the results are more consistent when expressed this way, the CI will be narrower, as indicated in the following table.

Experiment No.	Control	Hormone	Ratio
Experiment 1	123	209	1.70
Experiment 2	64	103	1.61
Experiment 3	189	343	1.81
Experiment 4	265	485	1.83

The mean ratio is 1.74 with a SE of 0.05. The 95% CI for the ratio is 1.58 to 1.90. You can be 95% sure that hormone treatment increases receptor number 1.58- to 1.90-fold.

Chapter 25 explains this approach in more detail. It is better to calculate the logarithm of the ratios and then calculate the mean and 95% CI for the logarithms. You can then take the antilogarithms of each confidence limit. With this example, you get almost the same results this way. The 95% CI for the mean ratio ranges from 1.55 to 1.91.

4. Why is the SE of a difference between two means larger than the SE of either mean?

The SE of a difference quantifies how well you know the difference between two population means. It depends on how well you know each mean as reflected in the two SEMs. Uncertainty in each of the two means contributes to uncertainty in the difference.

CHAPTER 8

1. The relative risk of death from lung cancers in smokers (compared with nonsmokers) is about 10. The relative risk of death from coronary artery disease in smokers is about 1.7. Does smoking cause more deaths from lung cancer or from coronary artery disease in a population where the mortality rate in nonsmokers is 5/100,000 for lung cancer and 170/100,000 for coronary artery disease?

The table analyzes the data.

Disease	Number of Deaths in 100,000 Nomsmokers	Number of Deaths in 100,000 Smokers	Excess Deaths Among 100,000 Smokers
Lung cancer	5	50	45
Coronary disease	170	289	119

The first column shows the number of deaths among 100,000 nomsmokers, calculated from the mortality rate. The second column shows the number of deaths among the smokers. It is calculated by multiplying the number of deaths among nonsmokers times the relative risk. The final column shows the number of excess deaths, calculated as the difference between the number of deaths among smokers and among nonsmokers.

Even though the relative risk is much higher for lung cancer, there are far more excess deaths for coronary disease. This is because coronary heart disease is so much more common than lung cancer. This question points out how overemphasizing relative risks can be misleading.

2. Goran-Larsson et al. wondered whether hypermobile joints caused symptoms in musicians. They sent questionnaires to many musicians and asked about hypermobility of the joints and about symptoms of pain and stiffness. They asked about all joints, but this problem concerns only the data they collected about the wrists. Of 96 musicians with hypermobile wrists, 5% had pain and stiffness of the wrists. In contrast, 18% of 564 musicians without hypermobility had such symptoms.

   A. Is this a prospective, retrospective, or cross-sectional study?
   B. Analyze these data as fully as possible.

A. This is a cross-sectional study. The subjects were selected because they were musicians. The investigators did not select subjects according to pain and stiffness of their joints, not because they do or don't have hypermobile joints. The investigators sampled subjects from the population of musicians and determined whether the joints were hypermobile and whether the joints were painful and stiff at the same time.

A retrospective study could also address this question. The investigators chose one group of musicians with painful and stiff joints and another group of musicians without painful or stiff joints. They would then ask both groups about hypermobility.

   This would be a prospective study if the investigators chose one group of subjects with hypermobile joints and another group of musicians without hypermobile joints. The investigators would then wait and see how many musicians in each group developed pain and stiff joints.

B. Before analyzing the data, it helps to display the data on a contingency table. To start, 5% of 96 musicians with hypermobile wrists had pain and stiffness. If you multiply 0.05 times 96, the result is 4.8. Since the result must be an integer, the investigators must have rounded off the percentage. So we place the number 5 into the upper left cell of the spreadsheet. The rest of the table is calculated in the same way.

	Pain and Stiffness	No Pain or Stiffness	Total
Hypermobile wrists	5	91	96
Normal wrists	102	462	564
Total	107	553	192

It is meaningful to calculate a relative risk from cross-sectional studies. The relative risk is 0.29. Musicians with hypermobile wrists are only 29% as likely to have painful or stiff wrists than are musicians without hypermobility. Using InStat or Equation 8.7, you can calculate that the approximate 95% CI ranges from 0.12 to 0.69. We are 95% sure that musicians with hypermobile joints have somewhere between 12% and 69% as high a prevalence of wrist pain and stiffness compared to musicians with normal wrists. Hypermobile wrists protect musicians from pain and stiffness.

You could also calculate the difference of prevalence rates. Among musicians with hypermobile wrists, the prevalence of pain and suffering is 5.2%. Among musicians without hypermobile wrists, the prevalence of pain and suffering is 18.1%. The difference is 12.9%, with an approximate 95% CI for the difference ranging from 7.4% to 18.4%.

To interpret these results, you have to assume that the musicians studied here are representative of musicians in other places. You also have to assume that the assessment of hypermobility is objective, and not influenced by the presence or absence of pain and stiffness.

3. The same number of cells (100,000 per ml) were placed into four flasks. Two cell lines were used. Some flasks were treated with drug, while the others were treated only with vehicle (control). The data in the following table are mean cell counts (thousands per milliliter) after 24 hours.

	No drug	Drug
Cell Line 1	145	198
Cell Line 2	256	356

Analyze these data as fully as possible. If you had access to all the original data, how else might you wish to summarize the findings?

This table is *not* a contingency table, and none of the methods presented in Chapter 8 are appropriate. The values in the table are mean cell counts, not counts of the number of subjects in each category. If this were a true contingency table, there would be 145 + 198 + 256 + 356 subjects or experiments. In fact, there is only one experiment reported here. A larger number in the table means that the cells grew faster, not that a certain outcome occurred in more subjects.

4. Cohen et al. investigated the use of active cardiopulmonary resuscitation (CPR).*
   In standard CPR the resuscitator compresses the victim's chest to force the heart to pump blood to the brain (and elsewhere) and then lets go to let the chest expand. Active CPR is done with a suction device. This enables the resuscitator to pull up on the chest to expand it as well as pressing down to compress it. These investigators randomly assigned cardiac arrest patients to receive either standard or active CPR. Eighteen of 29 patients treated with active CPR were resuscitated. In contrast, 10 of 33 patients treated with standard CPR were resuscitated.

   A. Is this a cross-sectional, prospective, retrospective, or experimental study?
   B. Analyze these data as you see fit.

A. This is an experimental study. The investigators sampled one group of subjects (victims of cardiac arrest) and gave them two different treatments.

---

*TJ Cohen, BG Goldner, PC Maccaro, AP Ardito, S Trazzera, MB Cohen, SR Dibs. A comparison of active compression-decompression cardiopulmonary resuscitation with standard cardiopulmonary resuscitation for cardiac arrests occurring in the hospital. *N Engl J Med* 329:1918–1921, 1993.

B. As always with these kind of data, it helps to arrange the data onto a contingency table. The results can be summarized with a relative risk which is $(18/29)/(10/33) =$ 2.05. Cardiac arrest victims given active CPR were about twice as likely to be resuscitated as cardiac arrest victims given standard CPR. The approximate 95% CI for the relative risk ranges from 1.14 to 3.70.

	Resuscitated	Not Resuscitated	Total
Active CPR	18	11	29
Standard CPR	10	23	33
Total	28	34	62

You could also calculate the results as the difference in success rates. Among victims given active CPR, 62.1% were resuscitated. Among victims given standard CPR, only 30.3% were resuscitated. The difference in resuscitation rates is 31.8%, with a 95% CI of 6.9% to 56.6%.

To interpret these results, you have to assume that these victims of cardiac arrest are representative of others in other places. You also have to assume that there was no difference between the two groups except for the kind of CPR. The results wouldn't be helpful, for example, if the group given active CPR happened to be much younger than the group given standard CPR or if the two groups of subjects lived in different areas or were treated by different paramedics.

To fully assess these results, you would want to look at more than just the resuscitation rate. You'd also want to know how many of the resuscitated patients survived long enough to be discharged from the hospital, and how many survived for one or more years after that. You'd also want to know whether the survivors were living active lives or whether they had lost mental function.

5. In a letter in *Nature* (356:992, 1992) it was suggested that artists had more sons than daughters. Searching *Who's Who in Art,* the author found that artists had 1834 sons and 1640 daughters, a ratio of 1.118. As a comparison group, the investigator looked at the first 4002 children of nonartists listed in *Who's Who* and found that the ratio of sons/daughters is 1.0460. Do you think the excess of sons among the offspring of artists is a coincidence?

Let's create a contingency table from these data.

	Sons	Daughters	Total
Artists in WW	1834	1640	3474
Nonartists in WW	2046	1956	4002
Total	3880	3596	7476

The only trick in creating the contingency table is filling in the numbers for the nonartists. We know that the total number in the sample is 4002 and that the ratio of boys/girls is 1.046. This means that the ratio of boys/(boys + girls) is 1.046/(1 + 1.046) or 51.1%. Therefore there are .511 * 4002 = 2046 boys, which leaves 1956 daughters.

You can calculate the relative risk. The word *risk* is a bit awkward here, and *relative proportion* is more appropriate. The relative proportion equals (1834/3474)/ (2046/4002) = 1.033. The artists were 1.033 times more likely to have a son than were the nonartists. The 95% CI ranges from 0.99 to 1.08.

Is the difference in sex ratio a coincidence in these samples, or is it a real difference between artists and nonartists? There is no way to answer this question for sure. You'll learn more about how to approach the question in Chapter 27. For now, you can see that the 95% CI contains the value 1.00, so it is certainly reasonable to suspect that the observed difference is a coincidence. The relative risk is not so far from 1.0 that you feel compelled to believe that the difference is real.

In assessing these data, you also need to consider the background of the study. Why did the investigators suspect that artists would have more sons? It doesn't seem to make any biological sense. And why study artists? Why not architects or plumbers or surgeons? If you look at enough occupations, you are sure to find some that have a higher or lower fraction of sons than average.

So there are two reasons to be dubious about the conclusions. First, the findings are not very striking, the relative risk is not far from 1.0 and the 95% CI contains 1.0 (barely). Second, the investigators do not appear to have a logical reason to think that the findings could be real. I conclude that the difference between this sample of artists and nonartists is almost certainly a coincidence.

CHAPTER 9

1. Can an odds ratio be greater than one? Negative? Zero?

The odds ratio ranges from 0 to 1. It cannot be negative.

2. Can the logarithm of the odds ratio be greater than one? Negative? Zero?

The logarithm of the odds ratio can be any value, including negative values.

3. Gessner and colleagues investigated an outbreak of illness in an Alaskan community. They suspected that one of the town's two water supplies delivered too much fluoride leading to fluoride poisoning. They compared 38 cases with 50 controls. Thirty-three of the cases recalled drinking water from water system 1, while only four of the controls had drunk water from that system.

Analyze these data as you think is appropriate. State your assumptions.

	Cases	Controls	Total
System 1	33	4	37
System 2	5	46	51
Totals	38	50	88

This is a retrospective study. The subjects were selected either because they had the illness or did not. Therefore a relative risk would be meaningless, and the results should be summarized with an odds ratio. The odds ratio is $(33/4)/(5/46) = 75.9$. The approximate 95% CI ranges from 18.9 to 304.4. If you assume that the disease was rare in the community, then you can conclude that people drinking from system 1 were 76 times more likely to have the illness than people drinking from system 2. These findings are striking and are strong evidence that the water system may be involved in causing the disease.

In interpreting these data, you must assume that the subjects and controls studied are representative of other people who did or didn't get the illness. You must also assume that the subjects and controls were selected independently. For example, the study wouldn't be valid if the investigators chose several cases or controls from the same household (of course, they all drink from the same water system). Finally, you must assume that there are no other differences between people who drank from the two water systems. You wouldn't be able to interpret the results if the two groups of people also bought their food at different sources, or worked in different villages, or were different in some other way.

How could these authors have conducted a prospective study to test their hypothesis?

To do a prospective study, the investigators should choose people based on which system provides their drinking water and then assess whether they got the illness. If the illness is rare, it will require more subjects to perform a prospective study than a retrospective study.

CHAPTER 10

1. You wish to test the hypothesis that a toin coss is fair. You toss the coin six times and it lands on heads each time. What is the null hypothesis? What is the P value? Is a one- or two-tailed test more appropriate? What do you conclude?

The null hypothesis is that the coin is flipped fairly, so the probability of landing on heads is exactly 0.5. If this were true, the chance that six consecutive flips would all land on heads is $0.5^6 = 1/64 = 0.016$. If the null hypothesis were true, the chance that six consecutive flips would all land on tails is also 0.016. These are one-tailed P values. The chance that six flips would be identical (either heads or tails) is twice 0.016 or 3.2%. This is the two-tailed P value.

It is impossible to evaluate these results without any context. Why did anyone suspect that the coin flipping was unfair? If you use a one-tailed P value, you must have decided before the data were collected that you expected to see too many heads. Otherwise you should choose the two-tailed P value of 0.032. How should the results be interpreted? Either there is a systematic problem with the coin flipping so heads comes up more than half the time or a coincidence occurred. This coincidence would occur with a fair coin 3.2% of the time. Unless I had some reason to suspect that the coin flipping was unfair (for example, if this were part of a magic show), I'd be inclined to believe that it is a coincidence.

2. You conduct an experiment and calculate that the two-tailed P value is 0.08. What is the one-tailed P value? What assumptions are you making?

You can only calculate a one-tailed P value if the direction of the expected change were predicted before the data were collected. If the change occurred in the predicted direction, then the one-tailed P value is half the two-tailed P value, or 0.04.

If the change occurred in the opposite direction, the situation is more tricky. The one-tailed P value answers this question: If the null hypothesis were true, what is the chance of randomly picking subjects such that the difference goes in the direction of the experimental hypothesis with a magnitude as large or larger than observed? The answer is a bit ambiguous when the result went in the direction opposite to the experimental hypothesis. One answer is that the one-tailed P value is 1.0 minus half the two-tailed value. So the one-tailed P value is 0.96.

It really doesn't help to calculate a one-tailed P value when the observed difference is opposite in direction to the experimental hypothesis. You should conclude that the P value is very high and the difference was due to chance.

3. (Same data as Problem 4 in Chapter 5 and Problem 1 in Chapter 7.) The serum levels of a hormone (the Y factor) was measured to be 93 ± 1.5 (mean $\pm$ SEM) in 100 nonpregnant women and 110 ± 2.3 (mean $\pm$ SEM) in women in the first trimester of pregnancy. When the two groups are compared with a t test, the P value is less than 0.0001.

A. Explain what the P value means in plain language.

The null hypothesis is that the mean levels of the Y factor are identical in pregnant and nonpregnant women. If this null hypothesis were true, there is less than a 0.01% chance that the difference between the means of two randomly selected samples (N = 100 each) would be 17 units/ml or greater.

B. Review your answers to Problem 4 in Chapter 5 and Problem 1 in Chapter 7. Explain how the results complement one another.

The low P value tells us that the results are very unlikely to be due to coincidence. And it certainly is not surprising to find that the mean value of the hormone is elevated in pregnant women (it might be surprising to find a hormone whose level does not change in pregnancy!). The question in Chapter 5 asked whether measurements of this hormone would be useful for diagnosing pregnancy. That is a very different question. To be useful for diagnosis, it is not enough that the means differ. The two distributions must also differ, with very little (if any) overlap. As explained in the answer in Chapter 5, the two distributions in this example overlap substantially, so the test is worthless for diagnosis.

4. (Same data as Problem 4 in Chapter 8.) Cohen et al. investigated the use of active cardiopulmonary resuscitation (CPR). In standard CPR the resuscitator compresses the victim's chest to force the heart to pump blood to the brain (and elsewhere) and then lets go to let the chest expand. Active CPR is done with a suction device. This enables the resuscitator to pull up on the chest to expand it as well as pressing down to compress it. These investigators randomly assigned cardiac arrest patients to receive either standard or active CPR. Eighteen of 29 patients treated with active CPR were resuscitated. In contrast, 10 of 33 patients treated with standard CPR were resuscitated.

Using Fisher's test, the two-sided P value is 0.0207. Explain what this means in plain language.

The null hypothesis is that the resuscitation rate is the same for active and standard CPR. If that hypothesis were true, the chance of observing such a large difference (or larger) in a study this size is 2.1%.

CHAPTER 11

1. For Problems 3 and 4 of the last chapter, explain what a Type I and Type II error would mean.

   Problem 3. (hormone Y in pregnancy). Since the P value is low, you would conclude that the mean levels of hormone are significantly different. You'd be making a Type I error if the mean levels in the overall population were really identical. It is only possible to make a Type II error if you conclude that there is no statistically significant difference. Since the P value is so low in this example (below any reasonable value for α), you'll definitely conclude that there is a significant difference, so it is impossible to make a Type II error.

   Problem 4. (CPR). If you set $\alpha = 0.05$, you would conclude that the resuscitation rates are significantly different. You'd be making a Type I error if the resuscitation rates in the overall populations were really identical. If you set $\alpha = 0.01$, would you conclude that the resuscitation rates are not significantly different? If you did this, you'd have made a Type II error if the rates are really different.

2. Which of the following are inconsistent?

   A. Mean difference = 10. The 95% CI: −20 to 40. P = 0.45.

   Consistent.

   B. Mean difference = 10. The 95% CI: −5 to 15. P = 0.02.

   Inconsistent. The 95% CI includes 0, so the P value must be greater than 0.05.

   C. Relative risk = 1.56. The 95% CI: 1.23 to 1.89. P = 0.013.

   Consistent.

   D. Relative risk = 1.56. The 95% CI: 0.81 to 2.12. P = 0.04.

   Inconsistent. The P value is less than 0.05, so the 95% CI for the relative risk cannot include 1.0.

   E. Relative risk = 2.03. The 95% CI: 1.01 to 3.13. P < 0.001.

Inconsistent. The P value is way less than 0.05, so the 95% CI must not include 1.0 and shouldn't even come close. But the 95% CI starts at a value just barely greater than 1.00.

CHAPTER 13

1. Assume that you are reviewing a manuscript prior to publication. The authors have tested many drugs to see if they increase enzyme activity. All P values were calculated using a paired t test. Each day they also perform a negative control, using just vehicle with no drug. After discussing significant changes caused by some drugs, they state that "in contrast, the vehicle control has consistently given a

nonsignificant response (P > 0.05) in each of 250 experiments.'' Comment on these negative results. Why are they surprising?

In the negative control experiments, you expect the null hypothesis to be true. Therefore you expect 1 out of 20 control experiments to have a P value less than 0.05. In 250 experiments, therefore, you expect to see about 0.05 * 250 = 12.5 ''significant'' responses. You'd be very surprised to see zero significant responses results among the 250 control experiments.

CHAPTER 14

1. A test has a specificity of 92% and a sensitivity of 99%. Calculate the predictive values of positive and negative tests in a population in which 5% of the individuals have the disease.

Create this table following the same steps as used for the examples in the book.

	Disease Present	Disease Absent	Total
Positive test	495	760	1255
Negative test	5	8740	8745
Totals	500	9500	10000

The predictive value of a positive test is 495/1255 = 39.4%. When a patient has a positive test, there is a 39.4% chance that he or she actually has the disease.

The predictive value of a negative test is 8740/8745 = 99.9%. When a patient has a negative test, you can be 99.9% sure that he or she does not have the disease.

2. A test has a specificity of 92% and a sensitivity of 99%. Calculate the predictive values of positive and negative tests in a population in which 0.1% of the individuals have the disease.

	Disease Present	Disease Absent	Total
Positive test	99	7992	8091
Negative test	1	91908	91909
Totals	100	99,900	100,000

The predictive value of a positive test is 99/8091 = 1.2%. When a patient has a positive test, there is a 1.2% chance that he or she actually has the disease. Virtually all (98.8%) of the positive tests are false positives.

The predictive value of a negative test is 91908/91909 = 99.9%. When a patient has a negative test, you can be 99.9% sure that he or she does not have the disease.

3. A woman wants to know if her only son is color blind. Her father is color blind, so she has a 50% chance of being a carrier (color blindness is a sex-lined trait).

This means that, on average, half her sons will be color blind (she has no other sons). Her son is a smart toddler. But if you ask him the color of an object, his response seems random. He simply does not grasp the concept of color. Is he color blind? Or has he not yet figured out what people mean when they ask him about color? From your experience with other kids that age, you estimate that 75% of kids that age can answer correctly when asked about colors. Combine the genetic history and your estimate to determine the chance that this kid is color blind.

The sensitivity is 100%, the specificity is 75%, and the prior probability is 50%. If you were to see the infants of 1000 women whose fathers are color blind, the results you'd expect to see are given in the following table.

	Color Blind	Not Color Blind	Total
No sense of color	500	125	625
Knows colors	0	375	375
Total	500	500	1000

Of the kids who have no sense of color, 500/625 = 80% are color blind, while the other 20% simply are slow to learn colors. You conclude that there is an 80% chance that this kid is color blind. Of course, the accuracy of this conclusion depends on how accurate the number "75%" is for the fraction of non-color blind kids who can answer questions about color blindness at this age.

4. For patient C in the porphyria example, calculate the predictive value of the test if your clinical intuition told you that the prior probability was 75%.

You can create this table.

	Disease Present	Disease Absent	Total
<99 units	615	9	624
>99 units	135	241	376
Total	750	250	1000

The predictive value is 615/624 = 98.6%. Since the prior probability is higher, the predictive value must be higher too. Before obtaining the test results, you already thought the chance was higher. The test results increase the probability further.

CHAPTER 15

1. A student wants to determine whether treatment of cells with a particular hormone increases the number of a particular kind of receptors. She and her advisor agree that an increase of less than 100 receptors per cell is too small to care about. Based on the standard deviation of results you have observed in similar studies, she calculates the necessary sample size to have 90% power to detect an increase of

100 receptors per cell. She performs the experiment that number of times, pools the data, and obtains a P value of 0.04.

The student thinks that the experiment makes a lot of sense and thought that the prior probability that her hypothesis was true was 60%. Her advisor is more skeptical and thought that the prior probability was only 5%.

A. Combining the prior probability and the P value, what is the chance that these results are due to chance? Answer from both the student's perspective and that of the advisor.

The student's perspective is illustrated by the following table which shows the results of 1000 hypothetical experiments.

Prior Probability = 0.60	Receptor Number Really Increases by More than 100 Sites/ Cell	Null Hypothesis Is True: Receptor Number Doesn't Change	Total
P < 0.04	540	16	556
P > 0.04	60	384	444
Total	600	400	1000

Of 556 experiments with a P value less than 0.04, receptor number really increases in 540. So the chance that the receptor number really increases is 540/556 = 97.1%, leaving a 2.9% chance that the results are due to coincidence.

The advisor's perspective is different, as illustrated in the following table.

Prior Probability = 0.05	Receptor Number Really Increases by More than 100 Sites/ Cell	Null Hypothesis Is True: Receptor Number Doesn't Change	Total
P < 0.04	45	38	83
P > 0.04	5	912	917
Total	50	950	1000

Of all the hypothetical experiments with P < 0.04, 45/83 = 54.2% occurred when receptor number really increased. The other 45.8% of the low P values were Type I errors, due to random sampling. So the new evidence is not convincing to the advisor, as she still thinks there is only about a 50% chance that the effect is real.

B. Explain why two people can interpret the same data differently.

The results could have been obtained in two ways:

First possibility: The treatment really does not change receptor number, and the change we observed is simply a matter of coincidence. The statistical calculations tell you how unlikely this coincidence would be. For this example, you would see a difference as large or larger than you observed in 4% of experiments, even if the null hypothesis were true.

Second possibility: The hormone really does increase receptor number.

In interpreting the results, you have to decide which of these possibilities is more likely. The student in this example thought that the second possibility was probably true even before collecting any data. She is far happier believing this hypothesis is true rather than believing in a coincidence that happens only 1 in every 25 experiments.

The advisor has reasons to think that this experiment shouldn't work. Perhaps she knows that the cells used do not have receptors for the hormone used, so it seems very unlikely that the hormone treatment could alter receptor number. So the advisor has to choose between believing that an "impossible" experiment worked or believing that a 4% coincidence has occurred. It is a toss up, and the conclusion is not clear.

To interpret experimental results, you need to integrate the results from one particular experiment (summarized with a P value) with your prior opinions about the experiment. If different people have different prior opinions, they may reach different conclusions when the P value is only moderately low.

    C. How would the advisor's perspective be different if the P value were 0.001 (and the power were still 90%)?

Now the advisor has to choose between a result that makes no sense to her or a coincidence that happens only 1 time in 1000. She believes the result, rather than such an unlikely coincidence. Although she probably makes this judgment intuitively, you can explain it with this table. Given the assumptions of this problem, only 1 in 46 results like this would be due to chance, while 45 out of 46 reflect a true difference. (To maintain the power with a smaller threshold value of p requires increasing the sample size.)

Prior Probability = 0.05	Receptor Number Really Increases by More than 100 Sites/Cell	Null Hypothesis Is True: Receptor Number Doesn't Change	Total
P < 0.001	45	1	46
P > 0.001	5	949	954
Total	50	950	1000

2. You go to Las Vegas on your 25th birthday, so bet on the number 25 in roulette. You win. You bet a second time, again on 25, and win again! A roulette wheel has 38 slots (1 to 36, 0, and 00), so there is a 1 in 38 chance that a particular spin will land on 25.

    A. Assuming that the roulette wheel is not biased, what is that chance that two consecutive spins will land on 25?

The chance is $(1/38) * (1/38) = 1/1444 = 0.069\%$.

    B. If you were to spend a great deal of time watching roulette wheels, you would note two consecutive spins landing on 25 many times. What fraction of those times would be caused by chance? What fraction would be caused by an unfair roulette wheel?

There are two possibilities. Either it is a coincidence or the roulette wheels are biased in your favor. Deciding between the two depends on the context. If you are in a commercial casino, the chance that the roulette wheel is unfair is pretty small, and the chance that it is biased in your favor is very very small. I find it easier to believe that a 1 in 1444 coincidence has occurred than that the casino has a roulette wheel biased in my favor. If you are watching a magic show, then you expect the roulette wheel to be rigged and your conclusion would be different. You can't answer the question without knowing the context (the prior probability).

CHAPTER 16

1. In Example 16.1, assume that the woman had three unaffected sons. What is the probability that she is a carrier?

	Carrier	Not Carrier	Total
All sons without disease	62	500	562
At least one son with disease	438	0	438
Total	500	500	1000

Follow the same steps as Example 16.1. In step 3 substitute 1/8 for 1/4. The chance she is a carrier is 62/562 = 11.0%).

2. If the lod score is -3, what is the probability that the marker is linked to the disease?

The post-test odds equals the pretest odds times the antilogarithm of the lod score. Assuming a randomly selected marker, the pretest odds are about 0.02. So the post-test odds equal $0.02 * 0.001 = 0.00002$. With such low values, the probability and the odds are almost identical. So the probability of linkage is only 0.002%. A lod score of -3 is strong evidence that the marker is not linked to the gene.

3. It would be possible to calculate a P value from linkage data. Explain in plain language what it would mean.

The P value would answer this question: If a marker and gene are truly not linked, what is the chance of randomly observing as few or fewer recombination events than were observed in an experiment this size (number and size of families studied). P values are rarely calculated from linkage data.

4. You perform a t test and obtain a P value less than 0.05. You used enough subjects to ensure that the experiment had a 80% power to detect a specified difference between population means with $P < 0.05$. Does it make sense to calculate a likelihood ratio? If so, calculate the ratio and explain what it means.

The likelihood ratio is a ratio. The numerator is the probability of obtaining a P value less than 0.05 if the specified difference really exists. The denominator is the probability of obtaining a P value less than 0.05 if the null hypothesis is true. In other words, the likelihood ratio is the power divided by α. For this example, the likelihood ratio is $0.80/0.05 = 16$. You are 16 times more likely to obtain a statistically significant result if the experimental hypothesis is true than if the null hypothesis is true.

You will rarely, if ever, see a likelihood ratio calculated for a statistical test. This is partly because it is hard to define the experimental hypotheses in many situations and thus hard to calculate power. But it is also just a tradition.

CHAPTER 17

1. In Example 17.1, how should the investigators have analyzed the data if they had measured the insulin sensitivity and %C20–22 twice in each subject?

The straightforward approach would be to consider each separate pair of measurements as a separate point on the graph. If you did this, you would have $N = 26$ instead of $N = 13$. But this is *not* a legitimate way to analyze the data, as you would be violating the assumption that the points are obtained independently. The second measurement from each subject is likely to be closer to the first measurement on that subject than to measurements from other subjects.

You should first average the two insulin sensitivity measurements for each subject and then average the two %C20–22 measurements in each subject. Then calculate the correlation coefficient between the average insulin measurement and the average %C20–22 measurement.

2. The P value in Example 17.1 was two tailed. What is the one-tailed P value? What does it mean?

The two-tailed P value is 0.0021. Recall what this means. If there really were no correlation between the X and Y variables in the overall population, there is a 0.21% chance of observing such a strong correlation (or stronger) in a study of this size. If the null hypothesis were true, there is a 0.1% chance of randomly picking 13 subjects and finding that $r \geq 0.77$ and a 0.1% chance of randomly picking 13 subjects and finding that $r \leq 0.77$.

You can only calculate a one-tailed P value if you decided to do so, and specified the direction of the experimental hypothesis, before collecting the data. If you predicted that the correlation was positive (increasing insulin sensitivity with increasing %C20–22), then the one-tailed P value is 0.001.

3. Do X and Y have to be measured in the same units to calculate a correlation coefficient? Can they be measured in the same units?

As Example 17.1 shows, X and Y do not have to be measured in the same units. However, they can be measured in the same units.

4. What is the P value if $r = 0.5$ with $N = 10$? What is the P value if $r = 0.5$ with $N = 100$?

Use Equation 17.3. When $N = 10$, $z = 1.45$. When $N = 100$, $z = 5.41$. Using the last column of Table A5.2 in the Appendix, the P values are 0.147 for $N = 10$ and $P < 0.0001$ for $N = 100$.

It makes sense that the P value is smaller with the bigger sample. You are far less likely to get $r = 0.5$ by coincidence with $N = 100$ than with $N = 10$.

344 APPENDICES

5. Can you calculate the correlation coefficient if all X values are the same? If all Y values are the same?

You can't calculate the correlation coefficient if all X or all Y values are identical.

You can see this mathematically by looking at Equation 17.1. If all X values are identical, then the SD of X is 0, and you'd be dividing by 0 in the equation. If all Y values were identical, the SD of Y would be 0, again leading to a division by 0.

You can see this intuitively as well. The correlation coefficient quantifies the degree to which the variation in one variable can be explained by variability in the other. If one of the variables has no variation, then the whole idea of correlation makes no sense.

6. Golino et al. investigated the effects of serotonin released during coronary angioplasty. After angioplasty (inflating a balloon positioned inside to coronary artery to force open a blocked artery) they measured the degree of vasoconstriction in the next few minutes as the percent change in cross sectional area (monitored by angiograms). They also measured the amount of serotonin in blood sampled from the coronary sinus. The data for eight patients are shown (I read these values off the graph in the publication, so they may not be exactly correct). To make the serotonin levels follow a distribution closer to a Gaussian distribution, the authors calculated the logarithm of serotonin. Calculate the correlation between the logarithm of serotonin levels and the percent change in cross sectional area.

Serotonin (ng/ml)	% Change in Cross-Sectional Area
2.0	4.0
5.0	7.0
6.0	28.0
10.0	26.0
15.0	30.0
60.0	34.0
65.0	35.0
165.0	42.0

First calculate the logarithm of the X values. Then use Equation 18.1. The correlation coefficient, r, equals 0.89 and r^2 equals 0.79. This means that 79% of the variance in cross-sectional area can be explained by differences in serotonin levels.

CHAPTER 19

1. Will the regression line be the same if you exchange X and Y? How about the correlation coefficient?

The regression line is the best line to predict Y from X. If you swap the definitions of X and Y, the regression line will be different unless the data points line up perfectly so every point is on the line.

The correlation coefficient measures the correlation between two variables. It makes no difference which variable is called X and which is called Y. Swapping the two will not change the value of r.

2. Why are the 95% CIs of a regression line curved?

Within 95% confidence, there are many different regression lines that could be the best-fit line for the entire population. This spectrum of lines is obtained by combining a wobbly slope and an uncertain intercept. The outer limits of the possible regression line is defined by a curve. This is because there is more uncertainty about the position of the regression line at the fringes of the data than in the middle.

3. Do the X and Y axes have to have the same units to perform linear regression?

No.

4. How many P values can be generated from a simple linear regression?

You can generate a P value for any null hypothesis. For linear regression, the standard null hypothesis is that the slope equals 0. You could also test the null hypothesis that the intercept equals 0. You don't have to stop there. You could also test the null hypothesis that the slope = 3.56, if that was of interest. Or the null hypothesis that the Y-intercept equals 2.0. Or 3.92. Or 11.28. There are an infinite number of possible null hypotheses, so there are an infinite number of P values. You can't interpret a P value unless you know what null hypothesis it is testing.

5. The results of a protein assay are presented in the following table. Chemicals are added to tubes that contain various amounts of protein. The reaction forms a blue color. Tubes with higher concentrations of protein become a darker blue. The darkness of the blue color is measured as optical density.

Concentration (Micrograms)	Optical Density
0	0
4	0.017
8	0.087
12	0.116
16	0.172
Unknown 1	0.097
Unknown 2	0.123

A. Calculate the best-fit linear regression line through the standard curve (five known concentrations).

The slope is 0.011 optical density units per microgram. The Y-intercept is −0.0102. The graph is shown, with 95% CIs (Figure A4.4).

B. Read the unknown concentrations from the standard curve.

You can approximate the values by reading graphically off the curve. But you can do it exactly by rearranging the equation. The linear equation is $Y = 0.011 * X - 0.0102$.

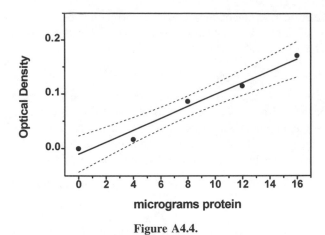

Figure A4.4.

Simple algebra puts X in front: X = (Y + 0.0102)/0.011. Now enter in the two known Y values and determine the unknown X values. Unknown 1 has X = 9.68. Unknown 2 has X = 12.03. Both of these values are well within the range of the standard curve, which is good. You shouldn't trust values extrapolated from a regression line outside the range of the known values.

6. What is r^2 if all points have the same X value? What about if all points have the same Y value?

If all points have the same X values or the same Y values, you can't calculate r^2. This makes sense intuitively. If all X values are the same, or all Y values are the same, there is no need to calculate linear regression. The value of r^2 is the fraction of the variation in Y that is explained by the regression line. If all Y values are the same, there is no variation in Y, so it makes no sense to partition the variation into various components. If all X values are the same it makes no sense to create a regression line, because the point of regression is to find the line that best predicts Y from X. If all X values are identical, then there is no way to create a regression line.

7. Sketch some examples of residual plots for data that do not meet the assumptions of linear regression.

See Figure A4.5. The first graph shows data that appear to follow all the assumptions of linear regression. The residuals (shown on the top right of the figure) are randomly scattered above and below the line at Y = 0, and the distance of the residuals from that line is not related to the value of X. The data in the second graph have increasing scatter with increasing X. The distance of the residuals from the Y = 0 line increases as X increases. The data in the third graph are not linear. If you look at the graph carefully, you can see that the points with low and high values of X tend to be above the linear regression line and the points in the middle tend to be below the line. This is easier to see in the residual graph.

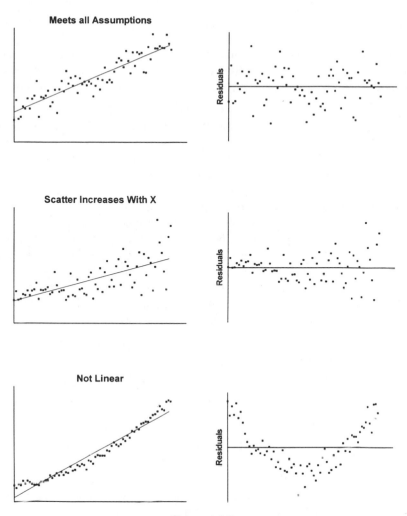

Figure A4.5.

8. Can r^2 ever be 0? Negative?

It will equal 0 if there is no trend whatsoever between X and Y. The best-fit line is exactly horizontal. Its value cannot be negative.

9. Do you need more than one Y value for each X value to calculate linear regression? Does it help?

You do not need more than one Y value for each X value to calculate linear regression. If you do have replicate Y values, you can perform some extra calculations (not explained in this book) to test whether the data are best fit by a straight line.

CHAPTER 22

1. You are preparing a grant for a study that will test whether a new drug treatment lowers blood pressure substantially. For previous experience, you know that 15 rats in each group is enough. Prepare a convincing power analysis to justify that sample size.

We need to use Equation 22.5. We know the answer we are looking for, N = 15, and need to find values for the other values to justify this. There are many combinations of variables that would work. Here is one. Set α = 0.05 and β = 0.20. This means (from Table 22.1) that the square of the sum of z_α and z_β equals 7.9. We know from prior work, that we expect the SD to be about 10 mmHg. Now we can solve for Δ, the minimum difference we wish to detect as significant. Using the numbers of this example, Δ = 10.2 mmHg. So you could write in a paper: "From previous work, we expected the SD of blood pressure to be 10 mmHg. Setting the significance level to 0.05, we calculated that a study with 15 subjects in each group would have 80% power to detect a mean difference in blood pressures of 10 mmHg."

   As this example shows, sample size calculations do not always proceed from the assumptions to the sample size. Sometimes, the sample size is chosen first and then justified. This is OK, so long as the assumptions are reasonable.

2. The overall incidence of a disease is 1 in 10,000. You think that a risk factor increases the risk. How many subjects do you need if you want to have 95% power to detect a relative risk as small as 1.1 in a prospective study?

You need to use Equation 22.6. From the problem, p_1 equals 0.00010. p_2 equals 1.1 times p1 or .00011. Since we want 95% power, β = 0.05. The problem doesn't specify α, so we'll use the conventional value of 0.05 (two-tailed). Plugging all those values into Equation 22.6, N equals 22 million subjects in each group, obviously an impractical number.

   Why such a large sample size? First, the disease is rare, with an incidence of only 1 in 10,000. So it will take a large number of subjects to get a reasonable number with the disease. Since you are looking for a small change in incidence, and want high power, you will need even larger samples. Detecting small changes in rare diseases is virtually impossible in prospective studies. The required sample size is enormous. That is why case-control studies are so useful.

3. How large a sample would you need in an election poll to determine the percent voting for each candidate to within 1%? What assumptions are you making?

Here you need to use Equation 22.3. The precision is stated in the problem, 0.01. We don't know what number to expect for p, so we'll set p = 0.50. This is the worst case, as any other value would lead to a higher sample size. The sample size required is 10,000.

4. Can a study ever have 100% power?

Only if the study collects information on 100% of the population. But then there would be no need for statistical inference.

5. About 10% of patients die during a certain operation. A new technique may reduce the death rate. You are planning a study to compare the two procedures. You

will randomly assign the new and standard technique, and compare the operative death rate.

A. How many subjects do you need to have 95% power to detect a 10% reduction in death rate ($\alpha = 0.05$)?

$p_1 = 0.10$, $p_2 = 0.90 * p_1 = 0.09$. $\alpha = 0.05$, $\beta = .05$. Use Equation 22.6 to calculate that you require 22,354 subjects in each group. It takes an enormous study to have a high power to detect a small difference.

B. How many subjects do you need to have 60% power to detect a 50% reduction in death rate ($\alpha = 0.10$)?

$$p_1 = 0.10; \quad p_2 = 0.5 * p_1 = 0.05; \quad \alpha = 0.10; \quad \beta = 0.40.$$

You can't use the table in the chapter to find the sum of the square of the z values, because the table doesn't include $\beta = 0.40$. Look in the second column of Table A4.2 in the Appendix to find 40%, then read over to see that z_β is about 0.25 (40% of a Gaussian population has z > 0.25). For $\alpha = 0.10$ (two-tailed), use the fourth column of Table A4.2 to find that z is about 1.65 (10% of a Gaussian population has either z > 1.65 or z < 1.65).

Now, you can plug all the values into Equation 22.6. You need 200 subjects in each group. Finally, we have a reasonable number.

C. Is it ethical to compare a new technique (that you think is better) with a standard technique. Why shouldn't all the subjects get the new technique?

It you are really really sure that the new technique is superior, then it is not ethical to give the older technique to anyone. But how sure are you really? The history of medicine is full of examples of new techniques that turned out to be worthless, despite the enthusiasm of people who used the technique early. The way to really compare two techniques is to perform an experiment. The patients who get the standard treatment may turn out to be better off than the patients who received the new treatment.

6. Lymphocytes contain beta-adrenergic receptors. Epinephrine binds to these receptors and modulates immune responses. It is possible to count the average number of receptors on human lymphocytes using a small blood sample. You wish to test the hypothesis that people with asthma have fewer receptors. By reading various papers, you learn that there are about 1000 receptors per cell and that the coefficient of variation in a normal population is about 25%.

A. How many asthmatic subjects do you need to determine the receptor number to plus or minus 100 receptors per cell with 90% confidence?

You need to use Equation 22.1. Since you know that the coefficient of variation is 25%, you know that the SD is about 250 receptors/cell. Equation 22.1 calculates sample size for 95% CIs. To make 90% intervals (which are narrower) the value of t* must be replaced with a smaller number. From Table A5.3 in the Appendix, you can see that the value of t* for 90% confidence and many degrees of freedom is 1.65.

So $N = 1.65^2 * (250/100)^2 = 17$ subjects in each group.

B. You want to compare a group of normal subjects with a group of asthmatics.

How many subjects do you need in each group to have 80% power to detect a mean decrease of 10% of the receptors using $\alpha = 0.05$ (two-tailed)?

Use Equation 22.5. N = 99 in each group.

C. How many subjects do you need in each group to have 95% power to detect a mean difference in receptor number of 5% with $\alpha = 0.01$?

Here you need 890 subjects in each group. You are looking for a smaller difference and have set harsher criteria for α and β.

7. You read the following in a journal

> Before starting the study, we calculated that with a power of 80% and a significance level of 5%, 130 patients would be required in each group to demonstrate a 15-percentage-point reduction in mortality (from the expected rate of 33 percent to 18 percent).

A. Explain in plain language what this means.

If the null hypothesis were really true (the two treatments are identical in terms of mortality), then there is a 5% chance of obtaining a significant difference by coincidence of random sampling. (The statement that the "significant level is 5%" means that $\alpha = 0.05$.) If the alternative hypothesis is true (that the mortality drops from 33% to 15%), then there is a 80% chance that a study with 130 patients in each group will end up with a statistically significant difference and a 20% chance of ending up with a conclusion of not statistically significant.

B. Show the calculations, if possible.

It sounds like they are analyzing the results by comparing survival curves. The methods needed to calculate sample size for comparing survival curves are complicated, and not discussed in this book. However, we can probably come up with a similar answer by treating the data as comparison of proportions.

We know that $p_1 = .33$ and $p_1 = .18$. We also know α and β. So you can use Equation 22.6. With this equation, N = 134 in each group. The difference between N = 130 and N = 134 is not important, as all these calculations are based on assumptions that may not be exactly true.

CHAPTER 23

1. Calculate the t test for Example 7.1. The authors compared stool output between treated and untreated babies. The results were 260 ± 254 (SD) for 84 control babies and 182 ± 197 for 85 treated babies.

First calculate the two SEMs, 27.7 and 21.4 for the control and treated babies. Then calculate the SE of the difference between means using Equation 23.3. The SE of the difference is 34.8. The difference between the means is 78. So the t ratio is $88/34.8 = 2.24$. There are 167 degrees of freedom. If you use a computer program, you can determine that the P value (two-tailed) is 0.0269. Using Table A5.4 in the Appendix, you can see that the P value must be a bit more than 0.02.

2. Calculate the t test for Problem 2 in Chapter 7.

3. t = 6.3 with 63 degrees of freedom. P < 0.0001.

A. Explain why power goes up as Δ_H gets larger.

It is easier to find a big difference than a small one. Sampling error is more likely to produce a small difference than a large one. Since a large Δ_H is less likely to be due to chance, it is more likely to be due to a real difference. In other words, the power is higher.

B. Explain why power goes down as SD gets larger.

If the SD is large, the data are scattered more. With more scatter, sampling error is likely to produce larger differences. So an observed difference is more likely to be due to random sampling with a large SD than with a small one. So the power is lower.

C. Explain why power goes up as N gets larger.

If you collect more evidence, you have more power. With a larger sample size, sampling error is likely to produce small differences. So an observed difference is less likely to be due to random sampling with a large N than with a small N.

D. Explain why power goes down as α gets smaller.

When you make α smaller, you set a harsher criteria for finding a significant difference. This decreases your chance of making a Type I error. But you have made it harder to declare a difference significant even if the difference is real. By making α smaller, you increase the chance that a real difference will be declared not significant. You have decreased the power.

4. For the study in Example 23.2, calculate the following:

A. Calculate the t test and determine the t ratio and the P value.

$$t = .240 \quad P = .81$$

B. What was the power of that study to detect a mean difference of 25 sites/cell?

Use Equation 23.7. $\Delta_H = 25$, SE of difference = 25, and $t* = 2.04$. $z_{power} = -1.04$. From Table A5.6, the power is about 15%. From the example in the text, we know that the study had a 48% power to detect a mean difference of 50 sites/cell. Of course the study had less power to detect a smaller difference as statistically significant.

C. What was the power of that study to detect a mean difference of 100 sites/cell?

Use Equation 23.7. $\Delta_H = 100$, SE of difference = 25, and $t* = 2.04$. $z_{power} = 2.96$. From Table A5.6, the power is >99%. If the difference between population means was really huge, this study had plenty of power to find a significant difference.

D. If the study were repeated with 50 subjects in each group, what would the power be to detect a difference of 50 sites/cell?

If we were to repeat the study with more subjects, we can't be sure what the SD values will be. But our best guess is that the SD will be the same as observed in the first study. The problem gives SEM values, but these can be converted to SD values since you know the sample size. Then you can calculate the pooled SD, which is 73.85 receptors/cell (Equation 23.3). From this value and the new sample sizes, you can calculate that the SE of the difference will be about 14.8 receptors/cell. Now you can use Equation 23.7 to find that $z_{power} = 1.33$. The power is 91%. This is the same problem as the example (we haven't changed the SD of the data or the size of Δ_H) except that the samples are now larger. Increasing the sample size increased the power.

CHAPTER 24

1. The data compare the number of beta-adrenergic receptors on lymphocytes of control subjects and those taking a drug.

Control	Drug
1162	892
1095	903
1327	1164
1261	1002
1103	961
1235	875

A. Calculate the Mann-Whitney test.

The best thing to do is use a computer program. It tells you that the sum of ranks in the control group is 54 and that the sum of ranks in the drug-treated group is 24. The P value is 0.0152 (two tailed).

If you want to analyze the data manually, the first step is to rank the values, without respect to whether they are control or drug treated. Then sum the ranks in each group.

Control	Rank	Drug	Rank
1162	8	892	2
1095	6	903	3
1327	12	1164	9
1261	11	1002	5
1103	7	961	4
1235	10	875	1
Sum of ranks	54	Sum of ranks	24

Many books (not this one) give tables for the Mann-Whitney test, so you might look up the P value from the sum of ranks and the sample size.

To calculate it manually, you need to use Equations 24.2 and 24.3, which give approximate answers. Equation 24.2 calculates two values for U. U = 33 and U = 3. You will get the same value of Z no matter which U value you use. Using Equation

24.3 (and knowing that $N_a = N_b = 6$), $Z = 2.40$. From Table A5.2 in the Appendix, you can see that 1.64% of a Gaussian population has $z > 2.40$ or $z < -2.40$. So the two-tailed P value is approximately 0.0164.

B. Calculate the t test.

Again, you'll find it easiest to use a computer program. InStat tells us that $t = 3.978$ and $P = 0.0026$.

To calculate the t test manually, first calculate the mean and SEM of each group.

$$\text{Control: Mean} = 1197.2, \text{SEM} = 37.8$$

$$\text{Treated: Mean} = 966.2, \text{SEM} = 44.1$$

Then calculate the SE of the difference between means as the square root of the sum of the square of the two SEM values. It equals 58.1. The t ratio equals the difference between means divided by the SE of the difference $= (1197.2 - 966.2)/58.1 = 3.97$. To determine the degrees of freedom, sum the total number of subjects minus 2. DF = 10. Look in Table A5.7 of the Appendix. DF = 10, and P = 0.003.

CHAPTER 25

1. These data are the same as those presented in Problem 1 from Chapter 24. The experimenters compared the number of beta-adrenergic receptors on lymphocytes of control subjects and those taking a drug. Assume that each subject was first measured as a control and then while taking the drug.

   A. Calculate the paired t test.

First, calculate the difference for each subject.

Control	Drug	Difference	Rank of Difference
1162	892	270	−5
1095	903	192	−3
1327	1164	163	−2
1261	1002	259	−4
1103	961	142	−1
1235	875	360	−6

Then calculate the mean and SEM of the differences. Mean difference = 231 with a SE of 33.2. The t ratio is the ratio of the mean difference divided by its SE, which is $231/33.2 = 6.964$. The number of degrees of freedom is the number of pairs minus one, or five. Table A5.4 does not go down to five degrees of freedom, so use Table A5.5. With 5 df, the critical t ratio for $\alpha = 0.001$ is 6.869. Since the t ratio in this example is higher than this, the P value must be less than 0.001. A program would tell you that P = 0.0009.

   B. Calculate the Wilcoxon test.

First rank the differences, as shown in the last column above. Assign a negative rank when the value decreases and positive when it increases. In this problem, all values decrease so all ranks are negative. Then separately sum the ranks for the subjects that increase and decrease. The sum of the ranks for the increases is -21; the sum of the ranks for the decreases is zero. This book doesn't give a table to determine a P value from the sum of ranks, but a program would tell you that $P = 0.0313$ (two-tailed).

CHAPTER 26

1. You hypothesize that a disease is inherited as an autosomal dominant trait. That means that you expect that, on average, half of the children of people with the disease will have the disease and half will not. As a preliminary test of this hypothesis, we obtain data from 40 children of patients and find that only 14 have the disease. Is this discrepancy enough to make us reject the hypothesis?

With 40 children, you expect 20 to have the disease and 20 to not. You need to compare the observed and expected values with the chi-square test.

	Observed (O)	Expected (E)	$(O-E)^2/E$
Disease	14	20	1.8
No disease	26	20	1.8
Total	40	40	3.6

So $\chi^2 = 3.6$ with one degree of freedom (number of categories minus one).

From Table A5.7 in the Appendix, you can see that $P = 0.0578$. The null hypothesis is that the disease really is an autosomal dominant trait, and that all discrepancies between observed and expected are due to chance. If this hypothesis were true, you would see a discrepancy as large (or larger) than observed here in 5.78% of experiments. You have to decide if that probability is low enough to reject the hypothesis.

CHAPTER 27

1. Perform the chi-square test for the cat-scratch disease data for Example 9.1. State the null hypothesis, show the calculations, and interpret the results.

Here again are the observed data.

Observed Data	Cases	Controls	Total
Cats have fleas	32	4	36
Cats don't have fleas	24	52	76
Total	56	56	112

You can calculate the expected data directly from the experimental data. You don't need any special theory to calculate the expected values. Let's first figure out how many cases you expect to see where the cats have fleas, if the cases and controls were equally likely to have cats with fleas. Altogether 36 of the cases and controls had cats with fleas. Since there were equal number of cases and controls, we'd expect 18 cases with cats with fleas and 18 controls. So the expected values look like the following table. Of course, the row and column totals are the same in the expected and observed tables. The expected values were calculated from the row and column totals.

Expected Data	Cases	Controls	Total
Cats have fleas	18	18	36
Cats don't have fleas	38	38	76
Total	56	56	112

Now calculate χ^2 using Equation 27.2. $\chi^2 = 29.8$. There is only a single degree of freedom. With such an enormous value of χ^2, the P value is tiny. The P value is less than 0.0001.

The null hypothesis is that the cases and controls are equally likely to have cats with fleas. If this hypothesis were true, you'd find this big a discrepancy between observed and expected values in less than 0.01% of experiments of this size.

2. (Same as Problem 2 from Chapter 8.) Goran-Larsson et al. wondered whether hypermobile joints caused symptoms in musicians. They sent questionnaires to many musicians and asked about hypermobility of the joints and about symptoms of pain and stiffness. They asked about all joints, but this problem concerns only the data they collected about the wrists. Of 96 musicians with hypermobile wrists, 5% had pain and stiffness of the wrists. In contrast, 18% of 564 musicians without hypermobility had such symptoms. Analyze these data as fully as possible.

Observed Data	Pain and Stiffness	No Pain or Stiffness	Total
Hypermobile wrists	5	91	96
Normal wrists	102	462	564
Total	107	553	660

If you use a computer program, you can analyze the data with Fisher's exact test. The P value (two-tailed) is 0.0008. If there were no association between hypermobility of the wrist and pain and stiffness, there is only a 0.08% chance that you'd find an association this strong (and a relative risk this high) in a study this size.

If you do the calculations yourself, you need to first calculate the expected values. If the null hypothesis were true, how many subjects would you expect to find with both hypermobile wrists and pain/stiffness? Overall 107/660 = 16.2% of the subjects had pain/stiffness. Therefore, out of the 96 subjects with hypermobile wrists, you'd expect 96 * .162 = 15.6 would have pain and stiffness. You can work out the remainder of the expected values using similar logic, or by subtraction from the row and column totals. The expected results are given in the following table.

Expected Data	Pain and Stiffness	No Pain or Stiffness	Total
Hypermobile wrists	15.6	80.4	96
Normal wrists	91.4	472.6	564
Total	107	553	660

It's OK that the expected results include fractional components. The table really shows the average expected results if you were to repeat the experiments many times. To calculate the chi-square test, you need to combine the observed and expected values. If you use the Yates' correction, the result is $\chi^2 = 9.088$. The P value is 0.0026.

3. Will a study with N = 50 have greater power to detect a difference between $p_1 = 0.10$ and $p_2 = 0.20$, or between $p_1 = 0.20$ and $p_2 = 0.40$?

First answer intuitively. A chance difference of just one or two cases will change the proportions more when the proportions are low. Therefore, you'd expect to have more power to detect the difference between $p_1 = 0.20$ and $p_2 = 0.40$. To answer mathematically, calculate Equation 27.3. For the first example, H = 0.28. For the second example, H = 0.44. A bigger vlaue of H means more power.

4. Calculate a P value from the data in Problem 4 of Chapter 8. Interpret it.

First, show the data as a contingency table.

	Resuscitated	Not Resuscitated	Total
Active CPR	18	11	29
Standard CPR	10	23	33
Total	28	34	62

The best way to analyze the data is to perform the Fisher's test with a computer. The P value is 0.0207 (two sided).

To calculate a P value yourself, use the chi-square test. Using Yates' correction, chi-square = 5.072. The two-sided P value is 0.024 from Table A5.7.

5. In response to many case reports of connective tissue diseases after breast implants, the FDA called for a moratorium on breast implants in 1992. Gabriel and investigators did a prospective study to determine if there really was an association between breast implants and connective tissue (and other) diseases. They studied 749 women who had received a breast implant and twice that many control subjects. They analyzed their data using survival analysis to account for different times between implant and disease and to correct for differences in age. You can analyze the key

findings more simply. They found that 5 of the cases and 10 of the controls developed connective tissue disease.

A. What is the relative risk and P value? What is the null hypothesis?

The null hypothesis is that there is no association between breast implants and connective tissue diseases. If the null hypothesis is true, the overall incidence of connective tissue diseases among cases ought to equal the incidence among controls.

They used twice as many controls as patients with breast implants. Twice as many controls as cases developed connective tissue disease. This is exactly what you expect under the null hypothesis, so the relative risk is exactly 1. To calculate the relative risk: $(5/749)/(10/1494) = 1.0$. The data provide no evidence whatsoever of any association between breast implant and connective tissue disease.

The P value (two-tailed) must be 1.0. The P value answers this question: If the null hypothesis is true, what is the chance of randomly sampling subjects and obtaining such strong evidence (or stronger) that there is an association. Since there was absolutely no evidence of association in this study, 100% of all studies would produce evidence this strong or stronger. If you were to calculate the chi-square test, you'd find that the expected values equaled the observed values. Chi-square equals zero, so $P = 1.0$.

B. What is the 95% CI for the relative risk?

Before concluding that breast implants don't cause connective tissue disease, you should do further calculations. One is to calculate the 95% CI for the relative risk. This is best done by computer. The answer is 0.34 to 2.92. So the data are consistent with a reduction of risk by a factor of three or an increase in the risk by a factor of three. Even though the study shows no evidence of association, the study is too small to be convincingly negative.

C. If breast implants really doubled the risk of connective tissue disease, what is the power of a study this size to detect a statistically significant result with $P < 0.05$?

One way to think about the usefulness of the data from the study is to ask what power the study had to find hypothetical differences. For example, if patients with breast implants really got connective tissue disease at twice the rate of control patients, what was the power of this study to find a statistically significant difference?

We'll set p_1 to 0.0067, since this is the fraction of controls (and cases) who had connective tissue disease (10/1484). We'll set p_2 to half that value or 0.0034. To calculate power first use Equation 27.3 to calculate that $H = 0.0472$.

Figure 27.1 isn't very useful with such small values of H, so use Equation 27.4. Set $z_\alpha = 1.96$, since we are using the conventional definition of statistical significance ($P < 0.05$, two tailed). Since the sample sizes are different, you need to calculate the harmonic mean according to Equation 27.5. $N = 2 * 747 * 1484/(747 + 1484) = 993.7$. Now, use Equation 27.4 to calculate $z_{power} = 0.91$. From Table A4.14 in the Appendix, power is about 18%. If breast implants really increased the risk of connective tissue disease twofold, a study this size would observe a significant association less than one time in five. Since the study had such low power, you shouldn't interpret the findings too strongly.

CHAPTER 28

1. You use a hemocytometer to count white blood cells. When you look in the microscope you see lots of squares and 25 squares enclose 0.1 microliter. You count the
   number of cells in nine squares and find 50 white blood cells. Can you calculate
   the 95% CI for the number of cells per microliter? If so, calculate the interval. If
   not, what information do you need? What assumptions must you make?

You counted 50 cells in 9 squares. Since there are 250 squares per microliter, there
are 50 * (250/9) = 1389 cells per microliter. To calculate the CI, we must work with
the number actually counted, 50. From Table A5.9 in the Appendix, you can see that
the 95% CI for the average number of cells in 9 squares ranges from 37.1 to 65.9.
Multiply each confidence limit by (250/9) to determine the 95% CI for the number of
cells per microliter. The 95% CI ranges from 1031 to 1831 cells per microliter.

2. In 1988 a paper in *Nature* (333:816, 1988) caused a major stir in the popular and
   scientific press. The authors claimed that antibodies diluted with even 10^{-120} of the
   starting concentration stimulated basophils to degranulate. With that much dilution,
   the probability that even a single molecule of antibody remains in the tube is almost
   0. The investigators hypothesized that the water somehow "remembered" that it
   had seen antibody. These results purported to give credence to homeopathy, the
   theory that extremely low concentrations of drugs are therapeutic.
       The assay is simple. Add a test solution to basophils and incubate. Then stain.
   Count the number of stained cells in a certain volume. Compare the number of
   cells in control and treated tubes. A low number indicates that many cells had
   degranulated, since cells that had degranulated won't take up the stain.
       The authors present the "mean and standard error of basophil number actually
   counted in triplicate." In the first experiment, the three control tubes were reported
   as 81.3 ± 1.2, 81.6 ± 1.4, and 80.0 ± 1.5. A second experiment on another day
   gave very similar results. The tubes treated with the dilute antibody gave much
   lower numbers of stained cells, indicating that the dilute antibody had caused
   degranulation. For this problem, think only about the control tubes.

   A. Why are these control values surprising?

The authors counted only the cells that still had granules. In the results shown, the
average number of cells in the volume counted was about 80. If this is true, then the
Poisson distribution tells us what we expect to see in different experiments. From
Table A5.9 in the Appendix, the 95% CI ranges from 63 to 100. If you repeated the
experiment many times, you'd expect the number of cells counted to be within this
range 95% of the time and outside the range 5% of the time. The results reported by
the authors have far less variation than that. There is way too little scatter.

   B. How could these results have been obtained?

One possibility is that the investigators were lucky and just happened to have so little
variation. The chance of having so little variation is very small, so this is quite unlikely.
If the cells are well mixed and randomly sampled, the Poisson distribution tells us the
distribution of values we expect to see. This is the "best" you can do, assuming perfect
experimental technique. Any sloppiness in the experiment will lead to more variability.

No matter how carefully you do the experiment, you can't get less scatter than predicted by the Poisson distribution unless you are extremely lucky.

So how then could the results have been obtained? There is no way to be sure, but here is one possibility. It is difficult to count cells in a reliable way. Counting requires some judgment as you ask yourself these kinds of questions: How do you deal with cells on the border? How do you distinguish between real cells and debris? How do you deal with clumps? If you know the results you are "supposed" to get, your answers might be swayed to get the count you expect to get.

For an assay like this one (counting cells manually), the only way to avoid bias is to have the person counting the cells not know which tube is which, so he or she can't let the expected results bias the count. Even better, use a machine that counts in an unbiased way. The machines can also count a larger volume, thus narrowing the confidence interval.

You could also improve the experimental methodology in other ways. Why not count all cells, not just the ones that haven't degranulated. Why not use a more precise assay, such as measuring histamine release from the cells?

Appendix 5

Statistical Tables

This appendix contains tables that you will find useful when you do your own calculations. To keep this book compact, I have only tabulated certain values. You can easily calculate other values using the spreadsheet program Microsoft Excel, which I used to create all the tables (version 5 for Windows).* The spreadsheet formulas are listed below. While most are straightforward and documented in the Excel function manual, the equations for Tables 1 and 9 are not obvious. I obtained these equations from Lothar Sachs, *Applied Statistics. A Handbook of Techniques.* 2nd ed., New York, Springer-Verlag, 1984.

Table 1
Lower limit: N/(N+ (D−N+1) *FINV((100−P)/200, (2*D−2*N+2),2*N))
Upper limit: 1/(1+ (D−N) / ((N+1) *FINV((100−P)/200,2*N+2,2*D−2*N)))
[N=Numerator, D=Denominator, P=Percentage confidence, i.e. 95]

Table 2
First column: 2*NORMSDIST(Z)−1
Second and third columns: 1−NORMSDIST(Z)
Fourth column: Sum of the values in the second and third column.

Table 3
TINV([100−P]/100,DF)[P=Percent confidence,i.e.95, DF=degrees of freedom]

Table 4
TDIST(T,DF,2)

Table 5
TINV(ALPHA,DF)

Table 6
NORMSDIST(ZPOWER)−1

Table 7
CHIDIST(CHI2,1)

Table 8
CHIINV(ALPHA,DF).

---

*The Macintosh version is identical. Other spreadsheet programs have similar functions.

Table 9

Lower confidence limit: CHIINV((100+P)/200,2*C)*0.5

Upper confidence limit: CHIINV((100−P)/200,2*C+2)*0.5

[C=Counts, Percent confidence, i.e.95]

Table 10

FINV(ALPHA,DFN,DFD)

TABLE A5.1. 95% CONFIDENCE INTERVAL OF A PROPORTION

Find the column corresponding to the numerator (N) and the row corresponding to the denominator (D). The values are the lower and upper 95% CI for the proportion. For example if you observe that 4 of 20 patients have a certain complication, the proportion is 0.200 and the 95% CI ranges from 0.057 to 0.437.

Table A5.1. 95% Confidence Interval of a Proportion

Each cell gives the lower and upper limits (lower, upper) for the given Denominator (D) and Numerator.

D	0	1	2	3	4	5	6	7	8	9
1	.000, .975	.025, 1.00								
2	.000, .842	.013, .987	.158, 1.00							
3	.000, .708	.008, .906	.094, .992	.292, 1.00						
4	.000, .602	.006, .806	.068, .932	.194, .994	.398, 1.00					
5	.000, .522	.005, .716	.053, .853	.147, .947	.284, .995	.478, 1.00				
6	.000, .459	.004, .641	.043, .777	.118, .882	.223, .957	.359, .996	.541, 1.00			
7	.000, .410	.004, .579	.037, .710	.099, .816	.184, .901	.290, .963	.421, .996	.590, 1.00		
8	.000, .369	.003, .527	.032, .651	.085, .755	.157, .843	.245, .915	.349, .968	.473, .997	.631, 1.00	
9	.000, .336	.003, .482	.028, .600	.075, .701	.137, .788	.212, .863	.299, .925	.400, .972	.518, .997	.664, 1.00
10	.000, .308	.003, .445	.025, .556	.067, .652	.122, .738	.187, .813	.262, .878	.348, .933	.444, .975	.555, .997
11	.000, .285	.002, .413	.023, .518	.060, .610	.109, .692	.167, .766	.234, .833	.308, .891	.390, .940	.482, .977
12	.000, .265	.002, .385	.021, .484	.055, .572	.099, .651	.152, .723	.211, .789	.277, .848	.349, .901	.428, .945
13	.000, .247	.002, .360	.019, .454	.050, .538	.091, .614	.139, .684	.192, .749	.251, .808	.316, .861	.386, .909
14	.000, .232	.002, .339	.018, .428	.047, .508	.084, .581	.128, .649	.177, .711	.230, .770	.289, .823	.351, .872
15	.000, .218	.002, .319	.017, .405	.043, .481	.078, .551	.118, .616	.163, .677	.213, .734	.266, .787	.323, .837
16	.000, .206	.002, .302	.016, .383	.040, .456	.073, .524	.110, .587	.152, .646	.198, .701	.247, .753	.299, .802
17	.000, .195	.001, .287	.015, .364	.038, .434	.068, .499	.103, .560	.142, .617	.184, .671	.230, .722	.278, .770
18	.000, .185	.001, .273	.014, .347	.036, .414	.064, .476	.097, .535	.133, .590	.173, .643	.215, .692	.260, .740
19	.000, .176	.001, .260	.013, .331	.034, .396	.061, .456	.091, .512	.126, .566	.163, .616	.203, .665	.244, .711
20	.000, .168	.001, .249	.012, .317	.032, .379	.057, .437	.087, .491	.119, .543	.154, .592	.191, .639	.231, .685
21	.000, .161	.001, .238	.012, .304	.030, .363	.054, .419	.082, .472	.113, .522	.146, .570	.181, .616	.218, .660
22	.000, .154	.001, .228	.011, .292	.029, .349	.052, .403	.078, .454	.107, .502	.139, .549	.172, .593	.207, .636
23	.000, .148	.001, .219	.011, .280	.028, .336	.050, .388	.075, .437	.102, .484	.132, .529	.164, .573	.197, .615
24	.000, .142	.001, .211	.010, .270	.027, .324	.047, .374	.071, .422	.098, .467	.126, .511	.156, .553	.188, .594
25	.000, .137	.001, .204	.010, .260	.025, .312	.045, .361	.068, .407	.094, .451	.121, .494	.149, .535	.180, .575
26	.000, .132	.001, .196	.009, .251	.024, .302	.044, .349	.066, .394	.090, .436	.116, .478	.143, .518	.172, .557
27	.000, .128	.001, .190	.009, .243	.024, .292	.042, .337	.063, .381	.086, .423	.111, .463	.138, .502	.165, .540
28	.000, .123	.001, .183	.009, .235	.023, .282	.040, .327	.061, .369	.083, .410	.107, .449	.132, .487	.159, .524
29	.000, .119	.001, .178	.008, .228	.022, .274	.039, .317	.058, .358	.080, .397	.103, .435	.127, .472	.153, .508
30	.000, .116	.001, .172	.008, .221	.021, .265	.038, .307	.056, .347	.077, .386	.099, .423	.123, .459	.147, .494

(continued)

Table A5.1. *Continued*

	Numerator									
D	10	11	12	13	14	15	16	17	18	19
11	.587 .998	.715 1.00	.735	.753	.768	.782	.794	.805	.815	.824
12	.516 .979	.615 .998	.640 1.00	.661	.681	.698	.713	.727	.740	.751
13	.462 .950	.546 .981	.572 .998	.595 1.00	.617	.636	.653	.669	.683	.696
14	.419 .916	.492 .953	.519 .982	.544 .998	.566 1.00	.586	.604	.621	.637	.651
15	.384 .882	.449 .922	.476 .957	.501 .983	.524 .998	.544 1.00	.563	.581	.597	.612
16	.354 .848	.413 .890	.440 .927	.465 .960	.488 .984	.509 .998	.528 1.00	.546	.563	.578
17	.329 .816	.383 .858	.410 .897	.434 .932	.457 .962	.478 .985	.498 .999	.516 1.00	.533	.549
18	.308 .785	.357 .827	.384 .867	.408 .903	.430 .936	.451 .964	.471 .986	.489 .999	.506 1.00	.522
19	.289 .756	.335 .797	.361 .837	.384 .874	.407 .909	.427 .939	.447 .966	.465 .987	.482 .999	.498 1.00
20	.272 .728	.315 .769	.340 .809	.364 .846	.385 .881	.406 .913	.425 .943	.443 .968	.460 .988	.476 .999
21	.257 .702	.298 .743	.322 .782	.345 .819	.366 .854	.387 .887	.406 .918	.424 .946	.441 .970	.457 .988
22	.244 .678	.282 .718	.306 .756	.328 .793	.349 .828	.369 .861	.388 .893	.406 .922	.423 .948	.439 .971
23	.232 .655	.268 .694	.291 .732	.313 .768	.334 .803	.353 .836	.372 .868	.389 .898	.406 .925	.950
24	.221 .634	.256 .672	.278 .709	.299 .744	.319 .779	.339 .812	.357 .844	.374 .874	.902	.929
25	.211 .613	.244 .651	.266 .687	.287 .722	.306 .756	.325 .789	.343 .820	.851	.879	.906
26	.202 .594	.234 .631	.255 .666	.275 .701	.294 .734	.313 .766	.798	.828	.857	.884
27	.194 .576	.224 .612	.245 .647	.264 .681	.283 .713	.745	.776	.806	.835	.862
28	.186 .559	.215 .594	.235 .628	.255 .661	.694	.725	.755	.785	.814	.841
29	.179 .543	.207 .577	.227 .611	.643	.675	.706	.736	.765	.793	.821
30	.173 .528	.199 .561	.594	.626	.657	.687	.717	.745	.773	.801

Numerator

D	20	21	22	23	24	25	26	27	28	29
21	.381	.439 1.00								
22	.354	.404 .999	.461 1.00							
23	.332	.377 .989	.426 .999	.481 1.00						
24	.313	.354 .973	.398 .990	.446 .999	.500 1.00					
25	.296	.335 .955	.375 .975	.418 .990	.465 .999	.518 1.00				
26	.282	.318 .934	.355 .956	.395 .976	.437 .991	.482 .999				
27	.269	.302 .914	.338 .937	.375 .958	.413 .976	.454 .991	.499 .999	.549 1.00		
28	.257	.289 .893	.322 .917	.357 .939	.393 .960	.431 .977	.471 .991	.514 .999	.564 1.00	
29	.246	.276 .873	.308 .897	.341 .920	.375 .942	.410 .961	.447 .978	.486 .992	.529 .999	.577 1.00
30	.236	.265 .853	.295 .877	.326 .901	.358 .923	.392 .944	.426 .962	.463 .979	.501 .992	.542 .999

TABLE A5.2. THE GAUSSIAN DISTRIBUTION

For any value y from a Gaussian distribution, calculate

$$z = |Y - \text{mean}|/\text{standard deviation}.$$

z	Fraction of the population whose distance from the mean is			
	Within z SDs	>z SDs	<z SDs	Outside z SDs
0.00	0.00%	50.00%	50.00%	100.00%
0.05	3.99%	48.01%	48.01%	96.01%
0.10	7.97%	46.02%	46.02%	92.03%
0.15	11.92%	44.04%	44.04%	88.08%
0.20	15.85%	42.07%	42.07%	84.15%
0.25	19.74%	40.13%	40.13%	80.26%
0.30	23.58%	38.21%	38.21%	76.42%
0.35	27.37%	36.32%	36.32%	72.63%
0.40	31.08%	34.46%	34.46%	68.92%
0.45	34.73%	32.64%	32.64%	65.27%
0.50	38.29%	30.85%	30.85%	61.71%
0.55	41.77%	29.12%	29.12%	58.23%
0.60	45.15%	27.43%	27.43%	54.85%
0.65	48.43%	25.78%	25.78%	51.57%
0.70	51.61%	24.20%	24.20%	48.39%
0.75	54.67%	22.66%	22.66%	45.33%
0.80	57.63%	21.19%	21.19%	42.37%
0.85	60.47%	19.77%	19.77%	39.53%
0.90	63.19%	18.41%	18.41%	36.81%
0.95	65.79%	17.11%	17.11%	34.21%
1.00	68.27%	15.87%	15.87%	31.73%
1.05	70.63%	14.69%	14.69%	29.37%
1.10	72.87%	13.57%	13.57%	27.13%
1.15	74.99%	12.51%	12.51%	25.01%
1.20	76.99%	11.51%	11.51%	23.01%
1.25	78.87%	10.56%	10.56%	21.13%
1.30	80.64%	9.68%	9.68%	19.36%
1.35	82.30%	8.85%	8.85%	17.70%
1.40	83.85%	8.08%	8.08%	16.15%
1.45	85.29%	7.35%	7.35%	14.71%
1.50	86.64%	6.68%	6.68%	13.36%
1.55	87.89%	6.06%	6.06%	12.11%
1.60	89.04%	5.48%	5.48%	10.96%
1.65	90.11%	4.95%	4.95%	9.89%
1.70	91.09%	4.46%	4.46%	8.91%
1.75	91.99%	4.01%	4.01%	8.01%
1.80	92.81%	3.59%	3.59%	7.19%
1.85	93.57%	3.22%	3.22%	6.43%
1.90	94.26%	2.87%	2.87%	5.74%
1.95	94.88%	2.56%	2.56%	5.12%
2.00	95.45%	2.28%	2.28%	4.55%
2.05	95.96%	2.02%	2.02%	4.04%
2.10	96.43%	1.79%	1.79%	3.57%
2.15	96.84%	1.58%	1.58%	3.16%

(continued)

	Fraction of the population whose distance from the mean is			
z	Within z SDs	>z SDs	<z SDs	Outside z SDs
2.20	97.22%	1.39%	1.39%	2.78%
2.25	97.56%	1.22%	1.22%	2.44%
2.30	97.86%	1.07%	1.07%	2.14%
2.35	98.12%	0.94%	0.94%	1.88%
2.40	98.36%	0.82%	0.82%	1.64%
2.45	98.57%	0.71%	0.71%	1.43%
2.50	98.76%	0.62%	0.62%	1.24%
2.55	98.92%	0.54%	0.54%	1.08%
2.60	99.07%	0.47%	0.47%	0.93%
2.65	99.20%	0.40%	0.40%	0.80%
2.70	99.31%	0.35%	0.35%	0.69%
2.75	99.40%	0.30%	0.30%	0.60%
2.80	99.49%	0.26%	0.26%	0.51%
2.85	99.56%	0.22%	0.22%	0.44%
2.90	99.63%	0.19%	0.19%	0.37%
2.95	99.68%	0.16%	0.16%	0.32%
3.00	99.73%	0.13%	0.13%	0.27%
3.05	99.77%	0.11%	0.11%	0.23%
3.10	99.81%	0.10%	0.10%	0.19%
3.15	99.84%	0.08%	0.08%	0.16%
3.20	99.86%	0.07%	0.07%	0.14%
3.25	99.88%	0.06%	0.06%	0.12%
3.30	99.90%	0.05%	0.05%	0.10%
3.35	99.92%	0.04%	0.04%	0.08%
3.40	99.93%	0.03%	0.03%	0.07%
3.45	99.94%	0.03%	0.03%	0.06%
3.50	99.95%	0.02%	0.02%	0.05%
3.55	99.96%	0.02%	0.02%	0.04%
3.60	99.97%	0.02%	0.02%	0.03%
3.65	99.97%	0.01%	0.01%	0.03%
3.70	99.98%	0.01%	0.01%	0.02%
3.75	99.98%	0.01%	0.01%	0.02%
3.80	99.99%	0.01%	0.01%	0.01%
3.85	99.99%	0.01%	0.01%	0.01%
3.90	99.99%	0.00%	0.00%	0.01%
3.95	99.99%	0.00%	0.00%	0.01%
4.00	99.99%	0.00%	0.00%	0.01%

TABLE A5.3. VALUES OF t* USED TO CALCULATE CONFIDENCE INTERVALS

This table shows the values of t* needed to calculate confidence intervals. A confidence interval of a mean extends in both directions from the mean by a distance equal to the SEM times the critical t value tabulated here. Set df equal to one less than the number of data points. This book uses the symbol t* to denote the critical value of the t distribution.

df	Critical value of t for __ % CI		
	90% CI	95% CI	99% CI
1	6.3137	12.7062	63.6559
2	2.9200	4.3027	9.9250
3	2.3534	3.1824	5.8408
4	2.1318	2.7765	4.6041
5	2.0150	2.5706	4.0321
6	1.9432	2.4469	3.7074
7	1.8946	2.3646	3.4995
8	1.8585	2.3060	3.3554
9	1.8331	2.2622	3.2498
10	1.8125	2.2281	3.1693
11	1.7959	2.2010	3.1058
12	1.7823	2.1788	3.0545
13	1.7709	2.1604	3.0123
14	1.7613	2.1448	2.9768
15	1.7531	2.1315	2.9467
16	1.7459	2.1199	2.9208
17	1.7396	2.1098	2.8982
18	1.7341	2.1009	2.8784
19	1.7291	2.0930	2.8609
20	1.7247	2.0860	2.8453
21	1.7207	2.0796	2.8314
22	1.7171	2.0739	2.8188
23	1.7139	2.0687	2.8073
24	1.7109	2.0639	2.7970
25	1.7081	2.0595	2.7874
26	1.7056	2.0555	2.7787
27	1.7033	2.0518	2.7707
28	1.7011	2.0484	2.7633
29	1.6991	2.0452	2.7564
30	1.6973	2.0423	2.7500
35	1.6896	2.0301	2.7238
40	1.6839	2.0211	2.7045
45	1.6794	2.0141	2.6896
50	1.6759	2.0086	2.6778
100	1.6602	1.9840	2.6259
200	1.6525	1.9719	2.6006
500	1.6479	1.9647	2.5857
1000	1.6464	1.9623	2.5807

TABLE A5.4. DETERMINING A P VALUE FROM t

To determine a P value (two-tailed), find the row that corresponds to the value of t that you calculated from your data and the column that corresponds to df. For an unpaired t test, df equals the total number of values (both groups) minus two. For a paired t test, df equals the number of pairs minus one.

Table A5.4. Determining a P Value from t

Degrees of Freedom

t	6	7	8	9	10	11	12	13	14	15	16	18	20	25	50	1000
1.0	.356	.351	.347	.343	.341	.339	.337	.336	.334	.333	.332	.331	.329	.327	.322	.318
1.1	.313	.308	.303	.300	.297	.295	.293	.291	.290	.289	.288	.286	.284	.282	.277	.272
1.2	.275	.269	.264	.261	.258	.255	.253	.252	.250	.249	.248	.246	.244	.241	.236	.230
1.3	.241	.235	.230	.226	.223	.220	.218	.216	.215	.213	.212	.210	.208	.205	.200	.194
1.4	.211	.204	.199	.195	.192	.189	.187	.185	.183	.182	.181	.179	.177	.174	.168	.162
1.5	.184	.177	.172	.168	.165	.162	.159	.158	.156	.154	.153	.151	.149	.146	.140	.134
1.6	.161	.154	.148	.144	.141	.138	.136	.134	.132	.130	.129	.127	.125	.122	.116	.110
1.7	.140	.133	.128	.123	.120	.117	.115	.113	.111	.110	.108	.106	.105	.102	.095	.089
1.8	.122	.115	.110	.105	.102	.099	.097	.095	.093	.092	.091	.089	.087	.084	.078	.072
1.9	.106	.099	.094	.090	.087	.084	.082	.080	.078	.077	.076	.074	.072	.069	.063	.058
2.0	.092	.086	.081	.077	.073	.071	.069	.067	.065	.064	.063	.061	.059	.056	.051	.046
2.1	.080	.074	.069	.065	.062	.060	.058	.056	.054	.053	.052	.050	.049	.046	.041	.036
2.2	.070	.064	.059	.055	.052	.050	.048	.046	.045	.044	.043	.041	.040	.037	.032	.028
2.3	.065	.055	.050	.047	.044	.042	.040	.039	.037	.036	.035	.034	.032	.030	.026	.022
2.4	.053	.047	.043	.040	.037	.035	.034	.032	.031	.030	.029	.027	.026	.024	.020	.017
2.5	.047	.041	.037	.034	.031	.030	.028	.027	.025	.025	.024	.022	.021	.019	.016	.013
2.6	.041	.035	.032	.029	.026	.025	.023	.022	.021	.020	.019	.018	.017	.015	.012	.009
2.7	.036	.031	.027	.024	.022	.021	.019	.018	.017	.016	.016	.015	.014	.012	.009	.007
2.8	.031	.027	.023	.021	.019	.017	.016	.015	.014	.013	.013	.012	.011	.010	.007	.005
2.9	.027	.023	.020	.018	.016	.014	.013	.012	.012	.011	.010	.010	.009	.008	.006	.004
3.0	.024	.020	.017	.015	.013	.012	.011	.010	.010	.009	.008	.008	.007	.006	.004	.003
3.1	.021	.017	.015	.013	.011	.010	.009	.008	.008	.007	.007	.006	.006	.005	.003	.002
3.2	.019	.015	.013	.011	.009	.008	.008	.007	.006	.006	.006	.005	.004	.004	.002	.001
3.3	.016	.013	.011	.009	.008	.007	.006	.006	.005	.005	.005	.004	.004	.003	.002	.001
3.4	.014	.011	.009	.008	.007	.006	.005	.005	.004	.004	.004	.003	.003	.002	.001	.001
3.5	.013	.010	.008	.007	.006	.005	.004	.004	.004	.003	.003	.003	.002	.002	.001	<.001
3.6	.011	.009	.007	.006	.005	.004	.004	.003	.003	.003	.002	.002	.001	.001	.001	<.001

3.7	.010	.008	.006	.005	.004	.004	.003	.003	.002	.002	.002	.002	.001	.001	.001	<.001
3.8	.009	.007	.005	.004	.003	.003	.003	.002	.02	.02	.001	.001	.001	.001	<.001	<.001
3.9	.008	.006	.005	.004	.003	.002	.002	.002	.002	.002	.001	.001	.001	<.001	<.001	<.001
4.0	.007	.005	.004	.003	.003	.002	.002	.002	.001	.001	.001	.001	<.001	<.001	<.001	<.001
4.1	.006	.005	.003	.003	.002	.002	.001	.001	.01	.001	.001	.001	<.001	<.001	<.001	<.001
4.2	.006	.004	.003	.002	.002	.001	.001	.001	.001	.001	.001	<.001	<.001	<.001	<.001	<.001
4.3	.005	.004	.003	.002	.001	.001	.001	.001	.001	.001	.001	<.001	<.001	<.001	<.001	<.001
4.4	.005	.003	.002	.002	.001	.001	.001	.001	.001	.001	<.001	<.001	<.001	<.001	<.001	<.001
4.5	.004	.003	.002	.001	.001	.001	.001	.001	.001	.001	<.001	<.001	<.001	<.001	<.001	<.001
4.6	.004	.002	.002	.001	.001	.001	.001	.001	<.001	<.001	<.001	<.001	<.001	<.001	<.001	<.001
4.7	.003	.002	.001	.001	.001	.001	.001	<.001	<.001	<.001	<.001	<.001	<.001	<.001	<.001	<.001
4.8	.003	.002	.001	.001	.001	.001	<.001	<.001	<.001	<.001	<.001	<.001	<.001	<.001	<.001	<.001
4.9	.003	.002	.001	.001	.001	.001	<.001	<.001	<.001	<.001	<.001	<.001	<.001	<.001	<.001	<.001
5.0	.002	.002	.001	.001	<.001	<.001	<.001	<.001	<.001	<.001	<.001	<.001	<.001	<.001	<.001	<.001
5.1	.002	.001	.001	.001	<.001	<.001	<.001	<.001	<.001	<.001	<.001	<.001	<.001	<.001	<.001	<.001
5.2	.002	.001	.001	<.001	<.001	<.001	<.001	<.001	<.001	<.001	<.001	<.001	<.001	<.001	<.001	<.001
5.3	.002	.001	.001	<.001	<.001	<.001	<.001	<.001	<.001	<.001	<.001	<.001	<.001	<.001	<.001	<.001
5.4	.002	.001	.001	<.001	<.001	<.001	<.001	<.001	<.001	<.001	<.001	<.001	<.001	<.001	<.001	<.001
5.5	.002	.001	.001	<.001	<.001	<.001	<.001	<.001	<.001	<.001	<.001	<.001	<.001	<.001	<.001	<.001
5.6	.001	.001	<.001	<.001	<.001	<.001	<.001	<.001	<.001	<.001	<.001	<.001	<.001	<.001	<.001	<.001
5.7	.001	.001	<.001	<.001	<.001	<.001	<.001	<.001	<.001	<.001	<.001	<.001	<.001	<.001	<.001	<.001
5.8	.001	.001	<.001	<.001	<.001	<.001	<.001	<.001	<.001	<.001	<.001	<.001	<.001	<.001	<.001	<.001
5.9	.001	.001	<.001	<.001	<.001	<.001	<.001	<.001	<.001	<.001	<.001	<.001	<.001	<.001	<.001	<.001
6.0	.001	.001	<.001	<.001	<.001	<.001	<.001	<.001	<.001	<.001	<.001	<.001	<.001	<.001	<.001	<.001
6.1	.001	<.001	<.001	<.001	<.001	<.001	<.001	<.001	<.001	<.001	<.001	<.001	<.001	<.001	<.001	<.001
6.2	.001	<.001	<.001	<.001	<.001	<.001	<.001	<.001	<.001	<.001	<.001	<.001	<.001	<.001	<.001	<.001
6.3	.001	<.001	<.001	<.001	<.001	<.001	<.001	<.001	<.001	<.001	<.001	<.001	<.001	<.001	<.001	<.001
6.4	.001	<.001	<.001	<.001	<.001	<.001	<.001	<.001	<.001	<.001	<.001	<.001	<.001	<.001	<.001	<.001
6.5	.001	<.001	<.001	<.001	<.001	<.001	<.001	<.001	<.001	<.001	<.001	<.001	<.001	<.001	<.001	<.001
6.6	.001	<.001	<.001	<.001	<.001	<.001	<.001	<.001	<.001	<.001	<.001	<.001	<.001	<.001	<.001	<.001
6.7	.001	<.001	<.001	<.001	<.001	<.001	<.001	<.001	<.001	<.001	<.001	<.001	<.001	<.001	<.001	<.001

TABLE A5.5. CRITICAL VALUES OF t

The table shows critical values from the t distribution. Choose the row according to the number of degrees of freedom and the column depending on the two-tailed value for α. If your test finds a value for t greater than the critical value tabulated here, then your two-tailed P value is less than α.

Table A5.5. Critical Values of t

α

df	0.250	0.200	0.150	0.100	0.070	0.060	0.050	0.040	0.030	0.020	0.010	0.005	0.001
2	1.604	1.886	2.282	2.920	3.578	3.896	4.303	4.849	5.643	6.965	9.925	14.089	31.600
3	1.423	1.638	1.924	2.353	2.763	2.951	3.182	3.482	3.896	4.541	5.841	7.453	12.924
4	1.344	1.533	1.778	2.132	2.456	2.601	2.776	2.999	3.298	3.747	4.604	5.598	8.610
5	1.301	1.476	1.699	2.015	2.297	2.422	2.571	2.757	3.003	3.365	4.032	4.773	6.869
6	1.273	1.440	1.650	1.943	2.201	2.313	2.447	2.612	2.829	3.143	3.707	4.317	5.959
7	1.254	1.415	1.617	1.895	2.136	2.241	2.365	2.517	2.715	2.998	3.499	4.029	5.408
8	1.240	1.397	1.592	1.860	2.090	2.189	2.306	2.449	2.634	2.896	3.355	3.833	5.041
9	1.230	1.383	1.574	1.833	2.055	2.150	2.262	2.398	2.574	2.821	3.250	3.690	4.781
10	1.221	1.372	1.559	1.812	2.028	2.120	2.228	2.359	2.527	2.764	3.169	3.581	4.587
11	1.214	1.363	1.548	1.796	2.007	2.096	2.201	2.328	2.491	2.718	3.106	3.497	4.437
12	1.209	1.356	1.538	1.782	1.989	2.076	2.179	2.303	2.461	2.681	3.055	3.428	4.318
13	1.204	1.350	1.530	1.771	1.974	2.060	2.160	2.282	2.436	2.650	3.012	3.372	4.221
14	1.200	1.345	1.523	1.761	1.962	2.046	2.145	2.264	2.415	2.624	2.977	3.326	4.140
15	1.197	1.341	1.517	1.753	1.951	2.034	2.131	2.249	2.397	2.602	2.947	3.286	4.073
16	1.194	1.337	1.512	1.746	1.942	2.024	2.120	2.235	2.382	2.583	2.921	3.252	4.015
17	1.191	1.333	1.508	1.740	1.934	2.015	2.110	2.224	2.368	2.567	2.898	3.222	3.965
18	1.189	1.330	1.504	1.734	1.926	2.007	2.101	2.214	2.356	2.552	2.878	3.197	3.922
19	1.187	1.328	1.500	1.729	1.920	2.000	2.093	2.205	2.346	2.539	2.861	3.174	3.883
20	1.185	1.325	1.497	1.725	1.914	1.994	2.086	2.197	2.336	2.528	2.845	3.153	3.850
21	1.183	1.323	1.494	1.721	1.909	1.988	2.080	2.189	2.328	2.518	2.831	3.135	3.819
22	1.182	1.321	1.492	1.717	1.905	1.983	2.074	2.183	2.320	2.508	2.819	3.119	3.792
23	1.180	1.319	1.489	1.714	1.900	1.978	2.069	2.177	2.313	2.500	2.807	3.104	3.768
24	1.179	1.318	1.487	1.711	1.896	1.974	2.064	2.172	2.307	2.492	2.797	3.091	3.745
25	1.178	1.316	1.485	1.708	1.893	1.970	2.060	2.167	2.301	2.485	2.787	3.078	3.725
30	1.173	1.310	1.477	1.697	1.879	1.955	2.042	2.147	2.278	2.457	2.750	3.030	3.646
35	1.170	1.306	1.472	1.690	1.869	1.944	2.030	2.133	2.262	2.438	2.724	2.996	3.591
40	1.167	1.303	1.468	1.684	1.862	1.936	2.021	2.123	2.250	2.423	2.704	2.971	3.551
45	1.165	1.301	1.465	1.679	1.856	1.929	2.014	2.115	2.241	2.412	2.690	2.952	3.520
50	1.164	1.299	1.462	1.676	1.852	1.924	2.009	2.109	2.234	2.403	2.678	2.937	3.496
100	1.157	1.290	1.451	1.650	1.832	1.902	1.984	2.081	2.201	2.364	2.626	2.871	3.390
200	1.154	1.286	1.445	1.653	1.822	1.892	1.972	2.067	2.186	2.345	2.601	2.838	3.340
1000	1.151	1.282	1.441	1.646	1.814	1.883	1.962	2.056	2.173	2.330	2.581	2.813	3.300

TABLE A5.6. CONVERTING z_{power} TO POWER

Use this table when calculating sample size. Calculate z_{power} using Equations 23.6 or 27.4. Then use this table to determine the power.

z_{power}	power	z_{power}	power	z_{power}	power
4.00	>99.99%	2.00	97.72%	0.00	50.00%
3.95	>99.99%	1.95	97.44%	−0.05	48.01%
3.90	>99.99%	1.90	97.13%	−0.10	46.02%
3.85	99.99%	1.85	96.78%	−0.15	44.04%
3.80	99.99%	1.80	96.41%	−0.20	42.07%
3.75	99.99%	1.75	95.99%	−0.25	40.13%
3.70	99.99%	1.70	95.54%	−0.30	38.21%
3.65	99.99%	1.65	95.05%	−0.35	36.32%
3.60	99.98%	1.60	94.52%	−0.40	34.46%
3.55	99.98%	1.55	93.94%	−0.45	32.64%
3.50	99.98%	1.50	93.32%	−0.50	30.85%
3.45	99.97%	1.45	92.65%	−0.55	29.12%
3.40	99.97%	1.40	91.92%	−0.60	27.43%
3.35	99.96%	1.35	91.15%	−0.65	25.78%
3.30	99.95%	1.30	90.32%	−0.70	24.20%
3.25	99.94%	1.25	89.44%	−0.75	22.66%
3.20	99.93%	1.20	88.49%	−0.80	21.19%
3.15	99.92%	1.15	87.49%	−0.85	19.77%
3.10	99.90%	1.10	86.43%	−0.90	18.41%
3.05	99.89%	1.05	85.31%	−0.95	17.11%
3.00	99.87%	1.00	84.13%	−1.00	15.87%
2.95	99.84%	0.95	82.89%	−1.05	14.69%
2.90	99.81%	0.90	81.59%	−1.10	13.57%
2.85	99.78%	0.85	80.23%	−1.15	12.51%
2.80	99.74%	0.80	78.81%	−1.20	11.51%
2.75	99.70%	0.75	77.34%	−1.25	10.56%
2.70	99.65%	0.70	75.80%	−1.30	9.68%
2.65	99.60%	0.65	74.22%	−1.35	8.85%
2.60	99.53%	0.60	72.57%	−1.40	8.08%
2.55	99.46%	0.55	70.88%	−1.45	7.35%
2.50	99.38%	0.50	69.15%	−1.50	6.68%
2.45	99.29%	0.45	67.36%	−1.55	6.06%
2.40	99.18%	0.40	65.54%	−1.60	5.48%
2.35	99.06%	0.35	63.68%	−1.65	4.95%
2.30	98.93%	0.30	61.79%	−1.70	4.46%
2.25	98.78%	0.25	59.87%	−1.75	4.01%
2.20	98.61%	0.20	57.93%	−1.80	3.59%
2.15	98.42%	0.15	55.96%	−1.85	3.22%
2.10	98.21%	0.10	53.98%	−1.90	2.87%
2.05	97.98%	0.05	51.99%	−1.95	2.56%
2.00	97.72%	0.00	50.00%	−2.00	2.28%

TABLE A5.7. DETERMINING A P VALUE FROM χ^2 (CHI-SQUARE) WITH 1 DEGREE OF FREEDOM

This table assumes that you have one degree of freedom, which is the case when analyzing a 2 × 2 contingency table. The P values tabulated here are two sided.

χ^2	P	χ^2	P	χ^2	P	χ^2	P
0.0	1.000	4.1	0.0429	8.1	0.0044	12.1	0.0005
0.1	0.7518	4.2	0.0404	8.2	0.0042	12.2	0.0005
0.2	0.6547	4.3	0.0381	8.3	0.0040	12.3	0.0005
0.3	0.5839	4.4	0.0359	8.4	0.0038	12.4	0.0004
0.4	0.5271	4.5	0.0339	8.5	0.0036	12.5	0.0004
0.5	0.4795	4.6	0.0320	8.6	0.0034	12.6	0.0004
0.6	0.4386	4.7	0.0302	8.7	0.0032	12.7	0.0004
0.7	0.4028	4.8	0.0285	8.8	0.0030	12.8	0.0003
0.8	0.3711	4.9	0.0269	8.9	0.0029	12.9	0.0003
0.9	0.3428	5.0	0.0253	9.0	0.0027	13.0	0.0003
1.0	0.3173	5.1	0.0239	9.1	0.0026	13.1	0.0003
1.1	0.2943	5.2	0.0226	9.2	0.0024	13.2	0.0003
1.2	0.2733	5.3	0.0213	9.3	0.0023	13.3	0.0003
1.3	0.2542	5.4	0.0201	9.4	0.0022	13.4	0.0003
1.4	0.2367	5.5	0.0190	9.5	0.0021	13.5	0.0002
1.5	0.2207	5.6	0.0180	9.6	0.0019	13.6	0.0002
1.6	0.2059	5.7	0.0170	9.7	0.0018	13.7	0.0002
1.7	0.1923	5.8	0.0160	9.8	0.0017	13.8	0.0002
1.8	0.1797	5.9	0.0151	9.9	0.0017	13.9	0.0002
1.9	0.1681	6.0	0.0143	10.0	0.0016	14.0	0.0002
2.0	0.1573	6.1	0.0135	10.1	0.0015	14.1	0.0002
2.1	0.1473	6.2	0.0128	10.2	0.0014	14.2	0.0002
2.2	0.1380	6.3	0.0121	10.3	0.0013	14.3	0.0002
2.3	0.1294	6.4	0.0114	10.4	0.0013	14.4	0.0001
2.4	0.1213	6.5	0.0108	10.5	0.0012	14.5	0.0001
2.5	0.1138	6.6	0.0102	10.6	0.0011	14.6	0.0001
2.6	0.1069	6.7	0.0096	10.7	0.0011	14.7	0.0001
2.7	0.1003	6.8	0.0091	10.8	0.0010	14.8	0.0001
2.8	0.0943	6.9	0.0086	10.9	0.0010	14.9	0.0001
2.9	0.0886	7.0	0.0082	11.0	0.0009		
3.0	0.0833	7.1	0.0077	11.1	0.0009		
3.1	0.0783	7.2	0.0073	11.2	0.0008		
3.2	0.0736	7.3	0.0069	11.3	0.0008		
3.3	0.0693	7.4	0.0065	11.4	0.0007		
3.4	0.0652	7.5	0.0062	11.5	0.0007		
3.5	0.0614	7.6	0.0058	11.6	0.0007		
3.6	0.0578	7.7	0.0055	11.7	0.0006		
3.7	0.0544	7.8	0.0052	11.8	0.0006		
3.8	0.0513	7.9	0.0049	11.9	0.0006		
3.9	0.0483	8.0	0.0047	12.0	0.0005		
4.0	0.0455						

TABLE A5.8. CRITICAL VALUES OF THE CHI-SQUARE DISTRIBUTION

This table shows the critical values of χ^2. The rows denote degrees of freedom, and the columns denote the value of α. If your test results in a value of χ^2 larger than the tabulated value, then your P value is less than α.

Table A5.8. Critical Values of the Chi-Square Distribution

α

df	0.250	0.200	0.150	0.100	0.070	0.060	0.050	0.040	0.030	0.020	0.010	0.005	0.001
1	1.323	1.642	2.072	2.706	3.283	3.537	3.841	4.218	4.709	5.412	6.635	7.879	10.827
2	2.773	3.219	3.794	4.605	5.319	5.627	5.991	6.438	7.013	7.824	9.210	10.597	13.815
3	4.108	4.642	5.317	6.251	7.060	7.407	7.815	8.311	8.947	9.837	11.345	12.838	16.266
4	5.385	5.989	6.745	7.779	8.666	9.044	9.488	10.026	10.712	11.668	13.277	14.860	18.466
5	6.626	7.289	8.115	9.236	10.191	10.596	11.070	11.644	12.375	13.388	15.086	16.750	20.515
6	7.841	8.558	9.446	10.645	11.660	12.090	12.592	13.198	13.968	15.033	16.812	18.548	22.457
7	9.037	9.803	10.748	12.017	13.088	13.540	14.067	14.703	15.509	16.622	18.475	20.278	24.321
8	10.219	11.030	12.027	13.362	14.484	14.956	15.507	16.171	17.011	18.168	20.090	21.955	26.124
9	11.389	12.242	13.288	14.684	15.854	16.346	16.919	17.608	18.480	19.679	21.666	23.589	27.877
10	12.549	13.442	14.534	15.987	17.203	17.713	18.307	19.021	19.922	21.161	23.209	25.188	29.588
11	13.701	14.631	15.767	17.275	18.533	19.061	19.675	20.412	21.342	22.618	24.725	26.757	31.264
12	14.845	15.812	16.989	18.549	19.849	20.393	21.026	21.785	22.742	24.054	26.217	28.300	32.909
13	15.984	16.985	18.202	19.812	21.151	21.711	22.362	23.142	24.125	25.471	27.688	29.819	34.527
14	17.117	18.151	19.406	21.064	22.441	23.017	23.685	24.485	25.493	26.873	29.141	31.319	36.124
15	18.245	19.311	20.603	22.307	23.720	24.311	24.996	25.816	26.848	28.259	30.578	32.801	37.698
16	19.369	20.465	21.793	23.542	24.990	25.595	26.296	27.136	28.191	29.633	32.000	34.267	39.252
17	20.489	21.615	22.977	24.769	26.251	26.870	27.587	28.445	29.523	30.995	33.409	35.718	40.791
18	21.605	22.760	24.155	25.989	27.505	28.137	28.869	29.745	30.845	32.346	34.805	37.156	42.312
19	22.718	23.900	25.329	27.204	28.751	29.396	30.144	31.037	32.158	33.687	36.191	38.582	43.819
20	23.828	25.038	26.498	28.412	29.991	30.649	31.410	32.321	33.462	35.020	37.566	39.997	45.314
21	24.935	26.171	27.662	29.615	31.225	31.895	32.671	33.597	34.759	36.343	38.932	41.401	46.796
22	26.039	27.301	28.822	30.813	32.453	33.135	33.924	34.867	36.049	37.659	40.289	42.796	48.268
23	27.141	28.429	29.979	32.007	33.675	34.370	35.172	36.131	37.332	38.968	41.638	44.181	49.728
24	28.241	29.553	31.132	33.196	34.893	35.599	36.415	37.389	38.609	40.270	42.980	45.558	51.179
25	29.339	30.675	32.282	34.382	36.106	36.824	37.652	38.642	39.880	41.566	44.314	46.928	52.619
26	30.435	31.795	33.429	35.563	37.315	38.044	38.885	39.889	41.146	42.856	45.642	48.290	54.051
27	31.528	32.912	34.574	36.741	38.520	39.259	40.113	41.132	42.407	44.140	46.963	49.645	55.475
28	32.620	34.027	35.715	37.916	39.721	40.471	41.337	42.370	43.662	45.419	48.278	50.994	56.892
29	33.711	35.139	36.854	39.087	40.919	41.679	42.557	43.604	44.913	46.693	49.588	52.335	58.301
30	34.800	36.250	37.990	40.256	42.113	42.883	43.773	44.834	46.160	47.962	50.892	53.672	59.702

TABLE A5.9. 95% CONFIDENCE INTERVALS FOR THE POISSON DISTRIBUTION

C is the number of events per unit time or number of objects per unit of space. The table shows the 95% confidence interval for the average number of events per unit time or number of objects per unit of space. For larger C, use this equation: $C - 1.96\sqrt{C}$ to $C + 1.96\sqrt{C}$. Don't normalize C to some standard unit; C should be the number of events or objects actually counted.

C	Lower	Upper	C	Lower	Upper	C	Lower	Upper
0	0.00	3.69	41	0.00	55.62	82	0.00	101.78
1	0.03	5.57	42	30.27	56.77	83	66.11	102.89
2	0.24	7.22	43	31.12	57.92	84	67.00	104.00
3	0.62	8.77	44	31.97	59.07	85	67.89	105.10
4	1.09	10.24	45	32.82	60.21	86	68.79	106.21
5	1.62	11.67	46	33.68	61.36	87	69.68	107.31
6	2.20	13.06	47	34.53	62.50	88	70.58	108.42
7	2.81	14.42	48	35.39	63.64	89	71.47	109.52
8	3.45	15.76	49	36.25	64.78	90	72.37	110.63
9	4.12	17.08	50	37.11	65.92	91	73.27	111.73
10	4.80	18.39	51	37.97	67.06	92	74.16	112.83
11	5.49	19.68	52	38.84	68.19	93	75.06	113.93
12	6.20	20.96	53	39.70	69.33	94	75.96	115.03
13	6.92	22.23	54	40.57	70.46	95	76.86	116.13
14	7.65	23.49	55	41.43	71.59	96	77.76	117.23
15	8.40	24.74	56	42.30	72.72	97	78.66	118.33
16	9.15	25.98	57	43.17	73.85	98	79.56	119.43
17	9.90	27.22	58	44.04	74.98	99	80.46	120.53
18	10.67	28.45	59	44.91	76.11	100	81.36	121.63
19	11.44	29.67	60	45.79	77.23	101	82.27	122.72
20	12.22	30.89	61	46.66	78.36	102	83.17	123.82
21	13.00	32.10	62	47.54	79.48	103	84.07	124.92
22	13.79	33.31	63	48.41	80.60	104	84.98	126.01
23	14.58	34.51	64	49.29	81.73	105	85.88	127.11
24	15.38	35.71	65	50.17	82.85	106	86.78	128.20
25	16.18	36.90	66	51.04	83.97	107	87.69	129.30
26	16.98	38.10	67	51.92	85.09	108	88.59	130.39
27	17.79	39.28	68	52.80	86.21	109	89.50	131.49
28	18.61	40.47	69	53.69	87.32	110	90.41	132.58
29	19.42	41.65	70	54.57	88.44	111	91.31	133.67
30	20.24	42.83	71	55.45	89.56	112	92.22	134.77
31	21.06	44.00	72	56.34	90.67	113	93.13	135.86
32	21.89	45.17	73	57.22	91.79	114	94.04	136.95
33	22.72	46.34	74	58.11	92.90	115	94.94	138.04
34	23.55	47.51	75	58.99	94.01	116	95.85	139.13
35	24.38	48.68	76	59.88	95.13	117	96.76	140.22
36	25.21	49.84	77	60.77	96.24	118	97.67	141.31
37	26.05	51.00	78	61.66	97.35	119	98.58	142.40
38	26.89	52.16	79	62.55	98.46	120	99.49	143.49
39	27.73	53.31	80	63.44	99.57	121	100.40	144.58
40	28.58	54.47	81	64.33	100.68	122	101.31	145.67

TABLE A5.10. CRITICAL VALUES OF THE F DISTRIBUTION

Find the column corresponding to the number of degrees of freedom in the numerator and the row corresponding to the number of degrees of freedom in the denominator. Be very sure that you are not mixing up the two df values, as it matters a lot. Find the critical value of F in the table. If you have obtained a value for F greater than that, then your P value is less than α. The next page gives values for $\alpha = 0.05$ and the following page gives values for $\alpha = 0.01$.

Table A5.10. Critical Values of the F Distribution. $\alpha = 0.05$

	DFn, degrees of freedom in the numerator															
DFd	2	3	4	5	6	7	8	9	10	12	15	20	50	100	500	1000
2	19.00	19.16	19.25	19.30	19.33	19.35	19.37	19.38	19.40	19.41	19.43	19.45	19.48	19.49	19.49	19.49
3	9.55	9.28	9.12	9.01	8.94	8.89	8.85	8.81	8.79	8.74	8.70	8.66	8.58	8.55	8.53	8.53
4	6.94	6.59	6.39	6.26	6.16	6.09	6.04	6.00	5.96	5.91	5.86	5.80	5.70	5.66	5.64	5.63
5	5.79	5.41	5.19	5.05	4.95	4.88	4.82	4.77	4.74	4.68	4.62	4.56	4.44	4.41	4.37	4.37
6	5.14	4.76	4.53	4.39	4.28	4.21	4.15	4.10	4.06	4.00	3.94	3.87	3.75	3.71	3.68	3.67
7	4.74	4.35	4.12	3.97	3.87	3.79	3.73	3.68	3.64	3.57	3.51	3.44	3.32	3.27	3.24	3.23
8	4.46	4.07	3.84	3.69	3.58	3.50	3.44	3.39	3.35	3.28	3.22	3.15	3.02	2.97	2.94	2.93
9	4.26	3.86	3.63	3.48	3.37	3.29	3.23	3.18	3.14	3.07	3.01	2.94	2.80	2.76	2.72	2.71
10	4.10	3.71	3.48	3.33	3.22	3.14	3.07	3.02	2.98	2.91	2.85	2.77	2.64	2.59	2.55	2.54
12	3.89	3.49	3.26	3.11	3.00	2.91	2.85	2.80	2.75	2.69	2.62	2.54	2.40	2.35	2.31	2.30
14	3.74	3.34	3.11	2.96	2.85	2.76	2.70	2.65	2.60	2.53	2.46	2.39	2.24	2.19	2.14	2.14
16	3.63	3.24	3.01	2.85	2.74	2.66	2.59	2.54	2.49	2.42	2.35	2.28	2.12	2.07	2.02	2.02
20	3.49	3.10	2.87	2.71	2.60	2.51	2.45	2.39	2.35	2.28	2.20	2.12	1.97	1.91	1.86	1.85
25	3.39	2.99	2.76	2.60	2.49	2.40	2.34	2.28	2.24	2.16	2.09	2.01	1.84	1.78	1.73	1.72
30	3.32	2.92	2.69	2.53	2.42	2.33	2.27	2.21	2.16	2.09	2.01	1.93	1.76	1.70	1.64	1.63
40	3.23	2.84	2.61	2.45	2.34	2.25	2.18	2.12	2.08	2.00	1.92	1.84	1.66	1.59	1.53	1.52
50	3.18	2.79	2.56	2.40	2.29	2.20	2.13	2.07	2.03	1.95	1.87	1.78	1.60	1.52	1.46	1.45
75	3.12	2.73	2.49	2.34	2.22	2.13	2.06	2.01	1.96	1.88	1.80	1.71	1.52	1.44	1.36	1.35
100	3.09	2.70	2.46	2.31	2.19	2.10	2.03	1.97	1.93	1.85	1.77	1.68	1.48	1.39	1.31	1.30
250	3.03	2.64	2.41	2.25	2.13	2.05	1.98	1.92	1.87	1.79	1.71	1.61	1.40	1.31	1.20	1.18
500	3.01	2.62	2.39	2.23	2.12	2.03	1.96	1.90	1.85	1.77	1.69	1.59	1.38	1.28	1.16	1.14
1000	3.00	2.61	2.38	2.22	2.11	2.02	1.95	1.89	1.84	1.76	1.68	1.58	1.36	1.26	1.13	1.11

DFn, degrees of freedom in the numerator

DFd	2	3	4	5	6	7	8	9	10	12	15	20	50	100	500	1000
2	99.00	99.16	99.25	99.30	99.33	99.36	99.38	99.39	99.40	99.42	99.43	99.45	99.48	99.49	99.50	99.50
3	30.82	29.46	28.71	28.24	27.91	27.67	27.49	27.34	27.23	27.05	26.87	26.69	26.35	26.24	26.15	26.14
4	18.00	16.69	15.98	15.52	15.21	14.98	14.80	14.66	14.55	14.37	14.20	14.02	13.69	13.58	13.49	13.47
5	13.27	12.06	11.39	10.97	10.67	10.46	10.29	10.16	10.05	9.89	9.72	9.55	9.24	9.13	9.04	9.03
6	10.92	9.78	9.15	8.75	8.47	8.26	8.10	7.98	7.87	7.72	7.56	7.40	7.09	6.99	6.90	6.89
7	9.55	8.45	7.85	7.46	7.19	6.99	6.84	6.72	6.62	6.47	6.31	6.16	5.86	5.75	5.67	5.66
8	8.65	7.59	7.01	6.63	6.37	6.18	6.03	5.91	5.81	5.67	5.52	5.36	5.07	4.96	4.88	4.87
9	8.02	6.99	6.42	6.06	5.80	5.61	5.47	5.35	5.26	5.11	4.96	4.81	4.52	4.41	4.33	4.32
10	7.56	6.55	5.99	5.64	5.39	5.20	5.06	4.94	4.85	4.71	4.56	4.41	4.12	4.01	3.93	3.92
12	6.93	5.95	5.41	5.06	4.82	4.64	4.50	4.39	4.30	4.16	4.01	3.86	3.57	3.47	3.38	3.37
14	6.51	5.56	5.04	4.69	4.46	4.28	4.14	4.03	3.94	3.80	3.66	3.51	3.22	3.11	3.03	3.02
16	6.23	5.29	4.77	4.44	4.20	4.03	3.89	3.78	3.69	3.55	3.41	3.26	2.97	2.86	2.78	2.76
20	5.85	4.94	4.43	4.10	3.87	3.70	3.56	3.46	3.37	3.23	3.09	2.94	2.64	2.54	2.44	2.43
25	5.57	4.68	4.18	3.85	3.63	3.46	3.32	3.22	3.13	2.99	2.85	2.70	2.40	2.29	2.19	2.18
30	5.39	4.51	4.02	3.70	3.47	3.30	3.17	3.07	2.98	2.84	2.70	2.55	2.25	2.13	2.03	2.02
40	5.18	4.31	3.83	3.51	3.29	3.12	2.99	2.89	2.80	2.66	2.52	2.37	2.06	1.94	1.83	1.82
50	5.06	4.20	3.72	3.41	3.19	3.02	2.89	2.78	2.70	2.56	2.42	2.27	1.95	1.82	1.71	1.70
75	4.90	4.05	3.58	3.27	3.05	2.89	2.76	2.65	2.57	2.43	2.29	2.13	1.81	1.67	1.55	1.53
100	4.82	3.98	3.51	3.21	2.99	2.82	2.69	2.59	2.50	2.37	2.22	2.07	1.74	1.60	1.47	1.45
250	4.69	3.86	3.40	3.09	2.87	2.71	2.58	2.48	2.39	2.26	2.11	1.95	1.61	1.46	1.30	1.27
500	4.65	3.82	3.36	3.05	2.84	2.68	2.55	2.44	2.36	2.22	2.07	1.92	1.57	1.41	1.23	1.20
1000	4.63	3.80	3.34	3.04	2.82	2.66	2.53	2.43	2.34	2.20	2.06	1.90	1.54	1.38	1.19	1.16

INDEX

Adjusting data for confounding variables, 293–296
Alpha
  choosing an appropriate value, 109
  defined, 106
  relationship to confidence intervals, 111
  relationship to P value, 110
Alternative hypothesis, 96
Analysis of variance
  multiple comparison post tests, 258–259
  nonparametric, 260
  one-way, 255–258
  repeated measures, 260
  two-way, 261
Attributable risk, 73
Average. *See* Mean

Bayesian analysis
  controversy, 145
  equation, 137
  genetic counseling, 149
  genetic linkage, 150
  informal use, 145
  interpreting lab tests, 129–138
  interpreting statistical significance, 142
Beta. *See* Power
Bias, 22
Binomial distribution
  defined, 11
  equation, 19
  used to generate confidence interval, 12
Blind, double, 185
Bonferroni's method for multiple
  comparison, 259
Box and whiskers plot, 25
Case-control study. *See* Retrospective study
Calculating sample size. *See* Sample size

Central limit theorem, 42–44
Chi-square test
  analyzing 2 × 2 contingency tables, 233–235
  analyzing larger contingency tables, 251–253
  compared to Fisher's test, 236, 301
  comparing observed and expected counts, 230
  power of, 236 240
  table, 375–377
  for trend, 253
Choosing a test, 297–302
Clinical trial
  designing, 183–191
  example, 71
Coefficient of determination, 173
Coefficient of variation, 29
Coincidences, 118
Confidence interval
  of correlation coefficient, 161
  of counted variable, 245–247, 379
  of difference between two proportions, 78
  of difference of two means, 63, 65, 67
  of linear regression line, 178
  of mean, 39–41
  of odds ratio, 78
  of proportion, 12–18, 362–364
  of ratio of two means, 285
  of relative risk, 78
  of slope in linear regression, 177
  of survival curve, 55
  what it means to be 95% sure, 13
Contingency table, 77
Confounding variables, adjusting for, 293–296
Continuity correction. *See* Yates' continuity correction

Contrasts, in multiple comparison, 259
Correlation coefficient
  assumptions, 157
  calculating, 160
  comparison with linear regression, 169
  confidence interval, 161
  effect of outliers, 158
  example, 155
  interpreting, 157
  P value, 162
  r vs. r^2, 158
  Spearman correlation coefficient, 160
Cross sectional study, 70
Cross-over design in clinical trials, 186
Curve fitting, 277–282
CV (coefficient of variation), 29

Decision analysis, 286
Double blind, 185
Dunnett's method for multiple comparison,
  259

Excel, 313, 360
Error bars, 48–49
Errors, type I vs. type II
  compared to errors in lab tests, 140
  defined, 109
Exact randomization test, 217
Experimental hypothesis, 96

F distribution
  in ANOVA, 257
  table, 379–381
Fisher's exact test, 233, 236, 301
Friedman test, 261

Gaussian distribution
  basis of many statistical tests, 46–47
  central limit theorem, 42–44
  deciding when to use nonparametric tests,
    297–300
  defined, 31–33
  and normal limits, 36–37
  table, 366–367
Genetics
  Bayesian analysis in genetic counseling, 149
  Bayesian analysis of genetic linkage,
    151–153
Graphpad InStat, 311–312
Graphpad Prism, 311–312

Histogram, 23
Homoscedasticity, 171, 265
Human subjects committees, 189
Hypothesis testing, 106

Innocent until proven guilty, 108
InStat, 311–312
Intention to treat, in clinical trials, 186
Interim analyses, 191
Interquartile range, 25

Kruskal-Wallis test, 261

Life tables. See survival curves
Likelihood ratio
  in genetic counseling, 151
  and lod score, 152
  for lab tests, 137
Linear regression
  assumptions, 171
  calculating, 176–178
  comparison with correlation, 169
  fallacy, 175
  least squares, 172
  meaning of r^2, 173
  model, 170
Linkage, genetic, 151–153
Lod score, 152
Log-rank test to compare survival curves,
  273
Logistic regression, 268–271

Mann-Whitney test, 221
Mantel-Haenszel test
  comparing survival curves, 273
  pooling several contingency tables, 253
Matched subjects. See Paired subjects
Maximum likelihood, 173
Mean
  confidence interval of, 39–41
  defined, 24
Median, 24
Mcnemar's test of paired observations, 250
Meta analysis, 288
Microsoft Excel, 313, 360
Multiple comparisons
  following one-way ANOVA, 258–259
  principles, 118–126
Multiple regression, 263–267
Multivariate statistics, 267

Negative predictive value, 130
Newman-Keuls method for multiple
    comparison, 259
Nonlinear regression, 277–282
Nonparametric tests
    exact randomization test, 217
    Friedman's test, 261
    instead of ANOVA, 261
    introduction, 217
    Kruskal-Wallis test, 261
    Mann-Whitney test, 221
    Spearman correlation, 159
    when to use, 297–300
    Wilcoxon signed rank sum test, 228
Normal limits, 36
Null hypothesis, 96

Odds ratio
    calculating a confidence interval, 78, 88
    defined, 76
    interpreting from retrospective studies, 82
    logistic regression, 269–271
    relationship to relative risk, 83
Odds vs. probabilities, 75, 136

P values
    common misinterpretations, 97
    definition, 96
    one-tail vs. two-tails, 97, 300
Paired subjects
    CI of difference between means, 67
    in case-control studies, 87
    McNemar's chi-square test, 250
    need for special analyses, 66, 301
    paired t test, 225–227
    repeated measures ANOVA, 260
Percentiles, 25
Picking a statistical test, 297–302
Placebos, 189
Poisson distribution, 245, 379
Population vs. sample, 4
Post tests. *See* Multiple comparisons
Positive predictive value, 130
Power
    analysis to determine sample size. *See*
        Sample size
    defined, 140
    interpreting, 201
    power of chi-square test, 236–240
    of t test, 213–315

used to interpret "not significant" results,
    116
Predictive value of lab test, 130
Prediction interval
    of mean, 35
    of regression line, 178
Prism, 311–312
Probabilities vs. odds, 75, 136
Probability distribution
    defined, 31
    Gaussian, 31–38
    t, 44
Proportion, confidence interval of, 12–18,
    362–364
Proportional hazards regression, 274–276
Prospective study, 70
Protocols in clinical trials, 187

R. *See* Correlation coefficient
R^2. *See* Linear regression
Randomization in clinical trials, 185
Randomization test, 217
References, 309–310
Regression
    different kinds, 166
    linear. *See* Linear regression
    logistic, 268–271
    multiple, 262–267
    nonlinear, 277–282
    polynomial, 283
    proportional hazards, 274–276
Relative risk
    calculating a confidence interval, 78
    defined, 73
    from survival studies, 76
    why it can be misleading, 74
Repeated measures ANOVA, 260
Residuals, 174
Retrospective study
    advantages and disadvantages, 85–86
    defined, 70
    example, 81
    interpreting odds ratios, 83
Robust tests. *See* Nonparametric tests

Sample size
    comparing survival curves, 200
    difference between two means, 196, 198
    difference between two proportions, 197,
        199

Sample size (*cont.*)
  one mean, 195
  one proportion, 196
Sample vs. population, 4
Scheffe's method for multiple comparison, 259
SD. *See* Standard deviation
SEM. *See* Standard error of the mean
Sensitivity of lab test, 129
Significance level, 106
Significant, statistically, 107
Spearman correlation, 160
Specificity of lab test, 129
Standard deviation
  checking the accuracy of your calculator, 28
  deciding between SD and SEM error bars, 48
  equation, 27
  interpreting, 28
  sample SD vs. population SD, 27
Standard error of the mean
  calculating, 44
  central limit theorem, 42–44
  deciding between SD and SEM error bars, 48
Statistically significant
  defined, 107, 113
  "extremely" significant, 113
  "marginally" significant, 114
  relationship to scientific importance, 111
Stratified randomization, 185
Student's t test. *See* T test
Student-Newman-Keuls method for multiple
    comparison, 259
Survival curves
  assumptions, 58
  censored observations, 54
  comparing, 272–276
  confidence interval of, 56
  creating, 55

log-rank test, 273
Mantel-Haenszel test, 273
median survival, 57
  sample size for, 200

T distribution
  defined, 44
  relationship between confidence interval and
    hypothesis testing, 211–213
  table, 368–373
  used to calculate a confidence interval, 41
T test
  paired, 225–227
  power of, 213–215
  unpaired, 207–213
Tukey's method for multiple comparison,
    259
Type I vs. Type II errors
  compared to errors in lab tests, 140
  defined, 109

Variability
  display with histogram, 23
  sources of, 22
Variance, 26–27

Wilcoxon signed rank sum test, 228
Will Roger's phenomenon, 274

Yates' continuity correction
  in chi-square test to compare two
    proportions, 235–236, 301–302
  comparing observed and expected, 231

Z table
  uses, 34
  table, 366–367